Ranis Ibragimov
Lie Group Analysis of Differential Equations

De Gruyter Series in Nonlinear Analysis and Applications

Volume 43

Ranis Ibragimov

Lie Group Analysis of Differential Equations

Invariant Solutions of Nonlinear Physical Phenomena

DE GRUYTER

Mathematics Subject Classification 2020
Primary: 34M04, 35Q99, 76-10, 76M60; Secondary: 70H33

Author
Dr. Ranis Ibragimov
De Gruyter
Boston, MA 02110
USA
ranis.ibragimov@degruyter.com

ISBN 978-3-11-138699-7
e-ISBN (PDF) 978-3-11-138749-9
e-ISBN (EPUB) 978-3-11-138780-2
ISSN 0941-813X

Library of Congress Control Number: 2023951290

Bibliographic information published by the Deutsche Nationalbibliothek
The Deutsche Nationalbibliothek lists this publication in the Deutsche Nationalbibliografie; detailed bibliographic data are available on the Internet at http://dnb.dnb.de.

Typesetting: VTeX UAB, Lithuania
Printing and binding: CPI books GmbH, Leck

www.degruyter.com

Preface

For several centuries, mathematical modeling has been considered as a branch of applied mathematics that encompasses topics from physics, engineering, biology, and environmental and computer sciences. It involves giving an approximate description of real processes, frequently in terms of linear and nonlinear differential equations.

Before the eighteenth century, the best mathematicians were driven by a desire to develop approximate descriptions of processes underlying real natural phenomena. Mathematical models' complexity can significantly increase with improvements in approximations or their extension to more general situations. Acquaintance with Lie group analysis of differential equations is important for constructing exact solutions of nonlinear mathematical models. Roughly speaking, Lie group analysis of differential equations represents an algorithm for finding invariant (exact) solutions of nonlinear ordinary and partial differential equations and systems of equations. The algorithm is based on finding a "symmetry," which is a local group of transformations that maps every solution of a mathematical model to another solution of the same model. In other words, it maps the solution set of the given model to itself. So, we can think of symmetry as a set of transformations of dependent and independent variables that leave an equation in question invariant. The obtained symmetries lead ultimately to exact (invariant) solutions. Lie group analysis reduces a nonlinear ordinary/partial differential equation (or system of nonlinear equations) to an integrable form and constructs its solution in the simplest possible form. Rephrasing the famous French aphorism *cherchez la femme*, my father Nail Ibragimov always used to say: "*If you cannot solve a nonlinear differential equation, cherchez le group*" [41]. Thus, numerous physical phenomena can be investigated using Lie symmetries to discover various invariant solutions and conservation laws of a given problem. Differential equations, conservation laws, solutions to boundary value problems, and so forth, can be derived from the group invariance principle. However, mathematical modeling in terms of Lie group analysis of differential equations is not limited to the mathematical solution of the problem in question. Mathematical modeling can be thought of as a three-step process: physical situation ⇒ mathematical formulation ⇒ invariant solutions ⇒ physical interpretation of the obtained invariant solutions.

Unfortunately, the last step in this process is often missed and therefore it is not always clear how one can interpret the obtained invariant solutions in the context of a given physical problem. I believe that it was a mistake to isolate Lie group analysis from its natural applications and treat it as a branch of abstract mathematics. The main focus of this book is on providing a physical interpretation of the obtained invariant solutions.

As one particular application of Lie group analysis, the dispersion relation of wave motion, especially of oceanic internal gravity waves, will be analyzed from a group analysis point of view. It will be shown that a similar dispersion relation appears in nonlinear wave motion and it can be used to obtain the more general form of physically relevant solutions of nonlinear equations.

https://doi.org/10.1515/9783111387499-201

As another particular example in this book, it will be shown that Lie group analysis of nonlinear differential equations might be able to reveal the existence of singularities provided by invariant solutions and invisible from the form of some nonlinear differential equations. We call them *internal singularities*, in contrast to *external singularities* manifested by the form of the equation. It will be illustrated by way of examples that internal singularities are useful for analyzing the behavior of solutions of nonlinear differential equations near external singularities.

As another illustration, the nonlinear incompressible nonviscous and viscous fluid flows within a thin rotating atmospheric shell that serves as a simple mathematical description of an atmospheric circulation caused by the temperature difference between the equator and the poles will be considered. A particular stationary flow is superimposed on the model in question which, under the assumption of no friction and a temperature distribution dependent only upon latitude, models the zonal West-to-East flows in the upper atmosphere between the Ferrel and polar cells. Owing to the Coriolis effects, the resulting achievable meteorological flows correspond to the asymptotic stable flows that are being translated along the equatorial plane. The exact solutions of Navier–Stokes equations used to describe the model will be found in terms of elementary functions by using Lie group methods. Special attention will be given to analyzing and visualizing the conserved densities associated with known symmetries. Particularly, it will be shown that the asymptotic behavior of the conserved densities can differ greatly depending on the selection of symmetry and the choice of the invariant solution.

As another example of applications of Lie group analysis, the exact solutions of three-dimensional equations of motion for internal gravity waves in cylindrical coordinates in unbounded media will be found by means of exact and approximate transformation groups of equations with a small parameter. The introduction of the small parameter has been motivated by justifying the analogy of the Kelvin hypothesis on vanishing of the component of the velocity normal to the wall for the rectilinear motion. In the present case of the cylindrical domain, the component of the velocity normal to the wall is nonzero and achieves its maximum in the interior, which also agrees with analytical predictions in previous studies for a strongly linear case. As a particular application to ocean and atmospheric modeling, in terms of nonlinear modeling, the time series of the energy density associated with invariant solutions will be visualized as spinning patterns that appear to be rotating in a counterclockwise sense when looking from above the North Pole. Such spinning patterns will be compared with the flow around a low-pressure area that is usually linked with models of hurricanes. In terms of zero-order approximate transformations, the invariant solutions will be visualized as funnels having something in common with the geometric structure of oceanic whirlpools.

As another modeling scenario, we will analyze and visualize the exact invariant solution of the nonlinear simplified version of the shallow water equations, which are used to simulate equatorial atmospheric waves of planetary scales. The method of obtaining the exact solution will be based on the Lie group invariance principle. It will be shown that the obtained invariant solution has a Fibonacci spiral-like form and has two arbitrary parameters. It will be shown that we can define a new time-dependent model hyperparameter. The question of particular interest in that example will be: can we tune the hyperparameter in order to match the exact invariant solution to the actual Fibonacci spiral?

This book represents a unique blend of traditional analytical and numerical methods enriched with the author's developments and applications to ocean and atmospheric sciences. It is meant for undergraduate and graduate students interested in applied mathematics, differential equations, and mathematical modeling of real-world problems, especially in physics of the ocean and atmosphere.

The book is divided into two parts: the first part presents the method of Lie group analysis without mixing it with physical applications, while the second part presents physical applications without details of the method. I believe that presenting the method and applications separately, not in a mixed form, is crucial for readers to effectively follow the book. Based on my teaching experience, presenting the method and applications simultaneously can be confusing for readers, and they may lose track of the main idea behind the method while focusing on applications. Therefore, by presenting the method before applications, readers can gain a deeper understanding of the method and its application, making it easier for them to follow and comprehend the book.

Acknowledgments

It is a pleasure to explicitly acknowledge the singular influence of my former teachers and colleagues, Professor Nikolai Makarenko (Lavryentiev's Institute of Hydrodynamics, Siberian Branch of the Russian Academy of Science, Russia), Hisashi Okamoto (Research Institute for Mathematical Sciences, Kyoto University, Japan), Professor Tuncay Aktosun (University of Texas, Arlington), and Kevin Lamb (University of Waterloo, Canada). Their prodigious contributions to the study of the dynamics of the atmosphere and oceans, as well as their example of scholarly integrity, have been a continuing source of inspiration. I would also like to express my special gratitude and thanks to my friends and colleagues from De Gruyter – Daniel Tiemann, Steve Elliot, Nadja Schedensack, David Pallai, Jaya Dalal, Damiano Sacco, Ute Skambraks, Helene Chavaroche, Bettina Noto, Marie Hammerschmidt, Jennifer Blaney, Karin Sora, and Kristin Berber-Nerlinger for their invaluable support. I also gratefully acknowledge many fruitful discussions with one of our authors, Olga Olegovna from Biosynthetic Machines Inc, who has inspired me to write this book.

https://doi.org/10.1515/9783111387499-202

Contents

1 Lie group analysis of differential equations

This preparatory chapter is designed to meet the needs of beginners and provides a background in Lie group analysis of differential equations illustrated by examples of various complexity with detailed solutions and illustrations. Surprisingly, sometimes differential equations appear where, it would seem, they should not appear at all. For example, differential equations may appear in the calculations when evaluating improper integrals, which is illustrated in the following example.

Example. Let us evaluate the following improper integral:

$$F(\omega) = \int_{-\infty}^{+\infty} e^{-i\omega x} e^{-x^2}\,dx. \tag{1.1}$$

Solution. Since F is a function of ω, we can differentiate F to obtain

$$\frac{dF}{d\omega} = -i \int_{-\infty}^{+\infty} x e^{-i\omega x} e^{-x^2}\,dx. \tag{1.2}$$

Let us denote

$$u = e^{-i\omega x} \quad \text{and} \quad dv = x e^{-x^2}\,dx,$$

so that

$$du = -i\omega e^{-i\omega x}\,dx \quad \text{and} \quad v = -\frac{e^{-x^2}}{2}.$$

We next use the formula for integration by parts

$$\int u\,dv = uv - \int v\,du. \tag{1.3}$$

Applying formula (1.3) to the integral in (1.2), we have

$$\frac{dF}{d\omega} = -i\left\{\left[-\frac{e^{-i\omega x}e^{-x^2}}{2}\right]_{-\infty}^{+\infty} - \frac{i\omega}{2}\int_{-\infty}^{+\infty} e^{-x^2}e^{-i\omega x}\,dx\right\} = -\frac{\omega}{2}F(\omega), \tag{1.4}$$

since the expression in the square brackets vanishes. Thus, we arrive at the following ordinary differential equation (ODE):

$$\frac{dF}{d\omega} = -\frac{\omega}{2}F. \tag{1.5}$$

We also note that

https://doi.org/10.1515/9783111387499-001

$$F(0) = \int_{-\infty}^{+\infty} e^{-x^2} dx = \sqrt{\pi}. \tag{1.6}$$

Thus, the value $F(\omega)$ of the improper integral is determined from the following boundary value problem:

$$\frac{dF}{d\omega} = -\frac{\omega}{2}F, \quad F(0) = \sqrt{\pi}. \tag{1.7}$$

Separating the variables we have

$$\ln|F| = -\frac{\omega^2}{4} + c,$$

where $c = \text{const.}$, so that

$$F(\omega) = Ae^{-\frac{\omega^2}{4}}, \tag{1.8}$$

where $A = \pm e^c$ is determined from the boundary condition in (1.7).

Thus, we have

$$F(\omega) = \sqrt{\pi}e^{-\frac{\omega^2}{4}}.$$ □

First, let us start with dissertating the idea of transformations by some basic examples from elementary mathematics.

1.1 Basic concepts and examples from elementary mathematics

Problems of elementary mathematics can often be solved by the method of transformations. Here we are going to illustrate the idea behind transformations on the basis of two intuitive examples that can also be found in [41].

Example 1.1. As a first simple example, consider the quadratic equation

$$Ax^2 + Bx + C = 0, \quad A \neq 0. \tag{1.9}$$

Solution. In accordance with tradition, students learn from school to solve equation (1.9) by completing the square. Indeed, this method is simple, but it is applicable only to quadratic equations and it is not suitable for equations of higher degrees, e. g., cubic equations. The simplest transformation converting any algebraic equation of degree n into an equation of the same degree is the linear transformation of x,

$$\bar{x} = x + \varepsilon, \tag{1.10}$$

with the parameter ε. Substituting $x = \bar{x} - \varepsilon$ in equation (1.9) we obtain

$$A\bar{x}^2 + (B - 2A\varepsilon)\bar{x} + A\varepsilon^2 - B\varepsilon + C = 0.$$

Hence, the transformation (1.10) converts equation (1.9) into a new quadratic equation, namely,

$$\bar{A}\,\bar{x}^2 + \bar{B}\,\bar{x} + \bar{C} = 0,$$

where

$$\bar{A} = A, \quad \bar{B} = B - 2A\varepsilon, \quad \bar{C} = C + A\varepsilon^2 - B\varepsilon.$$

Since the parameter ε is arbitrary, let us define it by the equation $\bar{B} = B - 2A\varepsilon = 0$. Then we obtain

$$\bar{B} = 0, \quad \bar{C} = C - \frac{B^2}{4A}.$$

Hence, the transformation

$$\bar{x} = x + \frac{B}{2A} \tag{1.11}$$

converts equation (1.9) into the equation

$$A\bar{x}^2 - \frac{B^2 - 4AC}{4A} = 0.$$

Substituting its roots

$$\bar{x}_{1,2} = \pm\frac{\sqrt{B^2 - 4AC}}{2A}$$

in equation (1.11), one arrives at the roots

$$x_{1,2} = \frac{-B \pm \sqrt{B^2 - 4AC}}{2A}$$

of equation (1.9). □

The transformation (1.10) provides an example of a one-parameter group with the group parameter ε. The group operation is the product $T_{\varepsilon''}T_{\varepsilon'}$ of the transformations $T_{\varepsilon'}$ and $T_{\varepsilon''}$ obtained by assigning to the parameter ε in (1.10) particular values $\varepsilon = \varepsilon'$ and $\varepsilon = \varepsilon''$, respectively. The product $T_{\varepsilon''}T_{\varepsilon'}$ is defined as the consecutive application of the transformation $T_{\varepsilon'}$ and then $T_{\varepsilon''}$. It is manifest that (1.10) satisfies the group property:

$$T_{\varepsilon''}T_{\varepsilon'} = T_{\varepsilon'+\varepsilon''}. \tag{1.12}$$

The group given by (1.10) is called the *group of translations*.

Example 1.2. In elementary geometry, students deal with concepts such as equality and similarity of geometric figures. Let us consider geometric figures (triangles, rectangles, etc.) in the (x, y)-plane. Two figures are said to be *equal* if one can be mapped onto another by a combination of translations

$$\bar{x} = x + a_1, \quad \bar{y} = y + a_2 \tag{1.13}$$

and rotations

$$\bar{x} = x\cos\theta + y\sin\theta, \quad \bar{y} = y\cos\theta - x\sin\theta. \tag{1.14}$$

The transformations (1.13) and (1.14) do not alter distances between any points in the (x, y)-plane. Therefore they are termed *isometric* or *rigid* motions in the (x, y)-plane. Thus, two geometric figures are equal if they coincide after a certain isometric motion. If one deals with the geometry on a sphere (for example, in geodesy), one cannot use the translations in (1.13) as isometric motions. They should be replaced by appropriate transformations on the sphere. The rotations in (1.14), however, belong to the group of isometric motions on the sphere. Two figures in the (x, y)-plane are said to be *similar* if one can be mapped onto another by a combination of isometric motions (1.13)–(1.14) and *scaling transformations*

$$\overline{x} = ax, \quad \overline{y} = by, \tag{1.15}$$

also known as *similarity transformations* or *dilations*. The scaling transformations are *uniform* dilations (expansions or contractions) from the origin if $a = b$ and *nonuniform* dilations if $a \neq b$. It is manifest that the isometric motions (1.13) and (1.14) do not change the area of geometric figures, but the scaling transformations in (1.15) do. Namely, if a geometric figure with area S is mapped by the dilation (1.15) to another figure with area $\overline{S}$, then

$$\overline{S} = abS. \tag{1.16}$$

□

Example 1.3. The commonly used scaling transformations in school geometry are uniform dilations. However, nonuniform dilations are also useful in applications. For example, subjecting the unit circle

$$x^2 + y^2 = 1$$

to the scaling transformation (1.15), one obtains the ellipse

$$\frac{\bar{x}^2}{a^2} + \frac{\bar{y}^2}{b^2} = 1. \tag{1.17}$$

Since the area of the unit circle is $S = \pi$, equation (1.16) gives the area $\overline{S}$ of the ellipse (1.17):

$$\overline{S} = ab\pi. \tag{1.18}$$

In particular, taking the uniform dilation with $k = l = R$, one obtains from (1.18) the area πR^2 of the sphere with radius R. □

Basic concepts from Lie group analysis of differential equations that are used in the present chapter are assembled here. For further information regarding Lie groups and their applications to the theory of differential equations, the reader should consult the various classical and modern texts in the field, such as Refs. [16, 20, 58, 10].

1.2 One-parameter group

Let us consider a change of the variable z involving a parameter a:

$$T_a: \quad \overline{z} = f(z, a), \tag{1.19}$$

where the function f is such that

$$T_0 = I, \quad \text{i. e., } f(z, 0) = z. \tag{1.20}$$

The inverse transformation is defined as

$$T_a^{-1}: \quad z = f^{-1}(\overline{z}, a). \tag{1.21}$$

We say that the transformation (1.19) forms an invertible one-parameter group G if the successive action of two transformations is equivalent to the action of another transformation of the form (1.19), i. e.,

$$T_a T_b = T_{\varphi(a,b)}. \tag{1.22}$$

Definition 1.1. The set G of transformations T_a is a local one-parameter group if the following is true:

[1] $T_0 = I$ existence of unity,
[2] $T_a^{-1} = T_{a^{-1}}$ existence of an inverse element,
[3] $T_c(T_b T_a) = (T_c T_b)T_a$ multiplication is associative.

Example 1.4. Let us check if the transformation $\overline{x} = x + ax$ defines a group G.

Solution. We need to check properties [1]–[3] for $\overline{x} = f(ax)$ with $f = x + ax$:

[1] $T_0: \quad f(x, 0) = x$, i. e., $T_0 = I$,

$$[2]\quad x(z+a)=\overline{x},\ \text{i. e.,}\ x=\frac{\overline{x}}{1+a},$$
$$[3]\quad \overline{\overline{x}}=\overline{x}+b\overline{x}=x+ax+b(x+ax)=x+(a+a+ab)x,$$

i. e., $T_bT_a = T_{\varphi(a,b)}$, with $\varphi(a,b) = a+b+ab$. Thus, the given transformation defines the group. □

Example 1.5. Check if the transformation

$$\overline{x}=x+ax,\quad \overline{y}=\frac{xy}{x+a}$$

defines a group G.

Solution. Again, we need to check if properties [1]–[3] are satisfied. First, let us check property [1]: $T_0 = I$. We have

$$\overline{x}|_{a=0} = (x+a)|_{a=0} = x,$$
$$\overline{y}|_{a=0} = \left(\frac{xy}{x+a}\right)\bigg|_{a=0} = y.$$

Now, we check property [2]: $T_a^{-1} = T_{a^{-1}}$. We have

$$x=\overline{x}-a,$$
$$y=\frac{x+a}{x}\overline{y}=\frac{\overline{x}-a+a}{\overline{x}-a}\overline{y}=\frac{\overline{x}\overline{y}}{\overline{x}-a}.$$

Finally, we need to check property [3]. We have

$$\overline{\overline{x}}=\overline{x}+b=x+a+b=x+(a+b),$$
$$\overline{\overline{y}}=\frac{\overline{x}\overline{y}}{\overline{x}+b}=\frac{(x+a)(\frac{xy}{x+a})}{x+a+b}=\frac{xy}{x+(a+b)},$$

i. e., $T_bT_a = T_{\varphi(a,b)} = T_{(a+b)}$. Thus, the given transformation defines the group. □

More generally, let us consider the transformation of points $z = (z^1,\dots,z^N) \in \mathbb{R}^N$ into points $\overline{z} = (\overline{z}^1,\dots,\overline{z}^N) \in \mathbb{R}^N$. Here a is a real parameter from a neighborhood of $a = 0$, and we impose the condition that transformation (1.19) is an identity if and only if $a = 0$, i. e.,

$$f^i(z,0) = z^i,\quad i=1,\dots,N. \tag{1.23}$$

The set G of transformations (1.19) satisfying condition (1.23) is called a (local) one-parameter group of transformations in $\mathbb{R}^N$ if the successive action of two transformations is identical to the action of a third transformation from G, i. e., if the function $f = (f^1,\dots,f^N)$ satisfies the following group property:

$$f^i(f(z,a),b) = f^i(z,c), \quad i = 1,\dots,N, \tag{1.24}$$

where

$$c = \varphi(a,b) \tag{1.25}$$

with a smooth function $\varphi(a,b)$ defined for sufficiently small a and b. The group parameter a in the transformation (1.19) can be changed so that the function (1.25) becomes $c = a + b$. In other words, the group property (1.24) can be written upon choosing an appropriate parameter a (called a *canonical parameter*) in the form

$$f^i(f(z,a),b) = f^i(z,a+b). \tag{1.26}$$

1.3 Lie equations

Let the transformation (1.19), $\overline{z} = f(z,a)$, generate a group and let the property (1.22) $T_aT_b = T_{\varphi(a,b)}$ have the form $T_aT_b = T_{a+b}$, i. e.,

$$f(f(z,a),b) = f(z,a+b). \tag{1.27}$$

Let us denote this group by G and represent $f(z,a)$ by the Taylor series about $a = 0$. We denote

$$\xi(z) = \left.\frac{\partial f(z,a)}{\partial a}\right|_{a=0}. \tag{1.28}$$

Then expanding f into Taylor series near $a = 0$ and keeping only the linear terms in a, one obtains the infinitesimal transformation of the group G:

$$\overline{z} = f(z,0) + \xi(z)a + o(a), \tag{1.29}$$

or

$$\overline{z} \approx z + a\xi(z), \tag{1.30}$$

so that, from a geometric point of view, the group G is determined by its tangent vector field ξ.

1.3.1 Lie theorem

Let G be a local group with function f satisfying the group property (1.27) and let the expansion (1.29) be valid. Then the function $f(z,a)$ is a solution of the first-order initial value problem (known as the *Lie equations*)

$$\frac{df}{da} = \xi(f) \tag{1.31}$$

with the initial condition

$$f|_{a=0} = z, \tag{1.32}$$

or in compact form

$$\frac{d\bar{z}}{da} = \xi(\bar{z}), \quad \bar{z}|_{a=0} = z. \tag{1.33}$$

In other words, the G-orbit of a point $z \in \mathbf{R}^n$ is an integral curve of the Lie equation (1.33) passing through the point z.

Proof. Let the group property (1.27) hold and let Δa be the increment of a. Rewrite the property (1.27) as

$$f(z, a + \Delta a) = f[f(z, a), \Delta a]. \tag{1.34}$$

At the leading order of the left side in (1.34) we have

$$f(z, a + \Delta a) = f(z, a) + \frac{\partial f}{\partial a}\Delta a + o(\Delta a). \tag{1.35}$$

Similarly, at the leading order of the right side in (1.34) we have

$$f[f(z, a), \Delta a] = f(z, a) + \left.\frac{\partial f}{\partial(\Delta a)}\right|_{\Delta a=0} \Delta a + o(\Delta a). \tag{1.36}$$

Now, by the definition (1.28) we have

$$\left.\frac{\partial f[f(z, a), \Delta a]}{\partial(\Delta a)}\right|_{\Delta a=0} = \xi(f(z, a)),$$

so if we subtract the terms $o(\Delta a)$ from (1.35) and (1.36) and divide both parts by Δa, we obtain

$$\frac{\partial f(z, a)}{\partial a} = \xi[f(z, a)].$$ □

1.3.2 Inverse Lie theorem

For any smooth vector field $\xi(z)$, the solution of the Cauchy problem (1.33),

$$\frac{d\bar{f}}{da} = \xi(fz), \quad f|_{a=0} = z,$$

satisfies the property (1.27),

$$f(f(z,a),b) = f(z,a+b).$$

Proof. Let $f(z,a)$ be a solution of (1.33). Let us fix a (sufficiently close to $a = 0$) and consider two functions

$$u(b) = f(\overline{z},b) = f[f(z,a),b],$$
$$v(b) = f(z,a+b).$$

From the Lie equations we have

$$\frac{du}{db} = \frac{df(\overline{z},b)}{db} = \xi(u), \quad u|_{b=0} = f(z,a),$$
$$\frac{dv}{db} = \frac{df(z,a+b)}{db} = \xi(v), \quad v|_{b=0} = f(z,a),$$

which means that u and v satisfy the same equations with the same initial data. The uniqueness of the Cauchy problem implies that $u = v$, so it implies that the property (1.27) holds. □

In other words, the solution of Lie's equation provides a one-parameter local group with a given infinitesimal transformation (1.29), $\overline{z}^i \approx z^i + a\xi^i(z)$.

We can now define infinitesimal transformations via

$$\overline{x} \approx x + \xi(x,y)a, \tag{1.37}$$

$$\overline{y} \approx y + \eta(x,y)a. \tag{1.38}$$

Example 1.6. Let us recall Example 1.1 for the quadratic equation and find the tangent vector for the transformation $\overline{x} = x + \varepsilon$. By the definition (1.28), we have

$$\xi(x) = \frac{\partial f(x,\varepsilon)}{\partial \varepsilon}\bigg|_{\varepsilon=0} = 1,$$

i. e.,

$$\frac{df}{d\varepsilon} = \frac{d\overline{x}}{d\varepsilon} = 1,$$

so that $\overline{x} = \varepsilon + C$. The initial condition $\overline{x}|_{\varepsilon=0} = 1$ implies that $C = x$. So, $\overline{x} = x + \varepsilon$. To check the answer, we have to verify that if $\xi = 1$, then $\overline{x} = x + \varepsilon$. For this particular example, the Lie equations are

$$\frac{df}{d\varepsilon} = \xi(f), \quad f|_{\varepsilon=0} = x$$

and

$$\frac{d\overline{x}}{d\varepsilon} = 1, \quad \overline{x}|_{\varepsilon=0} = x, \tag{1.39}$$

and the solution of the initial value problem (1.39) is $\overline{x} = x + \varepsilon$. □

Example 1.7. Find the transformation corresponding to the given vector field $\xi(x) = x$.

Solution. We start with the Lie equations (1.33),

$$\frac{d\overline{x}}{da} = \overline{x}, \quad \overline{x}|_{a=0} = x.$$

The first equation is a separable ODE, i. e.,

$$\frac{d\overline{x}}{\overline{x}} = da,$$

and $\ln \overline{x} = a + \ln C$, so the general solution is $\overline{x} = Ce^a$. The initial condition implies $\overline{x}|_{a=0} = C$, so $C = x$. Thus, the transformation is $\overline{x} = xe^a$. □

Example 1.8. Find the one-parameter group of transformations given the vector field $\xi(x, y) = (x, 2y)$.

Solution. We start with the Lie equations (1.33),

$$\frac{d\overline{x}}{da} = \overline{x}, \quad \overline{x}|_{a=0} = x. \tag{1.40}$$

$$\frac{d\overline{y}}{da} = 2\overline{y}, \quad \overline{y}|_{a=0} = y. \tag{1.41}$$

From (1.40), we have $\overline{x} = xe^a$. Similarly, from (1.41), we have $\overline{y} = ye^{2a}$. □

Example 1.9. Find the one-parameter group by its infinitesimal transformation

$$\overline{x} \approx x + ax^2, \quad \overline{y} \approx y + axy.$$

Solution. Here we consider the transformations of the form $\overline{x} = \varphi(x, y, a)$ and $\overline{y} = \psi(x, y, a)$. The Lie equations (1.33) are

$$\frac{d\overline{x}}{da} = \xi(\overline{x}, \overline{y}), \quad \overline{x}|_{a=0} = x, \tag{1.42}$$

$$\frac{d\overline{y}}{da} = \eta(\overline{x}, \overline{y}), \quad \overline{y}|_{a=0} = y. \tag{1.43}$$

From the definition of the tangent vector field (1.28), we know that

$$\xi(x, y) = \frac{\partial \varphi(x, y, a)}{\partial a}\bigg|_{a=0} = x^2.$$

So, $\xi(\overline{x}, \overline{y}) = \overline{x}^2$. Similarly,

$$\eta(x,y) = \left.\frac{\partial\psi(x,y,a)}{\partial a}\right|_{a=0} = xy.$$

So, $\eta(\overline{x},\overline{y}) = \overline{xy}$. Thus, the Lie equations (1.42)–(1.43) are written as

$$\frac{d\overline{x}}{da} = \overline{x}^2, \quad \overline{x}|_{a=0} = x, \tag{1.44}$$

$$\frac{d\overline{y}}{da} = \overline{xy}, \quad \overline{y}|_{a=0} = y. \tag{1.45}$$

Solving (1.44), we have

$$\frac{d\overline{x}}{\overline{x}^2} = da \quad \Rightarrow \quad -\frac{1}{\overline{x}} = a + C_1 \quad \Rightarrow \quad \overline{x} = -\frac{1}{a + C_1},$$

$$\overline{x}|_{a=0} = x = -\frac{1}{C_1} \quad \Rightarrow \quad \overline{x} = -\frac{1}{a - \frac{1}{x}} = \frac{x}{1 - ax}.$$

Solving (1.45), we have

$$\frac{d\overline{y}}{\overline{y}} = \overline{x}da \quad \Rightarrow \quad \frac{d\overline{y}}{\overline{y}} = \frac{x}{1 - ax}da \quad \Rightarrow \quad \ln\overline{y} = \ln\frac{C_2}{1 - ax},$$

or

$$\overline{y} = \frac{C_2}{1 - ax},$$

$$\overline{y}|_{a=0} = y = C_2 \quad \Rightarrow \quad \overline{y} = -\frac{y}{1 - ax}.$$

So,

$$\overline{x} = \frac{x}{1 - ax}, \quad \overline{y} = -\frac{y}{1 - ax}.$$

To conclude, the solution of Lie's equation provides a one-parameter local group with a given infinitesimal transformation (1.30). □

1.4 Canonical variables

Let us start with the Lie equation (1.33)

$$\frac{df}{da} = \xi(f), \quad f|_{a=0} = z$$

corresponding to the transformation $\overline{z} = f(z, a)$, where f satisfies the property (1.27),

$$f(f(z, a), b) = f(z, a + b).$$

Now let us pose the following interesting question: Can we reduce the multiplication law (1.22)

$$T_b T_a = T_{\varphi(a,b)}$$

to the simpler form (1.27)?

The answer to this question can be formulated as follows.

Theorem 1.1. *Every one-parameter group of transformations $\overline{x} = \varphi(x, y, a)$, $\overline{y} = \psi(x, y, a)$ can be reduced, by a suitable change of variables $t = t(x, y)$, $u = u(x, y)$, to the group of translations $\overline{t} = t + a$, $\overline{u} = u$.*

In other words, the group property (1.22) can be written, upon choosing an appropriate parameter $\overline{a}$ (called a *canonical parameter*), in the form

$$f(f(z, a), b) = f(z, a + b).$$

The variables $\overline{t}$ and $\overline{u}$ are called *canonical variables.*

In fact, it can be shown that the canonical parameter can be found by the formula

$$\overline{a} = \int_0^a A(a)da, \tag{1.46}$$

where $A(a)$ is given by

$$\left.\frac{\partial\varphi(a,b)}{\partial b}\right|_{b=0} = \frac{1}{A(a)}. \tag{1.47}$$

Let us illustrate this claim by some examples.

Example 1.10. Find the canonical variable for the transformation $\overline{z} = z + az$.

Solution. We know that $\varphi(a, b) = a + b + ab$. Now, since

$$\overline{\overline{z}} = \overline{z} + b\overline{z} = z + az + b(z + az) = z + \varphi(a, b)z,$$

formula (1.47) yields

$$\left.\frac{\partial\varphi(a,b)}{\partial b}\right|_{b=0} = 1 + a = \frac{1}{A(a)}.$$

Then, from formula (1.46), we have

$$\overline{a} = \int_0^a A(a)da = \int_0^a \frac{1}{1+a}da = \ln(1 + a). \tag{1.48}$$

□

Example 1.11. Can you check that formula (3) provides us with the canonical variable for the transformation $\overline{z} = z + az$?

Solution. To answer this question, let us express the parameter a via $\overline{a}$ as

$$a = e^{\overline{a}} - 1$$

and substitute this into the transformation. We have

$$\overline{z} = z + (e^{\overline{a}} - 1)z = e^{\overline{a}}z.$$

Thus,

$$\overline{\overline{z}} = e^{\overline{b}}\overline{z} = e^{\overline{b}}e^{\overline{a}}z = e^{(\overline{a}+\overline{b})}z,$$

so $\varphi(\overline{a}, \overline{b}) = \overline{a} + \overline{b}$. □

Example 1.12. Find the canonical variable for the transformation

$$\overline{x} = \frac{x + a}{1 - ax}.$$

Solution. First, let us find $\varphi(a, b)$ corresponding to this transformation. We have

$$\overline{\overline{x}} = \frac{\overline{x} + b}{1 - b\overline{x}} = \frac{\frac{x+a}{1-ax} + b}{1 - b\frac{x+a}{1-ax}} = \frac{x + \frac{a+b}{1-ab}}{1 - x\frac{a+b}{1-ab}} = \frac{x + c}{1 - cx}.$$

So,

$$\varphi(a, b) = \frac{a + b}{1 - ab},$$

and by formula (1.47) we have

$$\left.\frac{\partial\varphi(a, b)}{\partial b}\right|_{b=0} = 1 + a^2 = \frac{1}{A(a)}.$$

Then, from formula (1.46), we have

$$\overline{a} = \int_0^a A(a)da = \int_0^a \frac{1}{1 + a^2}da. \tag{1.49}$$

□

1.5 Invariants and invariant equations

A function $F(z)$ is said to be an invariant of the group of transformations G

$$\overline{z} = f(z, a) = z + \xi(z)a + o(a) \tag{1.50}$$

or

$$\overline{z} \approx z + a\xi(z) \tag{1.51}$$

if for each point $z = (z^1, \ldots, z^N) \in \mathbb{R}^N$ it is constant along the trajectory determined by the totality of transformed points $\overline{z}$: $F(\overline{z}) = F(z)$, i. e., if

$$F[f(z,a)] = F(z). \tag{1.52}$$

Theorem 1.2 (Invariance criterion). *A function $F(z)$ is an invariant of the group G iff it solves the first-order partial differential equation (PDE)*

$$\xi^i(z)\frac{\partial F(z)}{\partial z^i} = 0, \tag{1.53}$$

in which the Einstein summation is assumed.

For example, if $z = (x, y)$ and the tangent vector $\boldsymbol{\xi} = (\xi, \eta)$, then equation (1.53) is written as

$$\xi^i(z)\frac{\partial F(z)}{\partial z^i} = \xi(x,y)\frac{\partial F(x,y)}{\partial x} + \eta(x,y)\frac{\partial F(x,y)}{\partial y} = 0. \tag{1.54}$$

Proof. (Necessity) If $F(z)$ satisfies the invariance criterion (1.52), then the condition (1.53) follows from the expansion

$$F[f(z,a)] = F(z + a\xi(z) + o(a)) = F(z) + a\xi^i(z)\frac{\partial F(z)}{\partial z^i} + o(a) = F(z). \tag{1.55}$$

(Sufficiency) Let $F(z)$ be a solution of (1.53). Since (1.53) is valid for any z, it is also true for the points $\overline{z} = f(z,a)$, at which

$$\xi^i(\overline{z})\frac{\partial F(\overline{z})}{\partial \overline{z}^i} = 0. \tag{1.56}$$

Let us use equation (1.56) and the Lie equation

$$\frac{df^i(z,a)}{da} = \xi^i(\overline{z})\frac{\partial F(\overline{z})}{\partial \overline{z}^i}$$

to get

$$\frac{dF[f(z,a)]}{da} = \frac{\partial F(\overline{z})}{\partial \overline{z}^i}\frac{df^i(z,a)}{da} = \xi^i(\overline{z})\frac{\partial F(\overline{z})}{\partial \overline{z}^i} = 0. \tag{1.57}$$

Equation (1.57) shows that the function $F[f(z,a)]$ satisfies the initial value problem

$$\frac{dF(\overline{z})}{da} = 0, \quad F|_{a=0} = F(z),$$

which means that $F(\overline{z}) = F(z)$. □

We next introduce the first-order linear differential operator

$$X = \xi^i(z)\frac{\partial}{\partial z^i}, \tag{1.58}$$

which is known as the generator of the group G, also known as symmetry of the group G.

Then the invariance criterion (1.53) is written as

$$XF = 0. \tag{1.59}$$

More generally, we say that a function $J(z)$ is an invariant of the group G with generator (1.58) if and only if

$$X(J) \equiv \xi^i(z)\frac{\partial J}{\partial z^i} = 0. \tag{1.60}$$

In fact, any one-parameter group has exactly $N - 1$ functionally independent invariants (basis of invariants). One can take them to be the left-hand sides of $N - 1$ first integrals $J_1(z) = C_1, \dots, J_{N-1}(z) = C_{N-1}$ of the characteristic equations for the linear PDE (1.60). Then any other invariant is a function of $J_1(z), \dots, J_{N-1}(z)$.

1.5.1 Invariant equations

Let us consider an $(N - s)$-dimensional surface $M \in \mathbb{R}^N$ given by the system of equations

$$F_1(z) = 0, \dots, F_s(z) = 0, \quad s \leqslant N. \tag{1.61}$$

We say that the surface M is invariant under the group G of transformations (1.50) if each point z of the surface M moves along the surface under these transformations. In other words, if z is a solution of the system (1.61), then $\overline{z}$ is a solution of the system as well, i. e.,

$$F_k(\overline{z}) = 0, \quad k = 1, \dots, s. \tag{1.62}$$

In this case we say that the system of equations (1.61) is invariant under the group G or the system (1.61) *admits* the group G. Let X given by (1.58) be the generator of the group.

Theorem 1.3 (Invariant equations). *The system of equations (1.61) is invariant under the group G iff*

$$XF|_M = 0, \quad k = 1, \dots, s. \tag{1.63}$$

As an illustration of Theorem 1.3, let us show that the paraboloid

$$z^1 - \sum_{i=2}^{N} (z^i)^2 = 0 \tag{1.64}$$

is invariant under the nonuniform dilation group

$$\overline{z}^1 = z^1 e^{2a}, \quad \overline{z}^2 = z^2 e^a, \quad \dots, \quad \overline{z}^N = z^N e^{2a}. \tag{1.65}$$

First, we observe that the tangent vector is given by

$$\xi^1 = 2z^1, \quad \xi^i = z^i, \quad i = 2, \dots, N,$$

so the generator of the group is the differential operator

$$X = 2z^1 \frac{\partial}{\partial z^1} + z^2 \frac{\partial}{\partial z^2} + \dots + z^N \frac{\partial}{\partial z^N}. \tag{1.66}$$

The invariance criterion (1.63) can be written as

$$XF|_M = \left(2z^1 \frac{\partial}{\partial z^1} - z^2 \frac{\partial}{\partial z^2} - \dots - z^N \frac{\partial}{\partial z^N}\right)\bigg|_M = 2\left(z^1 - \sum_{i=2}^{N} (z^i)\right)\bigg|_{z^1 - \sum_{i=2}^{N}(z^i)^2 = 0} = 0$$

due to (1.64)

Remark. We note that this problem can be solved differently. We know that the surface (1.64) is invariant under the group (1.65) if each point z of the surface (1.64) moves along the surface under the transformations (1.65). In other words, if z is a solution of the system $F(z) = 0$, then $\overline{z}$ is a solution of the system $F(\overline{z}) = 0$ as well. Let us check if this is true. We have

$$\begin{aligned} \overline{z} &= f(z, a) = z + \xi(z)a + o(a), \\ \xi^1 &= 2z^1, \quad \xi^i = z^i, \quad i = 2, \dots, N, \end{aligned} \tag{1.67}$$

so we can write (1.67) as

$$\begin{aligned} \overline{z}^1 &= z^1 + 2z^1 a = z^1(1 + 2a) = z^1 e^{2a}, \\ \overline{z}^2 &= z^2 + z^2 a = z^2(1 + a) = z^2 e^a, \\ &\dots \\ \overline{z}^N &= z^N + z^N a = z^N(1 + a) = z^N e^a. \end{aligned}$$

Now, we have

$$F(z) = z^1 - \sum_{i=2}^{N} (z^i)^2 = 0,$$

$$F(\overline{z}) = e^{2a} z^1 - (z^2 e^a)^2 - \cdots - (z^N e^a)^2 = \left(z^1 - \sum_{i=2}^{N} (z^i)^2 \right) = 0$$

due to (1.64).

Very often, when we are dealing with differential equations admitting group of transformations, the following theorem can be useful.

Theorem 1.4. *A surface M invariant under a group of transformations G can be defined by a system of equations of the form*

$$\Phi_k [J_1(z), \ldots, J_{N-1}(z)] = 0, \quad k = 1, \ldots, s, \tag{1.68}$$

where the functions $J_1(z), \ldots, J_{N-1}(z)$ form a basis of invariants of the group G if the generator X of the group is not zero on the surface M.

As an illustration of Theorem 1.4, let us write equation (1.64) of the paraboloid

$$z^1 - \sum_{i=2}^{N} (z^i)^2 = 0$$

in an invariant form.

Since the surface is given by a single equation, we have $s = 1$. We also know that the surface is invariant under the group with the generator X given by (1.66). We can choose

$$J_1(z) = \frac{(z^2)^2}{z^1}, \quad J_2(z) = \frac{(z^3)^2}{z^1}, \quad \ldots, \quad J_{N-1}(z) = \frac{(z^N)^2}{z^1} \tag{1.69}$$

as a basis of invariants. Let us use the change of variables

$$\overline{z}^1 = z^1 - \sum_{i=2}^{N} (z^i)^2, \quad y^1 = z^2, \quad \ldots, \quad y^{N-1} = z^N. \tag{1.70}$$

Now, since $\overline{z}^1 = 0$ due to (1.64), we can denote

$$r^2 = z^1 = \overline{z}^1 + \sum_{i=2}^{N} (z^i)^2 = \sum_{l=1}^{N-1} (y^l)^2.$$

Then we can rewrite (1.69) as

$$\tilde{J}_1(y) = \frac{(y^1)^2}{r^2}, \quad \tilde{J}_2(y) = \frac{(y^2)^2}{r^2}, \quad \ldots, \quad \tilde{J}_{N-1}(y) = \frac{(y^{N-1})^2}{r^2}.$$

In order to find a functional relation of the form (1.68), we note that $(y^1)^2 + (y^2)^2 + \cdots. + (y^{N-1})^2 = (z^1)^2 + \cdots + (z^N)^2 = z^1$, in which

$$z^1 = \sum_{i=2}^{N} (z^i)^2 = r^2.$$

Hence,

$$\tilde{J}_1(y) + \cdots + \tilde{J}_{N-1}(y) = 1.$$

Thus, the equation $\Phi_k[J_1(z), \ldots, J_{N-1}(z)] = 0$ can be written as

$$J_1(z) + \cdots + J_{N-1}(z) = \frac{(z^1)^2 + \cdots + (z^N)^2}{r^2} = \frac{r^2}{r^2} = 1, \tag{1.71}$$

or, using (1.69), we can rewrite (1.71) as

$$J_1(z) + \cdots + J_{N-1}(z) = \frac{1}{z^1} \sum_{i=1}^{N} (z^i)^2 = 1,$$

i. e., for the given invariant surface we have the functional dependence of the form (1.68).

Question. Can you show that the y-axis is invariant under the group $G : \overline{x} = xe^a, \overline{y} = y$ but cannot be represented in the invariant form (1.68)? Would it contradict Theorem 1.4?

This question might be interesting to think about since the y-axis can be given by the equation $x = 0$. So $F(x) = x = 0$. The invariance condition (1.63) can be written as

$$XF|_M = x\frac{\partial}{\partial x}x\bigg|_{x=0} = x|_{x=0} = 0,$$

so the y-axis is invariant under the given group. Since

$$X = x\frac{\partial}{\partial x},$$

$J = y$. However, since $x = 0$, y cannot be expressed in terms of x.

Example 1.13. Find the invariant of the group G:

$$\overline{x} = xe^a, \quad \overline{y} = ye^{2a}, \quad \overline{z} = ze^{-2a}.$$

Solution. First, let us find the tangent vector field:

$$\begin{aligned}
\xi^1(x,y) &= \frac{\partial \overline{x}}{\partial a}\bigg|_{a=0} = xe^a\big|_{a=0} = x, \\
\xi^2(x,y) &= \frac{\partial \overline{y}}{\partial a}\bigg|_{a=0} = 2ye^{2a}\big|_{a=0} = 2y, \\
\xi^3(z) &= \frac{\partial \overline{z}}{\partial a}\bigg|_{a=0} = -2ze^{-2a}\big|_{a=0} = -2z.
\end{aligned}$$

The invariance criterion (1.59) (see also (1.60)) is $XJ = 0$, where

$$X = x\frac{\partial}{\partial x} + 2y\frac{\partial}{\partial y} - 2z\frac{\partial}{\partial z}.$$

Thus, we must have

$$x\frac{\partial J}{\partial x} + 2y\frac{\partial J}{\partial y} - 2z\frac{\partial J}{\partial z} = 0.$$

The corresponding characteristic equations are

$$\frac{dx}{x} = \frac{dy}{2y} = -\frac{dz}{2z}. \tag{1.72}$$

We can consider two pairs of these equations:

$$\frac{dx}{x} = \frac{dy}{2y} \tag{1.73}$$

and

$$\frac{dx}{x} = -\frac{dz}{2z}. \tag{1.74}$$

From (1.73) we have

$$\ln x^2 - \ln y = \ln C_1 \quad \Rightarrow \quad \frac{x^2}{y} = C_1. \tag{1.75}$$

From (1.74) we have

$$\ln x^2 + \ln z = \ln C_2 \quad \Rightarrow \quad x^2 z = C_2. \tag{1.76}$$

Expressions in (1.75) and (1.76) represent the first integrals of the system (1.72) and thus the basis of invariants is

$$J_1 = \frac{x^2}{y}, \quad J_2 = x^2 z.$$

Therefore, the invariant is

$$J = J\left(\frac{x^2}{y}, x^2 z\right).$$

For example, we can check that

$$J = J(\overline{x}, \overline{y}) = \frac{\overline{x}^2}{\overline{y}} = \frac{x^2 e^{2a}}{y e^{2a}} = \frac{x^2}{y}.$$ □

Example 1.14. Let us find the transformation group admitted by the operator

$$X = y\frac{\partial}{\partial x} - x\frac{\partial}{\partial y}. \tag{1.77}$$

Solution. First, let us find the tangent vector field $\xi = (\xi^1, \xi^2)$. We solve the Lie equations (1.33):

$$\xi^1 = \frac{d\overline{x}}{da} = \overline{y}, \quad \overline{x}|_{a=0} = x, \tag{1.78}$$

$$\xi^2 = \frac{d\overline{y}}{da} = -\overline{x}, \quad \overline{y}|_{a=0} = y. \tag{1.79}$$

First, we substitute $\overline{y}$ from (1.78) into (1.79) to obtain a second-order ODE:

$$\frac{d^2\overline{x}}{da^2} = -\overline{x},$$

so that

$$\overline{x} = C_1 \cos a + C_2 \sin a,$$

so

$$\overline{y} = \frac{d\overline{x}}{da} = -C_1 \sin a + C_2 \cos a.$$

From the initial conditions we have

$$\overline{x}|_{a=0} = x = C_1, \quad \overline{y}|_{a=0} = C_2.$$

Thus,

$$\overline{x} = x \cos a + y \sin a, \tag{1.80}$$

$$\overline{y} = -x \sin a + y \cos a. \tag{1.81}$$

The transformation (1.80)–(1.81) represents the group of rotations. □

Example 1.15. Let us find the transformation group and invariant admitted by the operator

$$X = y\frac{\partial}{\partial x} + x\frac{\partial}{\partial y}. \tag{1.82}$$

Solution. Comparison with formula (1.58) shows that the tangent vector field of the transformation group in question is given by $\xi = (y, x)$, i. e.,

$$X = \xi^1 \frac{\partial}{\partial x} + \xi^2 \frac{\partial}{\partial y} = y \frac{\partial}{\partial x} + x \frac{\partial}{\partial y}. \tag{1.83}$$

The Lie equations (1.33) are written as

$$\xi^1 = \frac{d\overline{x}}{da} = \overline{y}, \quad \overline{x}|_{a=0} = x, \tag{1.84}$$

$$\xi^2 = \frac{d\overline{y}}{da} = \overline{x}, \quad \overline{y}|_{a=0} = y. \tag{1.85}$$

Solving this system, we obtain

$$\overline{x} = c_1 e^a + c_2 e^{-a},$$

where the arbitrary constants c_1 and c_2 are determined from the initial condition in (1.84), which gives $c_1 = x - c_2$. Correspondingly, $\overline{y}$ is determined as

$$\overline{y} = \frac{d\overline{x}}{da} = c_1 e^a - c_2 e^{-a} \tag{1.86}$$

and the initial condition in (1.85) yields

$$c_2 = \frac{x - y}{2}. \tag{1.87}$$

Thus, the transformation group is given by

$$\overline{x} = x\left(\frac{e^a + e^{-a}}{2}\right) + y\left(\frac{e^a - e^{-a}}{2}\right), \tag{1.88}$$

$$\overline{y} = x\left(\frac{e^a - e^{-a}}{2}\right) + y\left(\frac{e^a + e^{-a}}{2}\right),$$

or in compact form,

$$\overline{x} = x \cosh a + y \sinh a, \tag{1.89}$$

$$\overline{y} = x \sinh a + y \cosh a, \tag{1.90}$$

which is known as a Lorentz transformation.

In order to find the invariant of the Lorentz transformation, we can write the corresponding characteristic equation in the form

$$\frac{dx}{y} = \frac{dy}{x}, \tag{1.91}$$

which leads to the separable ODE

$$\frac{dx}{dy} = \frac{y}{x}, \tag{1.92}$$

which can be integrated to give the solution in the form

$$\frac{x^2}{2} - \frac{y^2}{2} = \frac{c}{2}, \tag{1.93}$$

where c is an arbitrary constant, which can serve as an invariant, i. e., we can write the invariant of the Lorentz transformation (1.89)–(1.90) as $J = x^2 - y^2$. □

Example 1.16. Let us find the transformation group and invariant admitted by the operator

$$X = \frac{\partial}{\partial t} + t\frac{\partial}{\partial x} + x\frac{\partial}{\partial u}. \tag{1.94}$$

Solution. First, let us find the tangent vector field $\xi = (\xi^1, \xi^2, \xi^3)$. We solve the Lie equations (1.33):

$$\xi^1 = \frac{d\bar{t}}{da} = 1, \quad \bar{t}|_{a=0} = t, \tag{1.95}$$

$$\xi^2 = \frac{d\bar{x}}{da} = \bar{t}, \quad \bar{x}|_{a=0} = x, \tag{1.96}$$

$$\xi^3 = \frac{d\bar{u}}{da} = \bar{x}, \quad \bar{u}|_{a=0} = u. \tag{1.97}$$

Equation (1.95) implies that

$$\bar{t} = a + t. \tag{1.98}$$

Substituting the expression for $\bar{t}$ in (1.98) into (1.96), we obtain

$$\bar{x} = \frac{a^2}{2} + ta + x. \tag{1.99}$$

Finally, substituting the expression for $\bar{x}$ in (1.99) into (1.97), we obtain

$$\frac{d\bar{u}}{da} = \bar{x} = \frac{a^2}{2} + ta + x,$$

i. e.,

$$\bar{u} = \frac{a^3}{6} + \frac{a^2}{2}t + ax.$$

Let us check the answer by direct substitution:

$$\xi^1 = 1, \tag{1.100}$$

$$\xi^2 = \frac{\partial \overline{x}}{\partial a}\bigg|_{a=0} = (a + t)|_{a=0} = t, \tag{1.101}$$

$$\xi^3 = \frac{\partial \overline{u}}{\partial a}\bigg|_{a=0} = \left(\frac{a^2}{2} + ta + x\right)\bigg|_{a=0} = x. \tag{1.102}$$

□

Example 1.17. Let us find the two-dimensional transformation group and invariant admitted by the operator

$$X = \frac{\partial}{\partial x}. \tag{1.103}$$

Solution. First, let us find the tangent vector field $\xi = (\xi^1, \xi^2)$. We solve the Lie equations (1.33):

$$\xi^1 = \frac{d\overline{x}}{da} = 1, \quad \overline{x}|_{a=0} = x, \tag{1.104}$$

$$\xi^2 = \frac{d\overline{y}}{da} = 0, \quad \overline{y}|_{a=0} = y. \tag{1.105}$$

From (1.104) we have

$$\overline{x} = a + x, \tag{1.106}$$

and from (1.105) we have

$$\overline{y} = y. \tag{1.107}$$

The transformation (1.106)–(1.107) represents translation on the x-axis. □

Example 1.18. Let us find the transformation group admitted by the operator

$$X = l\frac{\partial}{\partial x} - k\frac{\partial}{\partial y}, \tag{1.108}$$

where k and l are constants.

Solution. First, let us find the tangent vector field $\xi = (\xi^1, \xi^2)$. We solve the Lie equations (1.33):

$$\xi^1 = \frac{d\overline{x}}{da} = l, \quad \overline{x}|_{a=0} = x, \tag{1.109}$$

$$\xi^2 = \frac{d\overline{y}}{da} = k, \quad \overline{y}|_{a=0} = y. \tag{1.110}$$

From (1.109) we have

$$\overline{x} = x + al, \tag{1.111}$$

and from (1.110) we have

$$\bar{y} = y + ak. \tag{1.112}$$

The transformation (1.111)–(1.112) represents translation along the line $kx + ly = 0$. □

Example 1.19. Let us find the invariant J for the transformation group

$$\bar{x} = xe^{a}, \quad \bar{y} = ye^{2a}. \tag{1.113}$$

Solution. First, let us find the generator of the group in the form of the differential operator (1.58). We start with finding the tangent vector $\xi = (\xi^1, \xi^2)$:

$$\xi^1 = \left.\frac{\partial \bar{x}}{\partial a}\right|_{a=0} = \left. xe^{a}\right|_{a=0} = x, \tag{1.114}$$

$$\xi^2 = \left.\frac{\partial \bar{y}}{\partial a}\right|_{a=0} = \left. 2ye^{2a}\right|_{a=0} = 2y. \tag{1.115}$$

Hence, the generator of the group is

$$X = x\frac{\partial}{\partial x} + 2y\frac{\partial}{\partial y}, \tag{1.116}$$

and thus we need to find such $J(z) = J[f(z,a)]$ that equation (1.60) is satisfied, i. e.,

$$XJ = x\frac{\partial J}{\partial x} + 2y\frac{\partial J}{\partial y} = 0.$$

The corresponding characteristic equations are

$$2\frac{dx}{x} = \frac{dy}{y},$$

so

$$\ln x^2 = \ln y = \ln C,$$

where C is a constant, which represents the first integral. So,

$$J = \frac{x^2}{y}.$$ □

Table 1.1 represents a summary of basic transformations.

Table 1.1: Tabulated basic transformations.

Basic transformations			
Transformation	**Group**	**Generator**	**Invariant**
Translation along the x-axis	$\bar{x} = x + a$ $\bar{y} = y$	$X = \frac{\partial}{\partial x}$	$J = y$
Translation along the y-axis	$\bar{x} = x$ $\bar{y} = y + a$	$X = \frac{\partial}{\partial y}$	$J = x$
Translation along the line $kx + ly = 0$	$\bar{x} = x + la$ $\bar{y} = y - ka$	$X = l\frac{\partial}{\partial x} - k\frac{\partial}{\partial y}$	$J = kx + ly$
Rotation	$\bar{x} = x\cos a + y\sin a$ $\bar{y} = y\cos a - x\sin a$	$X = y\frac{\partial}{\partial x} - x\frac{\partial}{\partial y}$	$J = x^2 + y^2$
Lorenz transformation	$\bar{x} = x\cosh a + y\sinh a$ $\bar{y} = y\cosh a + x\sinh a$	$X = y\frac{\partial}{\partial x} + x\frac{\partial}{\partial y}$	$J = x^2 - y^2$
Galilean transformation	$\bar{x} = x + ay,$ $\bar{y} = y$	$X = y\frac{\partial}{\partial x}$	$J = y$
Dilation	$\bar{x} = xe^a$ $\bar{y} = ye^a$	$X = x\frac{\partial}{\partial x} + y\frac{\partial}{\partial y}$	$J = \frac{x}{y}$
Projective transformation	$\bar{x} = \frac{x}{1-ax}$ $\bar{y} = \frac{y}{1-ax}$	$X = x^2\frac{\partial}{\partial x} + xy\frac{\partial}{\partial y}$	$J = \frac{x}{y}$

1.5.2 Universal invariant

The map $J : \mathbb{R}^N \to P$ is called a *universal invariant* of the group G if for any invariant $F : \mathbb{R}^N \to Z$ there exists a map $\Phi : P \to Z$:

$$F = \Phi(J).$$

As an illustration of this definition, let us find the universal invariant of the generator X given by (1.94).

First, let us find the invariant F by solving equation (1.53),

$$\frac{\partial F}{\partial t} + t\frac{\partial F}{\partial x} + x\frac{\partial F}{\partial u} = 0.$$

The corresponding characteristic equations are

$$\frac{dt}{1} = \frac{dx}{t} = \frac{du}{x}.$$

So,

$$\frac{dt}{1} = \frac{dx}{t} \quad \Rightarrow \quad \frac{dx}{dt} = t, \quad x = \frac{t^2}{2} + C \quad \Rightarrow \quad F_1 = x - \frac{t^2}{2},$$

$$\frac{du}{x} = dt \quad \Rightarrow \quad \frac{du}{dt} = x = F_1 + \frac{t^2}{2} \quad \Rightarrow \quad u = F_1 t + \frac{t^3}{6},$$

i. e.,

$$u = xt - \frac{t^3}{2} + \frac{t^3}{6} + C_1,$$

so

$$F_2 = u - xt + \frac{t^3}{3}.$$

Hence,

$$J = \Phi(F_1, F_2).$$

1.6 Transformations in terms of canonical variables

In fact, every one-parameter group of transformations $\overline{z} = f(z, a)$ with the corresponding operator

$$X = \xi^i(z)\frac{\partial}{\partial z^i} \tag{1.117}$$

can be reduced to a group of translations by introducing new variables

$$\overline{z}^i = \overline{z}^i(z). \tag{1.118}$$

Such variables are called canonical variables. How to find such canonical variables?

Suppose we are given the group G generated by the operator (1.117). Then, using the chain rule, we can write

$$\xi^i(z)\frac{\partial}{\partial z^i} = \xi^i(z)\frac{\partial \overline{z}^i}{\partial z^i}\frac{\partial}{\partial \overline{z}^i},$$

i. e., the generator X is written as

$$\overline{X} = X(\overline{z}^i)\frac{\partial}{\partial \overline{z}^i}. \tag{1.119}$$

Now, let us choose any $N-1$ independent invariants $J_1(z), \ldots, J_{N-1}(z)$ as new variables $\overline{z}^1, \ldots, \overline{z}^{N-1}$ and find $\overline{z}^N$ via

$$X(\overline{z}^N) = 1. \tag{1.120}$$

Then the functions

$$\bar{z}^1 = J_1(z), \ldots, \bar{z}^{N-1} = J_{N-1}(z), \bar{z}^N = \bar{z}^N(z)$$

are independent and they determine the change of variables (1.118). Indeed, in these variables, the operator (1.117) has the form

$$\overline{X} = X(\bar{z}^N)\frac{\partial}{\partial \bar{z}^N} = 1 \cdot \frac{\partial}{\partial \bar{z}^N} = \frac{\partial}{\partial \bar{z}^N}.$$

Example 1.20. Reduce the group of transformations

$$\bar{x} = x \cos a + y \sin a, \tag{1.121}$$

$$\bar{y} = y \cos a - x \sin a \tag{1.122}$$

to the group of translations.

Solution. We know from Example 1.14 that the generator of the group is given by

$$X = y\frac{\partial}{\partial x} - x\frac{\partial}{\partial y}. \tag{1.123}$$

The corresponding characteristic equations are

$$\frac{dx}{y} = -\frac{dy}{x} \quad \Leftrightarrow \quad xdx + ydy = 0,$$

and thus the invariant of the group is

$$J = x^2 + y^2.$$

According to (1.120), we choose the first variable as the invariant

$$r = \sqrt{x^2 + y^2}$$

and the second variable φ from the requirement

$$X(\bar{z}^N) = 1,$$

i. e.,

$$y\frac{\partial \varphi}{\partial x} - x\frac{\partial \varphi}{\partial y} = 1. \tag{1.124}$$

The corresponding characteristic equations are

$$\frac{dx}{y} = -\frac{dy}{x} = \frac{d\varphi}{1},$$

where $y = \sqrt{r^2 - x^2}$. Thus,

$$\varphi = \int \frac{dx}{\sqrt{r^2 - x^2}} = \arcsin\left(\frac{x}{r}\right).$$

Thus,

$$\overline{X} = X(\overline{z}^N)\frac{\partial}{\partial \overline{z}^N},$$

where $\overline{z}^N = \varphi$ and $X(\varphi) = 1$, so

$$\overline{X} = \frac{\partial}{\partial \varphi}.$$

Thus, the corresponding Lie equations are

$$\xi^1(r,\varphi) = \frac{d\overline{r}}{da} = 0, \quad \overline{r}|_{a=0} = r, \tag{1.125}$$

$$\xi^2(r,\varphi) = \frac{d\overline{\varphi}}{da} = 1, \quad \overline{\varphi}|_{a=0} = \varphi. \tag{1.126}$$

From (1.125) we have $\overline{r} = r$ and from (1.126) we have $\overline{\varphi} = \varphi + a$. □

1.7 Exponential map

Sometimes it is convenient to represent group transformations $\overline{z} = f(z,a)$ and also the functions $F(\overline{z}) = F[F(f(z,a))]$ in power series.

Let us start with the Lie equation (1.33),

$$\frac{df}{da} = \xi(f), \quad f|_{a=0} = z. \tag{1.127}$$

Let us try to solve (1.127) by power series. As is well known (see also (1.29)), expanding f into Taylor series near $a = 0$ and keeping only the linear terms in a, we can write the infinitesimal transformation of the group G as

$$\overline{z} = f(z,0) + \xi(z)a + o(a). \tag{1.128}$$

Also, let us rewrite (1.55) in the form

$$F(\overline{z}) = F(z) + a\xi^i(z)\frac{\partial F(z)}{\partial z^i} + o(a) \tag{1.129}$$

and introduce the following notation:

$$F(z) = F, \quad F(\overline{z}) = \overline{F}, \quad X = \xi^i(z)\frac{\partial}{\partial z^i}, \quad \overline{X} = \xi^i(\overline{z})\frac{\partial}{\partial \overline{z}^i}. \tag{1.130}$$

In terms of the notation (1.130) we can write (1.129) as follows:

$$\overline{F} = F + aXF + o(a). \tag{1.131}$$

Also, since

$$\frac{dF[f(z,a)]}{da} = \frac{\partial F(\overline{z})}{\partial \overline{z}^i}\frac{df^i(z,a)}{da},$$

the equation (see also (1.57))

$$\frac{dF[f(z,a)]}{da} = \xi^i(\overline{z})\frac{\partial F(\overline{z})}{\partial \overline{z}^i}$$

can be written as

$$\frac{d\overline{F}}{da} = \overline{XF}. \tag{1.132}$$

Thus, we can write the higher derivatives of $\overline{F}$ as

$$\frac{d^2\overline{F}}{da^2} = \frac{d}{da}(\overline{XF}) = \overline{X}(\overline{XF}) = \overline{X}^2\overline{F}, \quad \frac{d3\overline{F}}{da^2} = \overline{X}^3\overline{F}, \dots . \tag{1.133}$$

Substituting (1.133) into the Taylor series (1.129) we obtain

$$\overline{F} = [\overline{F}]_{a=0} + a\left[\frac{d\overline{F}}{da}\right]_{a=0} + \frac{a^2}{2}\left[\frac{d^2\overline{F}}{da^2}\right]_{a=0} + \cdots,$$

i. e.,

$$\overline{F} = F + aXF + \frac{a^2}{2}X^2F + \cdots,$$

or

$$F(\overline{z}) = F(z) + aXF(z) + \frac{(aX)^2}{2}F(z) + \cdots, \tag{1.134}$$

i. e.,

$$F(\overline{z}) = \left(1 + aX + \frac{(aX)^2}{2} + \cdots\right)F(z) = e^{aX}F(z). \tag{1.135}$$

In particular, evaluating the expansion (1.134) at $F = z$, we have

$$\overline{z} = e^{aX}z. \tag{1.136}$$

The expression (1.136) represents the exponential form of the transformation (1.128).

Example 1.21. Write the Taylor expansion for the group given by

$$X = x\frac{\partial}{\partial x}. \tag{1.137}$$

Solution. First, let us find the group generated by the given operator X:

$$\frac{d\overline{x}}{da} = \overline{x}, \quad \overline{x}|_{a=0} = x. \tag{1.138}$$

Solving the Lie equation (1.138) we find $\overline{x} = xa^a$. Next we find

$$X^2 = x\frac{\partial}{\partial x}\left(x\frac{\partial}{\partial x}\right) = x\frac{\partial}{\partial x} + x^2\frac{\partial^2}{\partial x^2},$$
$$X^3 = x\frac{\partial}{\partial x}\left(x\frac{\partial}{\partial x} + x^2\frac{\partial}{\partial x^2}\right) = x\frac{\partial}{\partial x} + 3x^2\frac{\partial^2}{\partial x^2} + x^3\frac{\partial^3}{\partial x^3}.$$

Thus,

$$aXF(x) = ax\frac{\partial}{\partial x}F(x) = axF'(x),$$
$$\frac{a^2}{2}X^2F(x) = \frac{a^2}{2}[xF'(x) + x^2F''(x)],$$
$$\frac{a^3}{3!}X^3F(x) = \frac{a^3}{3!}[xF'(x) + 3x^2F''(x) + x^3F''(x)], \ldots .$$

So,

$$F(xe^a) = F(x) + axF'(x) + \frac{a^2}{2}[xF'(x) + x^2F''(x)] + \cdots,$$

which we can rewrite as

$$F(\overline{x}) = \overline{F} = F + axF + \frac{a^2}{2}X^2F + \cdots = \left(1 + aX + \frac{(aX)^2}{2!} + \frac{(aX)^3}{3!}..\right)F(x). \tag{1.139}$$

□

Example 1.22. Find the series expansion of

$$F\left(\frac{x}{1-ax}\right) \tag{1.140}$$

with respect to a up to the order $o(a^3)$.

Solution. Using the expansion (8) we can write

$$F(\overline{x}) = \left(1 + aX + \frac{(aX)^2}{2!} + \frac{(aX)^3}{3!}..\right)F(x) = a^{aX}F(X).$$

So, we just need to find the generator $X = \xi^i \frac{\partial}{\partial x^i}$ of the projective transformation

$$\overline{x} = \frac{x}{1-ax}.$$

We have

$$\xi = \frac{\partial \overline{x}}{\partial a}\Big|_{a=0} = -\frac{x(-x)}{(1-ax)^2}\Big|_{a=0} = \frac{x^2}{(1-ax)^2}\Big|_{a=0} = x^2,$$

so

$$X = x^2 \frac{\partial}{\partial x}.$$

Thus,

$$X^2 = x^2\left(2x\frac{\partial}{\partial x} + x^2\frac{\partial^2}{\partial x^2}\right),$$

$$X^3 = x^2\left(6x^2\frac{\partial}{\partial x} + 6x^3\frac{\partial^2}{\partial x^2} + x^4\frac{\partial^3}{\partial x^3}\right),$$

and so on. Thus,

$$F(\overline{x}) = F(x) + ax^2F'(x) + \frac{a^2}{2}(2x^3F'(x) + x^4F''(x)) + \frac{a^3}{3!}(6x^4F'(x) + 6x^5F''(x) + x^6F'''(x)).$$

□

1.8 The invariance principle for differential equations

While there are many numerical packages that can be used for finding symmetries and corresponding invariant solutions of nonlinear differential equations, it is absolutely crucial to understand the process behind using numerical packages for solving nonlinear ODEs and PDEs. The book's objective is to teach readers the prolongation formulae and the significance of variable order in Lie group analysis of differential equations. Relying solely on numerical packages without understanding the underlying process can lead to incorrect results, which is why it is critical to comprehend the calculations' specifics. To ensure students' understanding, the book provides numerous worked examples, which are essential in comprehending the method's idea. Advanced students are strongly advised to solve the examples independently before comparing their solutions to those provided in the book.

Consider a change of variables x, y involving a parameter a:

$$T_a: \quad \overline{x} = \varphi(x,y,a) \approx x + a\xi(x,y), \quad \overline{y} = \psi(x,y,a) \approx y + a\eta(x,y). \tag{1.141}$$

We say that the group G of transformations (1.141) is called a symmetry group of a first-order ODE

$$\frac{dy}{dx} = f(x, y) \tag{1.142}$$

or that equation (1.142) *admits the group G* if the form of (1.142) remains the same after the change of variables (1.141), i. e.,

$$\frac{d\overline{y}}{d\overline{x}} = f(\overline{x}, \overline{y}).$$

Important fact: G is a symmetry group iff it converts any classical solution of (1.142) into a classical solution of the same equation.

Example 1.23. Show that any autonomous equation

$$\frac{dy}{dx} = f(y) \tag{1.143}$$

admits the group with operator

$$X = \frac{\partial}{\partial x}.$$

Solution. First, let us find the group associated with the given generator X. We solve the Lie equation

$$\frac{d\overline{x}}{da} = 1, \quad \overline{x}|_{a=0} = x$$

to obtain

$$\overline{x} = a + x,$$

which represents the group of translations along the x-axis. So, we substitute $\overline{x} = a + x$ and $\overline{y} = y$ into the original equation (1.143) and obtain

$$\frac{d\overline{y}}{d\overline{x}} = \frac{dy}{dx}, \quad f(\overline{y}) = f(y).$$

Thus, any solution of (1.143) is also a solution of

$$\frac{d\overline{y}}{d\overline{x}} = f(\overline{y}). \qquad \square$$

The idea of the invariance of first-order ODEs can be extended to higher-order nonlinear ODEs, nonlinear PDEs, and nonlinear systems of ODEs and PDEs, which will be done in the next subsection.

1.8.1 The prolongation formulae

Here we propose an algorithmic approach to solving nonlinear differential (ordinary and/or partial) equations. Let us say we have a collection of independent variables $x = (x^1, \dots, x^n)$, dependent variables $u = (u^1, \dots, u^m)$, and partial derivatives

$$\underset{1}{u} = \{u_i^\alpha\}, \quad \underset{2}{u} = \{u_{ij}^\alpha\}, \dots,$$

of u with respect to x up to certain order, where

$$u_i^\alpha = \frac{\partial u^\alpha}{\partial x^i}, \quad u_{ij}^\alpha = \frac{\partial^2 u^\alpha}{\partial x^i \partial x^j}, \dots .$$

Consider the group of transformations

$$\overline{x}^i = f^i(x, u, a), \quad f^i\big|_{a=0} = x^i, \tag{1.144}$$

$$\overline{u}^\alpha = \varphi^\alpha(x, u, a), \quad \varphi^\alpha\big|_{a=0} = u^\alpha \tag{1.145}$$

and the generator of the group

$$X = \xi^i(x, u)\frac{\partial}{\partial x^i} + \eta^\alpha(x, u)\frac{\partial}{\partial u^\alpha}, \tag{1.146}$$

where

$$\xi^i = \frac{\partial f^i}{\partial a}\bigg|_{a=0}, \quad \eta^\alpha = \frac{\partial \varphi^\alpha}{\partial a}\bigg|_{a=0}. \tag{1.147}$$

We next introduce the prolongation formulae.

First prolongation:

$$\underset{1}{X} = X + \zeta_i^\alpha \frac{\partial}{\partial u_i^\alpha}, \tag{1.148}$$

where the coordinates ξ^i, η^α are defined by (1.147) and ζ_i^α are given by

$$\zeta_i^\alpha = D_i(\eta^\alpha) - u_j^\alpha D_i(\xi^j). \tag{1.149}$$

We say that the operator $\underset{1}{X}$ extends the action of X to functions depending on $x, u, \underset{1}{u}$ and is called the first prolongation of generator (1.146). Equation (1.149) is called the first prolongation formula.

Second prolongation: The second prolongation for the generator X of the group of point transformations (1.144) and (1.145) is

$$\underset{2}{X} = \underset{1}{X} + \zeta_{ij}^\alpha \frac{\partial}{\partial u_{ij}^\alpha}, \tag{1.150}$$

where the new coordinates ζ_{ij}^{α} are defined by the following second prolongation formula:

$$\zeta_{ij}^{\alpha} = D_j(\zeta_i^{\alpha}) - u_{ik}^{\alpha} D_j(\xi^k). \tag{1.151}$$

Higher-order prolongation is defined recursively as

$$\zeta_{i_1,\dots,i_k}^{\alpha} = D_{i_k}(\zeta_{i_1,\dots,i_{k-1}}^{\alpha}) - u_{ji_1,\dots,i_{k-1}}^{\alpha} D_{i_k}(\xi^j). \tag{1.152}$$

We say that a system of equations

$$\begin{gathered} F_1(x, u, \dots, \underset{p}{u}) = 0, \\ \dots \\ F_s(x, u, \dots, \underset{s}{u}) = 0 \end{gathered} \tag{1.153}$$

admits the group G of the transformations (1.144) and (1.145) if the differential manifold $[F]$ determined by (1.153) is invariant under the prolonged group $\underset{p}{G}$.

Let $i = 1, \dots, n$ and $\alpha = 1, \dots, m$. We introduce some notation:

$$x = \{x^i\}, \quad u = \{u^{\alpha}\}, \quad \underset{1}{u} = \{u_i^{\alpha}\} = \frac{\partial u^{\alpha}}{\partial x^i}, \quad \underset{2}{u} = \{u_{ij}^{\alpha}\} = \frac{\partial^2 u^{\alpha}}{\partial x_i \partial x_j}, \dots,$$

$$D_x = \frac{\partial}{\partial x} + y' \frac{\partial}{\partial x} + y'' \frac{\partial}{\partial y'} + y''' \frac{\partial}{\partial y''} + \dots + y^{(s+1)} \frac{\partial}{\partial y^{(s)}}.$$

In general, the total differentiation with respect to x^i is given by

$$D_i = \frac{\partial}{\partial x^i} + u_i^{\alpha} \frac{\partial}{\partial u^{\alpha}} + u_{ij}^{\alpha} \frac{\partial}{\partial u_j^{\alpha}} + \cdots.$$

1.8.2 Generalization

The calculus of *differential algebra* furnishes us with a convenient language and effective tools for group analysis, calculation of conservation laws, and integration of differential equations. In classical mathematical analysis one deals with functions

$$u^{\alpha}(x), \quad \alpha = 1, \dots, m,$$

of independent variables $x = (x^1, \dots, x^n)$. The derivatives

$$u_i^{\alpha}(x) = \frac{\partial u^{\alpha}(x)}{\partial x^i}, \quad u_{ij}^{\alpha}(x) = \frac{\partial^2 u^{\alpha}(x)}{\partial x^i \partial x^j}, \dots$$

are also regarded as functions of x.

In differential algebra we treat the quantities

$$x = \{x^i\}, \quad u = \{u^\alpha\}, \quad u_{(1)} = \{u_i^\alpha\}, \quad u_{(2)} = \{u_{ij}^\alpha\}, \ldots$$

as independent variables and consider composite functions

$$f(x, u(x), \partial u(x)/\partial x, \ldots)$$

of x as functions $f(x, u, u_{(1)}, \ldots)$ of the variables $x, u, u_{(1)}, \ldots$.

Main variables. We consider the following infinite set of algebraically independent variables:

$$\begin{aligned} &x = \{x^i\}, \quad u = \{u^\alpha\}, \quad u_{(1)} = \{u_i^\alpha\}, \\ &u_{(2)} = \{u_{i_1 i_2}^\alpha\}, \quad u_{(3)} = \{u_{i_1 i_2 i_3}^\alpha\}, \ldots, \end{aligned} \tag{1.154}$$

where $\alpha = 1, \ldots, m$, $i, i_1, \ldots = 1, \ldots, n$. The variables $u_{i_1 i_2}^\alpha$, $u_{i_1 i_2 i_3}^\alpha, \ldots$ are supposed to be symmetric in the subscripts $i_1 i_2, i_1 i_2 i_3, \ldots$.

Total differentiations. The *total differentiations* D_i are defined by the following formal infinite sums:[1]

$$D_i = \frac{\partial}{\partial x^i} + u_i^\alpha \frac{\partial}{\partial u^\alpha} + u_{ii_1}^\alpha \frac{\partial}{\partial u_{i_1}^\alpha} + u_{ii_1 i_2}^\alpha \frac{\partial}{\partial u_{i_1 i_2}^\alpha} + \cdots, \quad i = 1, \ldots, n. \tag{1.155}$$

The action of the total differentiations is well defined on all functions involving any finite number of the variables (1.154) since in this case the infinite sum (1.155) is truncated and reduces to a finite sum. For example,

$$D_i(f(x, u, u_{(1)})) = \frac{\partial f}{\partial x^i} + u_i^\alpha \frac{\partial f}{\partial u^\alpha} + u_{ii_1}^\alpha \frac{\partial f}{\partial u_{i_1}^\alpha}.$$

Taking $f = u^\alpha, f = u_j^\alpha, \ldots$, one obtains

$$u_i^\alpha = D_i(u^\alpha), \quad u_{ij}^\alpha = D_i(u_j^\alpha) = D_i D_j(u^\alpha), \quad \ldots. \tag{1.156}$$

Thus, although the variables (1.154) are assumed to be *algebraically independent*, they are connected by the differential relations (1.156). Accordingly, x^i are called *independent variables* and u^α *differential variables* with the successive derivatives $u_{(1)}, u_{(2)}, \ldots$.

---

1 We use the usual *summation convention* of dropping the summation symbol for expressions containing repeated super- and subscripts. For example, the term $u_i^\alpha \frac{\partial}{\partial u^\alpha}$ in (1.155) stands for $\sum_{\alpha=1}^m u_i^\alpha \frac{\partial}{\partial u^\alpha}$, whereas $u_{ii_1}^\alpha \frac{\partial}{\partial u_{i_1}^\alpha}$ stands for $\sum_{\alpha=1}^m \sum_{i_1=1}^n u_{ii_1}^\alpha \frac{\partial}{\partial u_{i_1}^\alpha}$.

Theorem 1.5. *The system of differential equations (1.153) admits the group G with the generator X iff*

$$\underset{p}{X}F_k\Big|_{[F]} = 0, \tag{1.157}$$

where $k = 1, \dots, s$ and the symbol $|_{[F]}$ means "evaluated on the manifold $[F]$."

Equation (1.157) is called a *determining equation.*

The idea behind this algorithmic approach will be explained by means of examples in the next subsection.

1.8.3 Illustrative examples: prolonged generators

Here we will consider some illustrative examples showing how to use the determining equation (1.157) furnished with the prolongation formulae (1.149) and (1.151).

Example 1.24. Consider the following PDE:

$$\frac{\partial u}{\partial t} + t\frac{\partial u}{\partial x} = x. \tag{1.158}$$

Show that the PDE admits the operator (generator)

$$X = \frac{\partial}{\partial t} + t\frac{\partial}{\partial x} + x\frac{\partial}{\partial u}. \tag{1.159}$$

Solution. First, let us write the original equation (1.158) in "compact" form as

$$u_t + tu_x = x$$

and observe that the original space of the variables $R^3(t, x, u)$ is prolonged to the extended space $R^5(t, x, u, u_t, u_x)$, i. e., the space R^5 consists of the original variables t, x, u and all possible first-order derivatives. Again, let us use a "compact" notation and write

$$R^3(t, x, u) \longmapsto R^5(t, x, u, u_t, u_x),$$

where we use t as the first independent variable and x as the second independent variable. The order is not important, but it tells us what order of variables we are going to use in the prolongation formulae. In our case, since the original equation (1.158) is the equation of the first order and we do not need any mixed second-order derivatives, the total derivative operators will be

$$D_t = \frac{\partial}{\partial t} + u_t\frac{\partial}{\partial u}, \quad D_x = \frac{\partial}{\partial x} + u_x\frac{\partial}{\partial u},$$

and, as follows from the form of

$$X = \xi^1 \frac{\partial}{\partial t} + \xi^2 \frac{\partial}{\partial x} + \eta \frac{\partial}{\partial u}$$

given by (1.159), we have

$$\xi^1 = 1, \quad \xi^2 = t, \quad \eta = x.$$

Since we are dealing with a first-order PDE, we only need the first prolongation formula (1.148),

$$\underset{1}{X} = X + \zeta_i^\alpha \frac{\partial}{\partial u_i^\alpha} = \frac{\partial}{\partial t} + t\frac{\partial}{\partial x} + x\frac{\partial}{\partial u} + \zeta_1 \frac{\partial}{\partial u_t} + \zeta_2 \frac{\partial}{\partial u_x},$$

where, since we choose the order of independent variables t and x, we have from formula (1.149)

$$\zeta_i = D_i(\eta) - u_j D_i(\xi^j),$$

i. e.,

$$\begin{aligned}\zeta_1 = \zeta_t &= D_t(\eta) - u_t D_t(\xi^1) - u_x D_t(\xi^2)\\ &= \eta_t + u_t\eta_u - u_t(\xi_t^1 + u_t\xi_u^1) - u_x(\xi_t^2 + u_t\xi_u^2) = -u_x.\end{aligned}$$

Similarly,

$$\begin{aligned}\zeta_2 = \zeta_x &= D_x(\eta) - u_t D_x(\xi^1) - u_x D_x(\xi^2)\\ &= \eta_x + u_x\eta_u - u_t(\xi_x^1 + u_x\xi_u^1) - u_x(\xi_x^2 + u_x\xi_u^2) = 1 - u_t\cdot 0 - u_x \cdot 0 = 1.\end{aligned}$$

Thus,

$$\underset{1}{X} = X + \zeta_i^\alpha \frac{\partial}{\partial u_i^\alpha} = \frac{\partial}{\partial t} + t\frac{\partial}{\partial x} + x\frac{\partial}{\partial u} - u_x \frac{\partial}{\partial u_t} + \frac{\partial}{\partial u_x}. \tag{1.160}$$

Then we apply the extended generator (1.160) to the original equation (1.158) to obtain

$$\underset{1}{X}(u_t + tu_x - x)\Big|_{[F]} = u_x + t(-1) + x(0) - u_x(1) + t = u_x - u_t - u_x + t = 0,$$

i. e., equation (1.157) is satisfied, so the generator (1.159) is admitted by equation (1.158).

□

Remark about the order of independent variables. It is important to note that the order of independent variables in the original space is not significant. However, once the order is fixed, the same order should be used in the extended space. Additionally, the originally chosen order of variables should be used in the prolongation formulae for ζ_i^α and ζ_{ij}^α.

Example 1.25. Consider the following second-order PDE:

$$u_t = u_{xx}. \tag{1.161}$$

Show that the PDE admits the operator (generator)

$$X = u\frac{\partial}{\partial u}. \tag{1.162}$$

Solution. In this case, let us change the order of independent variables and consider t as the first independent variable and x as the second variable. In this case, since the equation is of second order, the original space of the variables $R^3(x, t, u)$ is prolonged to the extended space $R^8(x, t, u, u_x, u_t, u_{xt}, u_{xx}, u_{tt})$, i. e., the space R^8 consists of the original variables x, t, u and all possible second-order derivatives, i. e.,

$$R^3(x, t, u) \longmapsto R^5(x, t, u, u_x, u_t, u_{xt}, u_{xx}, u_{tt}),$$

where we use x as the first independent variable and t as the second independent variable. Again, we remark that the order is not important, but it tells us what order of variables we are going to use in the prolongation formulae. So we use the following order of variables and derivatives:

$$1 \Leftrightarrow x, \quad 2 \Leftrightarrow t, \quad 11 \Leftrightarrow xx, \quad 12 \Leftrightarrow xt, \quad 22 \Leftrightarrow tt.$$

In our case, since the original equation (1.161) is an equation of the first order and we should take into account the mixed derivatives as well (all possible derivatives), the prolonged operator is

$$\underset{2}{X} = u\frac{\partial}{\partial u} + \zeta_1\frac{\partial}{\partial u_x} + \zeta_2\frac{\partial}{\partial u_t} + \zeta_{11}\frac{\partial}{\partial u_{xx}} + \zeta_{12}\frac{\partial}{\partial u_{xt}} + \zeta_{22}\frac{\partial}{\partial u_{tt}}.$$

Since

$$X = \xi^1\frac{\partial}{\partial t} + \xi^2\frac{\partial}{\partial x} + \eta\frac{\partial}{\partial u},$$

comparison with (1.162) yields

$$\xi^1 = 0, \quad \xi^2 = 0, \quad \eta = u.$$

Since

$$D_x = \frac{\partial}{\partial x} + u_x \frac{\partial}{\partial u} + u_{xx} \frac{\partial}{\partial u_x} + u_{xt} \frac{\partial}{\partial u_t},$$

we have from the first prolongation formula (1.149)

$$\zeta_i = D_i(\eta) - u_j D_i(\xi^j),$$

i. e.,

$$\zeta_1 = \zeta_x = D_x(\eta) - u_x D_x(\xi^1) - u_t D_x(\xi^2) = D_x(u) - u_x D_x(0) - u_t D_x(0) = u_x.$$

Similarly,

$$\zeta_2 = \zeta_t = D_t(\eta) - u_x D_t(\xi^1) - u_t D_t(\xi^2) = D_t(u) - u_x D_t(0) - u_t D_t(0) = u_t.$$

Now, the second prolongation formula (1.151)

$$\zeta_{ij} = D_i(\zeta_i) - u_{ik} D_j(\xi^k)$$

yields

$$\zeta_{11} = D_x(\zeta_1) - u_{xx} D_x(\xi^1) - u_{xt} D_x(\xi^2) = D_x(u_x) = u_{xx}.$$

Since $\zeta_{12} = 0$ and $\zeta_{22} = 0$ (why?), we have

$$\underset{2}{X} = u\frac{\partial}{\partial u} + u_x \frac{\partial}{\partial u_x} + u_t \frac{\partial}{\partial u_t} + u_{xx} \frac{\partial}{\partial u_{xx}}. \tag{1.163}$$

Then we act by the extended generator (1.163) on the original equation (1.161) to obtain

$$\underset{2}{X}(u_t - u_{xx})\Big|_{[F]} = u(0) + u_x(0) + u_t(1) + u_{xx}(-1) = u_t - u_{xx} = 0,$$

i. e., equation (1.157) is satisfied. □

Example 1.26. Let us go back to the previous example (Example 1.25) and consider the following second-order PDE:

$$u_t - u_{xx} = 0. \tag{1.164}$$

Show that the PDE admits the operator (generator)

$$X = 2t\frac{\partial}{\partial x} - xu\frac{\partial}{\partial u}. \tag{1.165}$$

Solution. In this case, let us choose the following order of independent variables: x as the first independent variable and t as the second variable. In this case, since the equation is of second order, the original space of the variables $R^3(x,t,u)$ is prolonged to the extended space $R^8(x,t,u,u_x,u_t,u_{xt},u_{xx},u_{tt})$, i. e., the space R^8 consists of the original variables x,t,u and all possible second-order derivatives, i. e.,

$$R^3(x,t,u) \longmapsto R^8(x,t,u,u_x,u_t,u_{xt},u_{xx},u_{tt}).$$

We also have

$$\begin{aligned} D_x &= \frac{\partial}{\partial x} + u_x\frac{\partial}{\partial u} + u_{xx}\frac{\partial}{\partial u_x} + u_{xt}\frac{\partial}{\partial u_t}, \\ D_t &= \frac{\partial}{\partial t} + u_t\frac{\partial}{\partial u} + u_{xt}\frac{\partial}{\partial u_x} + u_{tt}\frac{\partial}{\partial u_t} \end{aligned}$$

and

$$\xi^1 = 2t, \quad \xi^2 = 0, \quad \eta = -xu.$$

The prolongation formulae are

$$\zeta_i = D_i(\eta) - u_j D_i(\xi^j), \tag{1.166}$$

$$\zeta_{ij} = D_i(\zeta_i) - u_{ik} D_j(\xi^k). \tag{1.167}$$

We read (1.166) as

$$\zeta_1 = \zeta_x = D_x(\eta) - u_x D_x(\xi^1) - u_t D_x(\xi^2) = D_x(-xu) - u_x D_x(2t) - u_t D_x(0) = -u - xu_x$$

and

$$\zeta_2 = \zeta_t = D_t(\eta) - u_x D_t(\xi^1) - u_t D_t(\xi^2) = D_t(-xu) - u_x D_t(2t) - u_t D_x(0) = -xu_t - 2u_x.$$

Similarly, we write (1.167) as

$$\zeta_{11} = D_x(\zeta_1) - u_{xx} D_x(\xi^1) - u_{xt} D_x(\xi^2) = D_x(-u - xu_x) - u_{xx} D_x(2t) - u_{xt} D_x(0) = -2u_x - xu_{xx}.$$

Thus, we have

$$\begin{aligned} \underset{2}{X}(u_t - u_{xx}) &= \Big[2t\frac{\partial}{\partial x} - xu\frac{\partial}{\partial u} - (u + xu_x)\frac{\partial}{\partial u_x} - (xu_t - 2u_x)\frac{\partial}{\partial u_t} \\ &\quad - (2u_x + xu_{xx})\frac{\partial}{\partial u_{xx}}\Big](u_t - u_{xx}) \\ &= x(u_{xx} - u_t). \end{aligned}$$

So the determining equation (1.157) is

$$\left.\underset{p}{X}F_k\right|_{[F]} = \left.\underset{2}{X}(u_t - u_{xx})\right|_{[F]} = xu_{xx} - u_t|_{u_t=u_{xx}} = 0.$$

Thus, the generator (1.165) is admitted by equation (1.164). □

Scaling transformations

Very often PDEs are invariant under scaling transformations:

$$\overline{x} = cx, \quad \overline{t} = bt, \quad \overline{u} = au. \tag{1.168}$$

Let us illustrate this by an example.

Example 1.27. Check if the PDE

$$u_t + 2u_x^2 + xu = 0 \tag{1.169}$$

is invariant under the scaling transformation (1.168).

Solution. In this case, we are looking for a generator of the form

$$X = cx\frac{\partial}{\partial x} + bt\frac{\partial}{\partial t} + au\frac{\partial}{\partial u}, \tag{1.170}$$

where, due to the original equation (1.169), the constants a, b, and c must be related by the equations

$$\frac{a}{b} = \frac{a^2}{c^2} = ac. \tag{1.171}$$

We can express a and b from (1.171) as

$$a = c^3, \quad \frac{1}{b} = c,$$

so we can rewrite the transformations (1.168) as

$$\overline{x} = cx, \quad \overline{t} = \frac{1}{c}t, \quad \overline{u} = c^3u,$$

and the corresponding tangent field is given by formula (1.28), in which we evaluate the derivatives at $c = 1$, i. e.,

$$\xi^1 = \left.\frac{\partial\overline{x}}{\partial c}\right|_{c=1} = x,$$
$$\xi^2 = \left.\frac{\partial\overline{t}}{\partial c}\right|_{c=1} = -t,$$
$$\eta = \left.\frac{\partial\overline{u}}{\partial c}\right|_{c=1} = 3u.$$

Correspondingly, the generator (1.170) is written

$$X = x\frac{\partial}{\partial x} - t\frac{\partial}{\partial t} + 3u\frac{\partial}{\partial u}. \tag{1.172}$$

Now we have to check if the generator (1.172) is admitted by the PDE (1.169). To this end, we are looking for the prolonged generator

$$\underset{1}{X} = X + \zeta_1\frac{\partial}{\partial u_x} + \zeta_2\frac{\partial}{\partial u_t},$$

where, since we randomly choose the order of independent variables x and t, from formula (1.149) we have

$$\zeta_i = D_i(\eta) - u_j D_i(\xi^j),$$

i. e.,

$$\zeta_1 = \zeta_x = D_x(\eta) - u_x D_x(\xi^1) - u_t D_x(\xi^2) = D_x(3u) - u_x D_x(x) - u_t D_x(t) = 2u_x,$$
$$\zeta_2 = \zeta_t = D_t(\eta) - u_x D_t(\xi^1) - u_t D_t(\xi^2) = D_t(3u) - u_x D_t(x) - u_t D_t(t) = 4u_t,$$

where

$$D_x = \frac{\partial}{\partial x} + u_x\frac{\partial}{\partial u} + u_{xx}\frac{\partial}{\partial u_x} + u_{xt}\frac{\partial}{\partial u_t},$$
$$D_t = \frac{\partial}{\partial t} + u_t\frac{\partial}{\partial u} + u_{xt}\frac{\partial}{\partial u_x} + u_{tt}\frac{\partial}{\partial u_t}.$$

Thus,

$$\underset{1}{X} = x\frac{\partial}{\partial x} - t\frac{\partial}{\partial t} + 3u\frac{\partial}{\partial u} + 2u_x\frac{\partial}{\partial u_x} + 4u_t\frac{\partial}{\partial u_t}.$$

Thus, the action of the prolonged generator on the given PDE yields

$$\underset{1}{X}(u_t + 2u_x^2 + xu)\Big|_{[u_t+2u_x^2+xu=0]} = 4u_t(1) + 2(2)u_x 2u_x + x(u) + 3u(x) = 4(u_t + 2u_x^2 + xu) = 0.$$

□

1.8.4 Invariant solutions (G/H factor system)

Consider a system of PDEs

$$F_k(x, u, u_{(1)}, \ldots) = 0, \quad k = 1, \ldots, s. \tag{1.173}$$

We already know that if the transformations $\overline{z}^i = f^i(z, a)$, $i = 1, \dots N$, are obtained by the transformations of the independent and dependent variables

$$\overline{x} = f(x, u, a), \quad \overline{u} = g(x, u, a) \tag{1.174}$$

and the extension of (1.174) to all derivatives $\underset{1}{u}$, etc., involved in the differential equations (1.173), then equations $F_k(\overline{z}) = 0$, $k = 1, \dots, s$, define a group G of transformations (1.174) admitted by the differential equations (1.173). In other words, an admitted group does not change the form of the system of differential equations (1.173). The generator of the admitted group G is termed an *infinitesimal symmetry* (or simply *symmetry*) of the differential equations (1.173). Let the differential equations (1.173) admit a multi-parameter group G and let H be a subgroup of G. A solution

$$u^\alpha = h^\alpha(x), \quad \alpha = 1, \dots, m, \tag{1.175}$$

of equations (1.173) is called an H-invariant solution (termed for brevity an *invariant solution*) if equations (1.175) are invariant under the subgroup H. If H is a one-parameter group and has the generator X, then the H-invariant solutions are constructed by calculating a basis of invariants $J_1, J_2, \dots$.

We note that the G/H factor system represents an ODE.

In this section, we will show how we can find invariant solutions by some examples.

Example 1.28. Show that the PDE

$$u_t - u_{xx} = 0 \tag{1.176}$$

is invariant under the group generated by the operator

$$X = 2t\frac{\partial}{\partial x} - xu\frac{\partial}{\partial u} \tag{1.177}$$

and find the solution invariant under this transformation.

Solution. We have already shown that the generator (1.177) is admitted by equation (1.176) in Example 1.26 by choosing x as the first independent variable and t as the second variable. In this example, as a small exercise, let us choose a different order of independent variables and choose t as the first independent variable and x as the second variable. In this case, we consider the following space extension:

$$R^3(t, x, u) \longmapsto R^5(t, x, u, u_t, u_t, u_{tx}, u_{tt}, u_{xx}).$$

We also have

$$D_t = \frac{\partial}{\partial t} + u_t\frac{\partial}{\partial u} + u_{tx}\frac{\partial}{\partial u_x} + u_{tt}\frac{\partial}{\partial u_t},$$

$$D_x = \frac{\partial}{\partial x} + u_x \frac{\partial}{\partial u} + u_{tx} \frac{\partial}{\partial u_t} + u_{xx} \frac{\partial}{\partial u_x},$$

and

$$\xi^1 = 0, \quad \xi^2 = 2t, \quad \eta = -xu.$$

The prolongation formulae are

$$\zeta_i = D_i(\eta) - u_j D_i(\xi^j), \tag{1.178}$$

$$\zeta_{ij} = D_i(\zeta_i) - u_{ik} D_j(\xi^k). \tag{1.179}$$

The extended generator is

$$\underset{2}{X} = 2t \frac{\partial}{\partial x} - xu \frac{\partial}{\partial u} + \zeta_1 \frac{\partial}{\partial u_t} + \zeta_2 \frac{\partial}{\partial u_x} + \zeta_{11} \frac{\partial}{\partial u_{tt}} + \zeta_{12} \frac{\partial}{\partial u_{tx}} + \zeta_{22} \frac{\partial}{\partial u_{xx}}$$

and

$$\underset{2}{X}(u_t - u_{xx}) = \zeta_1 - \zeta_{22}.$$

Thus, we only need to find ζ_1 and ζ_{22}.

We read (1.178) as

$$\zeta_1 = \zeta_t = D_t(\eta) - u_t D_t(\xi^1) - u_x D_t(\xi^2) = D_t(-xu) - u_t D_t(0) - u_x D_t(2t) = -xu_t - 2u_x$$

and

$$\zeta_2 = \zeta_x = D_x(\eta) - u_t D_x(\xi^1) - u_x D_x(\xi^2) = D_x(-xu) - u_t D_x(0) - u_x D_x(2t) = -u - xu_x.$$

Similarly, we write (1.179) as

$$\zeta_{22} = \zeta_{xx} = D_x(\zeta_2) - u_{xt} D_x(\xi^1) - u_{xx} D_x(\xi^2) = D_x(-u - xu_x) - u_{xx} D_x(2t) = -2u_x - xu_{xx}.$$

We can quickly check that

$$\zeta_1 - \zeta_{22} = -xu_t - 2u_x + 2u_x + xu_{xx} = -x(u_t - u_{xx}) = 0.$$

Now, how to find the invariant solution of equation (1.176)?

Step 1: *Invariants.* First, let us find the invariants from the characteristic system associated with the generator (1.177)

$$\frac{dx}{2t} = -\frac{du}{xu} = -\frac{dt}{0}. \tag{1.180}$$

One can readily see that one invariant is given by

$$J_1 = t. \tag{1.181}$$

Indeed,

$$X(t) = 0 \quad \Longleftrightarrow \quad 2t\frac{\partial t}{\partial x} - xu\frac{\partial t}{\partial u} = 0. \tag{1.182}$$

In order to find the second invariant, we shall consider the first equation of the characteristic system (1.180), which we can rewrite as

$$\frac{xdx}{2t} = -\frac{du}{u}. \tag{1.183}$$

Integrating (1.183), we obtain

$$\frac{x^2}{4t} + \ln u = \ln C,$$

where C is a constant of integration. Solving for C, we obtain the second invariant

$$J_2 = ue^{\frac{x^2}{4t}}. \tag{1.184}$$

Step 2: *Invariant solution.* A general presentation of the invariant solution is

$$J_2 = \varphi(J_1). \tag{1.185}$$

Substituting (1.181) and (1.184) into equation (1.185), we have

$$ue^{\frac{x^2}{4t}} = \varphi(t). \tag{1.186}$$

We solve (1.186) for u to obtain

$$u = \varphi(t)e^{-\frac{x^2}{4t}}. \tag{1.187}$$

We next calculate the derivatives

$$\frac{\partial u}{\partial t} = \frac{x^2}{4t}e^{-\frac{x^2}{4t}}\varphi(t) + e^{-\frac{x^2}{4t}}\frac{d\varphi}{dt}, \tag{1.188}$$

$$\frac{\partial u}{\partial x} = -\frac{x}{2t}e^{-\frac{x^2}{4t}}\varphi(t), \tag{1.189}$$

$$\frac{\partial^2 u}{\partial x^2} = \left(-\frac{1}{2t} + \frac{x^2}{4t^2}\right)e^{-\frac{x^2}{4t}}\varphi(t). \tag{1.190}$$

Substituting (1.188) and (1.190) into the original equation (1.176), we have

$$\frac{x^2}{4t}\varphi(t) + \frac{d\varphi}{dt} + \left(\frac{1}{2t} - \frac{x^2}{4t^2}\right)\varphi(t) = 0. \tag{1.191}$$

Simplifying equation (1.191), we obtain

$$\frac{d\varphi}{dt} + \frac{1}{2t}\varphi(t) = 0. \tag{1.192}$$

The resulting equation (1.192) is called a *G/H factor system.* We solve the G/H factor system (1.192),

$$\frac{d\varphi}{\varphi} = -\frac{1}{2t}dt,$$

which gives

$$\ln\varphi = -\frac{1}{2}\ln t + \ln C_1, \tag{1.193}$$

where C_1 is a constant of integration. Solving equation (1.193) for φ, we obtain

$$\varphi = ct^{-\frac{1}{2}}.$$

Substituting the obtained φ into the invariant solution presentation (1.187), we finally write the invariant solution of the original equation (1.176) in the form

$$u(x,t) = \frac{c}{\sqrt{t}}e^{-\frac{x^2}{4t}}. \tag{1.194}$$

Therefore, if, for example, we associate the variable t with time and u with the temperature distribution, the obtained solution shows that the temperature function (1.194) increases exponentially with increasing t. □

Example 1.29. Using scaling symmetries, derive the G/H factor system for the following nonlinear first-order PDE for $z(x,y)$:

$$\left(\frac{\partial z}{\partial x}\right)^2 + 2x\frac{\partial z}{\partial y} + z^2 = 0. \tag{1.195}$$

Solution. First, let us find scaling transformations

$$\overline{x} = ax, \quad \overline{y} = by, \quad \overline{z} = cz \tag{1.196}$$

generated by the operator

$$X = ax\frac{\partial}{\partial x} + by\frac{\partial}{\partial y} + cz\frac{\partial}{\partial z}, \tag{1.197}$$

where due to the original equation (1.195) the constants a, b, and c must be related by the equations

$$\frac{c^2}{a^2} = \frac{ac}{b} = c^2. \tag{1.198}$$

We can set $a = 1$ and express b and c from (1.198) as

$$c = \frac{1}{b}, \tag{1.199}$$

so we can rewrite the transformations (1.196) as

$$\overline{x} = x, \quad \overline{y} = by, \quad \overline{z} = \frac{z}{b} \tag{1.200}$$

and the corresponding tangent field is given by formula (1.28), in which we evaluate the derivatives at $b = 1$, i. e.,

$$\begin{aligned} \xi^1 &= \left.\frac{\partial \overline{x}}{\partial b}\right|_{b=1} = 0, \\ \xi^2 &= \left.\frac{\partial \overline{y}}{\partial b}\right|_{b=1} = y, \\ \eta &= \left.\frac{\partial \overline{z}}{\partial b}\right|_{b=1} = -z. \end{aligned}$$

Correspondingly, the generator (1.197) is written,

$$X = y\frac{\partial}{\partial y} - z\frac{\partial}{\partial z}. \tag{1.201}$$

Let us check if the generator (1.201) is admitted by the PDE (1.195). We can choose x as the first independent variable and y as the second variable. In this case, we consider the following space extension:

$$R^3(x, y, z) \longmapsto R^5(x, y, z, z_x, z_y).$$

We also have

$$D_x = \frac{\partial}{\partial x} + z_x\frac{\partial}{\partial z}, \quad D_y = \frac{\partial}{\partial y} + z_y\frac{\partial}{\partial z}.$$

We are looking for the prolonged generator

$$\underset{1}{X} = X + \zeta_1\frac{\partial}{\partial z_x} + \zeta_2\frac{\partial}{\partial z_y},$$

where, since we randomly choose the order of independent variables x and y, from formula (1.149) we have

$$\zeta_i = D_i(\eta) - u_j D_i(\xi^j),$$

i. e.,

$$\begin{aligned}
\zeta_1 = \zeta_x &= D_x(\eta) - z_x D_x(\xi^1) - z_t D_x(\xi^2) = D_x(-z) - z_x D_x(0) - z_y D_x(y) = z_y(-1) = -z_x,\\
\zeta_2 = \zeta_y &= D_y(\eta) - z_x D_y(\xi^1) - z_y D_y(\xi^2) = D_y(-z) - z_x D_y(0) - z_y D_y(y) = z_y(-1) - z_y(1)\\
&= -2z_y.
\end{aligned}$$

Thus, the action of the extended generator on equation (1.195) can be written as

$$\begin{aligned}
\underset{1}{X}(1.195)\Big|_{z_x^2+2xz_y+z^2=0} &= \left[y\frac{\partial}{\partial y} - z\frac{\partial}{\partial z} - z_x\frac{\partial}{\partial z_x} - 2z_y\frac{\partial}{\partial z_y}\right](z_x^2 + 2xz_y + z^2)\\
&= -z(2z) - z_x(2z_x) - 2z_y(2x) = -2(z_x^2 + 2xz_y + z^2) = 0.
\end{aligned}$$

Now, the characteristic equations associated with the generator (1.201) are written as

$$\frac{dy}{y} = -\frac{dz}{z} = \frac{dx}{0},$$

which gives us two functionally independent invariants $J_1 = x$ and $J_2 = yz$, so that writing $J_2 = \varphi(J_1)$ yields the relation

$$z = \frac{\varphi(x)}{y}. \tag{1.202}$$

Calculating the derivatives

$$z_x = \frac{\varphi'(x)}{y}, \quad z_y = -\frac{\varphi(x)}{y^2}$$

and substituting into (1.202), we have

$$\frac{1}{y^2}\varphi_x^2 + 2x\left(-\frac{1}{y^2}\right)\varphi + \frac{\varphi}{y^2} = 0,$$

which, after simplification, gives us the G/H factor system:

$$\left(\frac{d\varphi}{dx}\right)^2 = 2x\varphi - \varphi^2 = 0. \tag{1.203}$$

□

1.9 Group transformations admitted by nonlinear partial differential equations

Symmetry groups transform any solution of a given equation (or system of equations) (1.173) into a solution of the same equation. For example, let

$$\bar{t} = f(t, x, u, a), \quad \bar{x} = f(t, x, u, a), \quad \bar{u} = f(t, x, u, a) \tag{1.204}$$

be a symmetry group of the equation

$$u_t = F(t, x, u, u_x, u_{xx}), \quad \frac{\partial F}{\partial u_{xx}} \neq 0, \tag{1.205}$$

and let $u = \Phi(t, x)$ be a solution of equation (1.205). Since (1.204) represent a transformation group, $\Phi(t, x)$ can be written in new variables, so $\bar{u} = \Phi(\bar{t}, \bar{x})$ and thus the function h given by

$$h(t, x, u, a) = \Phi(f(t, x, u, a), g(t, x, u, a)) \tag{1.206}$$

solves equation (1.206) with respect to u, i. e., it represents a one-parameter family of solutions for equation (1.205).

Example 1.30. Let us show that the heat equation

$$u_t - u_{xx} = 0 \tag{1.207}$$

is invariant under transformations

$$\bar{t} = t, \quad \bar{x} = x + 2at, \quad \bar{u} = ue^{-(ax+a^2t)}. \tag{1.208}$$

Solution. First, let us find the generator of the group (1.208) in the form

$$X = \xi^1 \frac{\partial}{\partial t} + \xi^2 \frac{\partial}{\partial x} + \eta \frac{\partial}{\partial u}, \tag{1.209}$$

where the corresponding tangent field is given by formula (1.28),

$$\xi^1 = \left.\frac{\partial \bar{t}}{\partial a}\right|_{a=0} = 0,$$
$$\xi^2 = \left.\frac{\partial \bar{x}}{\partial a}\right|_{a=0} = 2t,$$
$$\eta = \left.\frac{\partial \bar{u}}{\partial a}\right|_{a=0} = -xu,$$

so

$$X = 2t\frac{\partial}{\partial x} - xu\frac{\partial}{\partial u}. \tag{1.210}$$

This means that any solution $u = \Phi(t, x)$ of (1.207) can be converted to a new solution by transformation (1.208). Since equation (1.207) is invariant under the transformation (1.208), we have

$$\bar{u}_{\bar{t}} - \bar{u}_{\bar{x}\bar{x}} = 0,$$

so the notation $\overline{u} = \Phi(\overline{t}, \overline{x})$ means that

$$ue^{-(ax+a^2t)} = \Phi(t, x + 2at),$$

which we can solve for u to obtain the invariant solution in the form

$$u(x,t) = \Phi(t, x + 2at)e^{-(ax+a^2t)}. \tag{1.211}$$

□

1.9.1 How to find a group admitted by a nonlinear PDE?

We will answer this question by considering a particular example.

Example 1.31. Find all possible transformations (which generate a group) that leave the second-order PDE

$$u_x u_{xx} + u_{yy} = 0 \tag{1.212}$$

invariant.

Solution. We are looking for the generator of the group in the form

$$X = \xi^1 \frac{\partial}{\partial x} + \xi^2 \frac{\partial}{\partial y} + \eta \frac{\partial}{\partial u}.$$

Now let us choose the following order of independent variables: x as the first independent variable and y as the second variable. In this case, since the equation is of second order, the original space of the variables $R^3(x, y, u)$ is prolonged to the extended space $R^8(x, y, u, u_x, u_y, u_{xy}, u_{xx}, u_{yy})$, i. e., the space R^8 consists of the original variables x, y, u and all possible second-order derivatives, i. e.,

$$R^3(x, y, u) \longmapsto R^8(x, y, u, u_x, u_y, u_{xy}, u_{xx}, u_{yy}).$$

Hence, the second prolongation of the generator X is written as

$$\underset{2}{X} = \xi^1 \frac{\partial}{\partial x} + \xi^2 \frac{\partial}{\partial y} + \eta \frac{\partial}{\partial u} + \zeta_1 \frac{\partial}{\partial u_x} + \zeta_2 \frac{\partial}{\partial u_y} + \zeta_{11} \frac{\partial}{\partial u_{xx}} + \zeta_{12} \frac{\partial}{\partial u_{xy}} + \zeta_{22} \frac{\partial}{\partial u_{yy}}. \tag{1.213}$$

Applying the prolonged generator (1.213) to equation (1.212) yields

$$\underset{2}{X}(u_x u_{xx} + u_{yy}) = \zeta_1 u_{xx} + \zeta_{11} u_x + \zeta_{22} = 0. \tag{1.214}$$

Equation (1.214) shows that the corresponding determining equation

$$\underset{p}{X} F_k \bigg|_{[F]} = 0$$

in our case is

$$\zeta_1 u_{xx} + \zeta_{11} u_x + \zeta_{22}|_{[F]} = 0, \tag{1.215}$$

where the transition to a manifold F consists in replacing u_{yy} on the left side of equation (1.212) with $u_x u_{xx}$. So, we need to find the coefficients $\zeta_1, \zeta_{11}, \zeta_{22}$ and substitute them back into (1.215).

The total derivative operators are

$$\begin{aligned} D_x &= \frac{\partial}{\partial x} + u_x \frac{\partial}{\partial u} + u_{xy} \frac{\partial}{\partial u_y} + u_{xx} \frac{\partial}{\partial u_x}, \\ D_y &= \frac{\partial}{\partial y} + u_y \frac{\partial}{\partial u} + u_{xy} \frac{\partial}{\partial u_x} + u_{yy} \frac{\partial}{\partial u_y}. \end{aligned}$$

The first prolongation formula (1.149)

$$\zeta_i = D_i(\eta) - u_j D_i(\xi^j)$$

gives

$$\begin{aligned} \zeta_1 = \zeta_x &= D_x(\eta) - u_x D_x(\xi^1) - u_y D_x(\xi^2) = \eta_x + u_x \eta_u - u_x[\xi_x^1 + u_x \xi_u^1] - u_y[\xi_x^2 - u_x \xi_u^2] \\ &= \eta_x + u_x(\eta_u - \xi_x^1) - u_x^2 \xi_u^1 - u_y \xi_x^2 - u_x u_y \xi_u^2, \\ \zeta_2 = \zeta_y &= D_y(\eta) - u_x D_y(\xi^1) - u_y D_y(\xi^2) = \eta_y + u_y \eta_u - u_x[\xi_y^1 + u_y \xi_u^1] - u_y[\xi_y^2 + u_y \xi_u^2] \\ &= \eta_y + u_y(\eta_u - \xi_y^2) - u_x \xi_y^1 - u_y^2 \xi_u^2 - u_x u_y \xi_u^1. \end{aligned}$$

Similarly, the second prolongation formula (1.151)

$$\zeta_{ij} = D_i(\zeta_i) - u_{ik} D_j(\xi^k)$$

gives

$$\begin{aligned} \zeta_{11} &= D_x(\zeta_1) - u_{xx} D_x(\xi^1) - u_{xy} D_x(\xi^2) \\ &= \eta_{xx} + u_x(\eta_{ux} - \xi_{xx}^1) - u_x^2 \xi_{ux}^1 - u_y \xi_{xx}^2 - u_x u_y \xi_{ux}^2 \\ &\quad + u_x[\eta_{xu} + u_x(\eta_{uu} - \xi_{xu}^1) - u_x^2 \xi_{uu}^1 - u_y \xi_{xu}^2 - u_x u_y \xi_{uu}^2] \\ &\quad + u_{xx}[(\eta_u - \xi_x^1) - 2u_x \xi_u^1 - u_y \xi_u^2] + u_{xy}[-\xi_x^2 - u_x \xi_u^2] \\ &\quad - u_{xx}[\xi_x^1 + u_x \xi_u^1] - u_{xy}[\xi_x^1 + u_x \xi_u^2]. \end{aligned}$$

Also,

$$\begin{aligned} \zeta_{22} &= D_y(\zeta_2) - u_{xy} D_x(\xi^1) - u_{yy} D_x(\xi^2) \\ &= \eta_{yy} + u_y(\eta_{uy} - \xi_{yy}^2) - u_x \xi_{yy}^1 - u_x u_y \xi_{uy}^1 - u_y^2 \xi_{uy}^2 \end{aligned}$$

$$+ u_y[\eta_{yu} + u_y(\eta_{uu} - \xi^2_{yu}) - u_x\xi^1_{yu} - -u_x u_y \xi^1_{uu} - u_y^2\xi^2_{uu}]$$
$$+ u_{yy}[(\eta_u - \xi^2_y) - u_x\xi^1_u - 2u_y\xi^2_u] + u_{xy}[-\xi^1_y - u_y\xi^1_u]$$
$$- u_{xy}[\xi^1_y + u_y\xi^1_u] - u_{yy}[\xi^2_y + u_y\xi^2_u].$$

Since all terms $x, y, u, u_x, u_y, u_{xx}$, and u_{xy} in the determining equation (1.215) are regarded as independent variables and the unknown functions ξ^1, ξ^2, and η depend on x, y, and u only, equation (1.215) holds only if the term at u_{xy} is zero, i. e., if

$$\xi^1_y + u_x\xi^2_x + u_y\xi^1_u + u_x^2\xi^2_u = 0. \tag{1.216}$$

Looking at equation (1.216) and thinking in the same way as above, we should equate all the coefficients at u_x^2, u_x, and u_y to zero, i. e., we should have

$$\xi^1_y = 0, \quad \xi^1_u = 0, \quad \xi^2_x = 0, \quad \xi^2_u = 0. \tag{1.217}$$

Now, taking into account the equations in (1.217), we collects all terms at u_{xx} in the determining equation (1.215) and equate them to zero. We have

$$\eta_x + u_x(\eta_u - 3\xi^1_x + 2\xi^2_y) = 0,$$

which means that

$$\eta_x = 0, \quad \eta_u = 3\xi^1_x - 2\xi^2_y. \tag{1.218}$$

If we differentiate η_u in (1.218) with respect to u, we obtain

$$\eta_{uu} = 3\xi^1_{xu} - 2\xi^2_{yu} = 3\xi^1_{ux} - 2\xi^2_{uy} = 0, \tag{1.219}$$

since $\xi^1_u = \xi^2_u = 0$ due to (1.217).

Similarly, if we differentiate η_u in (1.218) with respect to x, we obtain

$$\eta_{ux} = \eta_{xu} = 3\xi^1_{xx} - 2\xi^2_{xy} = 0, \tag{1.220}$$

since $\eta_{xu} = 0$ due to the first equation in (1.218) and $\xi^2_x = 0$ due to (1.217). Then it follows from (1.220) that

$$\xi^1_{xx} = 0.$$

So, the resulting determining equation (1.215) is

$$\eta_{yy} + u_y(2\eta_{uy} - \xi^2_{yy}) = 0. \tag{1.221}$$

Thus,

$$\eta_{yy} = 0, \quad 2\eta_{uy} = \xi^2_{yy}. \tag{1.222}$$

We remark that due to (1.218) we have

$$\eta_{uy} = 3\xi^1_{xy} - 2\xi^2_{yy} = 2\xi^2_{yy}$$

and due to (1.222) we have

$$\eta_{uy} = 0, \quad \xi^2_{yy} = 0. \tag{1.223}$$

So, we have

$$\xi^1_{xx} = 0 \quad \Rightarrow \quad \xi^1 = C_1 + C_2 x, \tag{1.224}$$

$$\xi^2_{yy} = 0 \quad \Rightarrow \quad \xi^2 = C_3 + C_4 y, \tag{1.225}$$

$$\eta_{yy} = 0 \quad \Rightarrow \quad \eta = C_5 + C_6 x. \tag{1.226}$$

We also have $\eta_{uy} = 0$. Then, as follows from (1.218),

$$\eta = (3\xi^1_x - 2\xi^2_y)u = (3C_2 - 2C_4)u. \tag{1.227}$$

Now, setting

$$C_1 = 1, \quad C_2 = 0, \quad C_3 = 0, \quad C_4 = 0, \quad C_5 = 0, \quad C_6 = 0,$$

we have

$$\xi^1 = 1, \quad \xi^2 = 0, \quad \eta = 0.$$

Then

$$X_1 = \frac{\partial}{\partial x}.$$

Setting

$$C_1 = 0, \quad C_2 = 1, \quad C_3 = 0, \quad C_4 = 0, \quad C_5 = 0, \quad C_6 = 0,$$

we have

$$\xi^1 = x, \quad \xi^2 = 0, \quad \eta = 3u.$$

Then

$$X_2 = \frac{\partial}{\partial x} + 3u\frac{\partial}{\partial u}.$$

Setting

$$C_1 = 0, \quad C_2 = 0, \quad C_3 = 1, \quad C_4 = 0, \quad C_5 = 0, \quad C_6 = 0,$$

we have

$$\xi^1 = 0, \quad \xi^2 = 1, \quad \eta = 0.$$

Then

$$X_3 = \frac{\partial}{\partial y}.$$

Setting

$$C_1 = 0, \quad C_2 = 0, \quad C_3 = 0, \quad C_4 = 1, \quad C_5 = 0, \quad C_6 = 0,$$

we have

$$\xi^1 = 0, \quad \xi^2 = y, \quad \eta = -2u.$$

Then

$$X_4 = y\frac{\partial}{\partial y} - 2u\frac{\partial}{\partial u}.$$

Setting

$$C_1 = 0, \quad C_2 = 0, \quad C_3 = 0, \quad C_4 = 0, \quad C_5 = 1, \quad C_6 = 0,$$

we have

$$\xi^1 = 0, \quad \xi^2 = 0, \quad \eta = 1.$$

Then

$$X_5 = \frac{\partial}{\partial u}.$$

Setting

$$C_1 = 0, \quad C_2 = 0, \quad C_3 = 0, \quad C_4 = 0, \quad C_5 = 0, \quad C_6 = 1,$$

we have

$$\xi^1 = 0, \quad \xi^2 = 0, \quad \eta = y.$$

Then

$$X_6 = y\frac{\partial}{\partial u}.$$

We also remark that, since the determining equations are linear, its solutions compose a vector space L_6, i. e., we obtain six one-parameter groups.

Let us find the transformation group associated with the generator X_1:

$$\xi^1: \quad \frac{d\overline{x}}{da} = 1, \quad \overline{x}|_a = x \quad \Rightarrow \quad \overline{x} = a + x, \quad \overline{y} = y, \quad \overline{u} = u.$$

Let us find the transformation group associated with the generator X_2:

$$\begin{aligned}
\xi^1: &\quad \frac{d\overline{x}}{da} = \overline{x}, \quad \overline{x}|_a = x \quad \Rightarrow \quad \overline{x} = xe^a,\\
\xi^2: &\quad \frac{d\overline{y}}{da} = 0, \quad \overline{y}|_a = y \quad \Rightarrow \quad \overline{y} = y,\\
\eta: &\quad \frac{d\overline{u}}{da} = 3\overline{u}, \quad \overline{u}|_a = u \quad \Rightarrow \quad \overline{u} = ue^{3a}.
\end{aligned}$$

Likewise, we can find all six generators. □

1.9.2 Lie algebras and multi-parameter groups

Let us consider the second-order PDE

$$u_x u_{xx} + u_{yy} = 0. \tag{1.228}$$

As we know from Example 1.31, equation (1.228) admits the following six symmetries:

$$X_1 = \frac{\partial}{\partial x}, \quad X_2 = \frac{\partial}{\partial y}, \quad X_3 = \frac{\partial}{\partial u}, \quad X_4 = y\frac{\partial}{\partial u}, \tag{1.229}$$

$$X_5 = x\frac{\partial}{\partial x} + 3u\frac{\partial}{\partial u}, \quad X_6 = y\frac{\partial}{\partial y} - 2u\frac{\partial}{\partial u}. \tag{1.230}$$

As an example, let us consider the operators X_3 and X_4. The corresponding transformations groups are

$$T_{a_3}: \quad \overline{u} = u + a_3$$

and

$$T_{a_4}: \quad \overline{u} = u + a_4 y.$$

The composition (product) of the transformations is $T_{a_3}T_{a_4}$, which produces the group

$$\overline{u} = u + a_4 y + a_3. \tag{1.231}$$

The group (1.231) depends on two parameters a_3 and a_4. Let us denote

$$T_{a_3}T_{a_4} = \{T_a\}, \quad a = (a_3, a_4), \quad b = (b_3, b_4). \tag{1.232}$$

Then we can show that

$$\bar{\bar{u}} = u + y(a_4 + b_4) + a_3 + b_3 \in \{T_a\}$$

with

$$c = (c_3, c_4) = (a_3 + b_3, a_4 + b_4).$$

This means that

$$T_b T_a = T_{\varphi(a,b)}$$

with $\varphi(a, b) = a + b$, i. e., the group property (1.24) is satisfied, which means that $\{T_a\}$ represents a two-parameter group G. Also, since $\varphi(a, b) = \varphi(b, a)$, the group G is commutative (i. e., G is a group in which the result of applying the group operation to two group elements does not depend on the order in which they are written). It is true that all one-parameter groups are commutative. However, it is not true that all multi-parameter groups are commutative.

Example 1.32. Repeat the above procedure for the operators X_4 and X_6.

Solution. We already know

$$T_{a_4} : \quad \bar{u} = u + ya_4.$$

We find the tangent field associated with X_6 by solving the initial value problems:

$$\begin{aligned} \xi^1 : \quad & \frac{d\bar{y}}{da} = \bar{y}, \quad \bar{x}|_a = y, \\ \xi^2 : \quad & \frac{d\bar{u}}{da} = 2\bar{u}, \quad \bar{u}|_a = u. \end{aligned}$$

Thus,

$$T_{a_6} : \quad \bar{y} = ye^{a_6}, \quad \bar{u} = ue^{-2a_6}.$$

The result of the consecutive superposition of T_{a_4} and T_{a_6} is

$$\bar{\bar{y}} = ye^{a_6+b_6}, \quad \bar{\bar{u}} = ue^{-2(a_6+b_6)} + y(a_4e^{-2(a_6+b_6)} + b_4e^{-2(a_6+b_6)})$$

and $T_{a_4}T_{a_6} = T_c$ with $c = (c_4, c_6)$, in which

$$c_4 = a_4 + b_4 e^{3a_6}, \quad c_6 = a_6 + b_6.$$

So,

$$\bar{\bar{y}} = ye^{c_6}, \quad \bar{\bar{u}} = ue^{-2c_6} + yc_4 e^{-2c_6},$$

which means that $\{T_a\}$ represents a two-parameter group G_2 (which is noncommutative since $\varphi(a,b) \neq \varphi(b,a)$). □

Main question: Can we get a six-parameter group using all the operators in (1.230)?

A very clear answer to this question is given by means of Lie algebra.

1.9.3 Commutator of the group

Consider the following two operators:

$$X_1 = \xi_1^i(z)\frac{\partial}{\partial z^i}, \quad X_2 = \xi_2^i(z)\frac{\partial}{\partial z^i}. \tag{1.233}$$

Define the commutator $[X_1, X_2]$ via

$$[X_1, X_2] = X_1 X_2 - X_2 X_1. \tag{1.234}$$

The result of such operation is an operator of the form (1.233). It can be seen from the formula

$$[X_1, X_2] = (X_1(\xi_2^i) - X_2(\xi_1^i))\frac{\partial}{\partial z^i}, \tag{1.235}$$

which follows directly from (1.234) and which can be used as the definition of the commutator instead of (1.234). The commutator has the following three basic properties:

1) It is *bilinear*, i. e.,

$$\begin{aligned} [cX_1, X_2] &= [X_1, cX_2] = c[X_1, X_2], \\ [X, X_1 + X_2] &= [X, X_1] + [X, X_2], \\ [X_1 + X_2, X] &= [X_1, X] + [X_2, X]. \end{aligned} \tag{1.236}$$

2) It is *antisymmetric*, i. e.,

$$[X_1, X_2] = -[X_2, X_1]. \tag{1.237}$$

3) It satisfies the *Jacobi* identity, i. e.,

$$[X_1, [X_2, X_3]] + [X_2, [X_3, X_1]] + [X_3, [X_1, X_2]] = 0. \tag{1.238}$$

Definition 1.2. An algebra is a vector space in which the bilinear multiplication law is defined.

So the Lie algebra of operators of the form (1.233) is a vector space L, where $[X_1, X_2] \in L$ for each $X_1, X_2 \in L$. The dimension of the vector space L is the dimension of the Lie algebra. Let $X_1, \dots, X_r$ be a basis of L. Consider the commutator $[X_\mu, X_\nu]$ of all possible pairs of these basis operators. Since any operator $X \in L$ can be written as a linear combination of basis operators, i. e.,

$$X = \sum_{\mu=1}^{r} e^{\mu} X_\mu, \quad e^{\mu} = \text{const.}, \tag{1.239}$$

all possible commutators $[X_\mu, X_\nu]$ can be uniquely determined from the algebra L by using the bilinear property. Thus, an r-dimensional space L_r with basis $X_1, \dots, X_r$ forms a Lie algebra if and only if the commutators of the basis operators belong to L_r, i. e.,

$$[X_\mu, X_\nu] = \sum_{\lambda=1}^{r} C_{\mu\nu}^{\lambda} X_\lambda, \quad \mu, \nu = 1, \dots, r. \tag{1.240}$$

The constants $C_{\mu\nu}^{\lambda}$ are called structure constants.

Example 1.33. Show that

$$X_1 = \frac{\partial}{\partial y}, \quad X_2 = y\frac{\partial}{\partial y}, \quad X_3 = \frac{\partial}{\partial u}$$

form a three-dimensional Lie algebra.

Solution. Consider all possible commutators of the given operators:

$$[X_1, X_2] = X_1 X_2 - X_2 X = \frac{\partial}{\partial y}\left(y\frac{\partial}{\partial y}\right) - y\frac{\partial}{\partial y}\frac{\partial}{\partial y} = \frac{\partial}{\partial y} + y\frac{\partial^2}{\partial y^2} - y\frac{\partial^2}{\partial y^2} = \frac{\partial}{\partial y} = X_1,$$

$$[X_2, X_3] = y\frac{\partial}{\partial y}\left(\frac{\partial}{\partial u}\right) - \frac{\partial}{\partial u}\left(y\frac{\partial}{\partial y}\right) = y\frac{\partial^2}{\partial y \partial u} - y\frac{\partial^2}{\partial y \partial u} = 0,$$

$$[X_1, X_3] = \frac{\partial}{\partial y}\frac{\partial}{\partial u} - \frac{\partial}{\partial u}\frac{\partial}{\partial y} = 0.$$

The result can be represented by the following table, also known as *commutator table*:

	X_1	X_2	X_3
X_1	0	$-X_1$	0
X_2	X_1	0	0
X_3	0	0	0

□

Example 1.34. Let $[u, v] = u \times v$, where $\times$ means the vector product. Show that the given pair forms a Lie algebra.

Solution. We need to check the properties (1.236)–(1.238) of the given commutator:

1) Is it *bilinear?* We have

$$[u + v, w] = [u \times w] + [v \times w],$$

so yes, it is.

2) It is *antisymmetric?* We have

$$[u \times v] = -[v \times u], \tag{1.241}$$

so yes, it is, by the properties of the vector product.

3) Does it satisfy the *Jacobi* identity? By the definition of the vector product $[u \times [v \times w]] = v(uw) - w(uv)$, we have

$$[u \times [v \times w]] + [v \times [w \times u]] + [w \times [u \times v]] = v(uw) - w(uv) + w(vu) - u(vw) - v(wu) = 0.$$

Yes, it does.

Thus, the pair u, w forms a Lie algebra. □

1.10 Summary

Basic concepts from Lie group analysis of differential equations used in the first part of the book are assembled here.

Definition of one-parameter groups. Let

$$\bar{z}^i = f^i(z, a), \quad i = 1, \ldots N, \tag{1.242}$$

be a one-parameter family of invertible transformations of points $z = (z^1, \ldots, z^N) \in \mathbb{R}^N$ into points $\bar{z} = (\bar{z}^1, \ldots, \bar{z}^N) \in \mathbb{R}^N$. Here a is a real parameter from a neighborhood of $a = 0$, and we impose the condition that the transformation (1.242) is an identity if and only if $a = 0$, i. e.,

$$f^i(z, 0) = z^i, \quad i = 1, \ldots, N. \tag{1.243}$$

The set G of transformations (1.242) satisfying condition (1.243) is called a (local) one-parameter group of transformations in $\mathbb{R}^N$ if the successive action of two transformations is identical to the action of a third transformation from G, i. e., if the function $f = (f^1, \ldots, f^N)$ satisfies the following group property:

$$f^i(f(z,a),b) = f^i(z,c), \quad i = 1,\dots,N, \tag{1.244}$$

where

$$c = \varphi(a,b) \tag{1.245}$$

with a smooth function $\varphi(a,b)$ defined for sufficiently small a and b. The group parameter a in the transformation (1.242) can be changed so that the function (1.245) becomes $c = a + b$. In other words, the group property (1.244) can be written, upon choosing an appropriate parameter a (called a *canonical parameter*), in the form

$$f^i(f(z,a),b) = f^i(z,a+b). \tag{1.246}$$

Group generator. Let G be a group of transformations (1.242) satisfying the condition (1.243) and the group property (1.246). Expanding the functions $f^i(z,a)$ into Taylor series near $a = 0$ and keeping only the linear terms in a, one obtains the infinitesimal transformation of the group G:

$$\overline{z}^i \approx z^i + a\xi^i(z), \tag{1.247}$$

where

$$\xi^i(z) = \left.\frac{\partial f^i(z,a)}{\partial a}\right|_{a=0}, \quad i = 1,\dots,N. \tag{1.248}$$

The first-order linear differential operator

$$X = \xi^i(z)\frac{\partial}{\partial z^i} \tag{1.249}$$

is known as the generator of the group G.

Invariants. A function $J(z)$ is said to be an invariant of the group G if for each point $z = (z^1,\dots,z^N) \in \mathbb{R}^N$ it is constant along the trajectory determined by the totality of transformed points $\overline{z}$: $J(\overline{z}) = J(z)$.

The function $J(z)$ is an invariant of the group G with generator (1.249) if and only if

$$X(J) \equiv \xi^i(z)\frac{\partial J}{\partial z^i} = 0. \tag{1.250}$$

Hence, any one-parameter group has exactly $N-1$ functionally independent invariants (basis of invariants). One can take them to be the left-hand sides of $N-1$ first integrals $J_1(z) = C_1,\dots,J_{N-1}(z) = C_{N-1}$ of the characteristic equations for the linear PDE (1.250). Then any other invariant is a function of $J_1(z),\dots,J_{N-1}(z)$.

Invariant equations. We say that a system of equations

$$F_k(z) = 0, \quad k = 1, \dots, s, \tag{1.251}$$

is invariant with respect to the group G (or *admits* the group G) if the transformations (1.242) of the group G map any solution of equations (1.251) into a solution of the same equations, i. e.,

$$F_k(\overline{z}) = 0, \quad k = 1, \dots, s, \tag{1.252}$$

whenever z solves equations (1.251). The group G with the generator (1.249) is admitted by equations (1.251) if and only if

$$X(F_k)|_{(1.251)} = 0, \quad k = 1, \dots, s, \tag{1.253}$$

where the symbol $|_{(1.251)}$ means evaluated on the solutions of equations (1.251).

If z is a collection of independent variables $x = (x^1, \dots, x^n)$, dependent variables $u = (u^1, \dots, u^m)$, and partial derivatives $u_{(1)} = \{u_i^\alpha\}$, $u_{(2)} = \{u_{ij}^\alpha\}, \dots$ of u with respect to x up to certain order, where

$$u_i^\alpha = \frac{\partial u^\alpha}{\partial x^i}, \quad u_{ij}^\alpha = \frac{\partial^2 u^\alpha}{\partial x^i \partial x^j}, \dots,$$

then (1.251) is a system of PDEs,

$$F_k(x, u, u_{(1)}, \dots) = 0, \quad k = 1, \dots, s. \tag{1.254}$$

Furthermore, if the transformations (1.242) are obtained by the transformations of the independent and dependent variables

$$\overline{x} = f(x, u, a), \quad \overline{u} = g(x, u, a) \tag{1.255}$$

and the extension of (1.255) to all derivatives $u_{(1)}$, etc., involved in the differential equations (1.254), then equations (1.252) define a group G of transformations (1.255) admitted by the differential equations (1.254). In other words, an admitted group does not change the form of the system of differential equations (1.254). The generator of the admitted group G is termed an *infinitesimal symmetry* (or simply *symmetry*) of the differential equations (1.254). Equations (1.253) serve for obtaining the infinitesimal symmetries and are known as the *determining equations*. These equations are linear and homogeneous and therefore the set L of solutions is a vector space. Integration of determining equations often provides several linearly independent infinitesimal symmetries. Moreover, the determining equations have a specific property that guarantees that the set L is closed with respect to the commutator $[X_1, X_2] = X_1X_2 - X_2X_1$. Due to this property, L is called a *Lie algebra*. If the dimension of the vector space L is equal to r, the space is denoted by L_r and is called an r-dimensional Lie algebra. An r-dimensional Lie algebra L_r generates a group depending on r parameters which is called an r-parameter group.

Invariant solutions. Let the differential equations (1.254) admit a multi-parameter group G and let H be a subgroup of G. A solution

$$u^\alpha = h^\alpha(x), \quad \alpha = 1, \dots, m, \tag{1.256}$$

of equations (1.254) is called an H-invariant solution (termed for brevity an *invariant solution*) if equations (1.256) are invariant with respect to the subgroup H. If H is a one-parameter group and has the generator X, then the H-invariant solutions are constructed by calculating a basis of invariants $J_1, J_2, \dots$.

1.11 Exercises

Exercise 1.1. Using the transformation (1.10) one can eliminate the term of degree $n-1$ in any algebraic equation of degree n. For example, using the transformation (1.10) one can convert the general cubic equation

$$C_0x^3 + C_1x^2 + C_2x + C_3 = 0$$

to the form

$$\bar{x}^3 + 3p\bar{x} + 2q = 0$$

and obtain the Cardano formula for its solution. Can you do it?

Exercise 1.2. Check if the transformation

$$\overline{x} = \frac{x + a}{1 - ax}$$

defines a group G.

Exercise 1.3. Check if the rotation by angle ε,

$$\begin{aligned} \overline{x} &= x \cos\varepsilon - y \sin\varepsilon, \\ \overline{y} &= x \sin\varepsilon + y \cos\varepsilon, \end{aligned}$$

defines a group G.

Exercise 1.4. Write and solve the Lie equations for the group of transformations $\overline{x} = x\cos a + y \sin a, \overline{y} = y\cos a - x\sin a$.

Find the transformation group admitted by the operator

$$X = x\frac{\partial}{\partial x} + ky\frac{\partial}{\partial y}, \tag{1.257}$$

where k is a constant.

Exercise 1.5. Find the transformation group and invariant admitted by the operator

$$X = x\frac{\partial}{\partial x} + x\frac{\partial}{\partial y}. \tag{1.258}$$

How would you classify this transformation?

Exercise 1.6. Can you show that the invariant for the projective transformation

$$\overline{x} = \frac{x}{1 - ax}, \quad \overline{y} = \frac{y}{1 - ax}$$

is given by $J = \frac{x}{y}$?

Exercise 1.7. Using exponential maps, show that the generator of the projective transformation

$$\overline{x} = \frac{x}{1 - ax}, \quad \overline{y} = \frac{y}{1 - ax}$$

is

$$X = x^2\frac{\partial}{\partial x} + xy\frac{\partial}{\partial y}.$$

Exercise 1.8. Revise Example 1.24 by choosing the order of variables x, t, u.

Exercise 1.9. Revise Example 1.25 by choosing the order of variables t, x, u.

Exercise 1.10. Revise Example 1.26 by choosing the order of variables t, x, u.

Exercise 1.11. Find scaling transformations for the PDE $u_t - u_{xx} = 0$.

Exercise 1.12. Show that the heat equation $u_t - u_{xx} = 0$ is invariant under the transformation $\overline{t} = t, \overline{x} = x + 2at, \overline{u} = ue^{-(ax+a^2t)}$.

Exercise 1.13. Find the generators X_3, X_4, X_5, and X_6 in Example 1.34.

2 Integration of ordinary differential equations

2.1 First-order equations

Let us consider an ODE of the first order for $y = y(x)$ written in the form

$$Q(x,y)dx + P(x,y)dy = 0. \tag{2.1}$$

Equation (2.1) is equivalent to the first-order PDE

$$P\frac{\partial F}{\partial x} + Q\frac{\partial F}{\partial y} = 0 \tag{2.2}$$

in the sense that the left part of any integral $F(x,y) = C$ of equation (2.1) is a solution of equation (2.2).

2.1.1 Integrating factor

Let us suppose that equation (2.1) admits a one-parameter transformation group $\overline{x} = f(x,y,a), \overline{y} = \varphi(x,y,a)$ generated by the operator

$$X = \xi(x,y)\frac{\partial}{\partial x} + \eta(x,y)\frac{\partial}{\partial y}. \tag{2.3}$$

This means that any solution of equation (2.1) is mapped to a solution of that equation under the symmetry generated by the operator (2.3). Hence, any integral $F(x,y) = C$ is mapped to the integral $F(\overline{x},\overline{y}) = \overline{C}$. Then, since, as shown by formula (1.135),

$$F(\overline{z}) = e^{aX}F(z),$$

the presentation $XF = C_1$ will be an integral as well. So, if $F(x,y)$ is a solution of (2.2), then XF is a solution as well. However, since equation (2.2) can have only one independent solution, there exists a functional dependence

$$XF = \Phi(F).$$

So the function $F(x,y)$ satisfies the following two conditions simultaneously:

$$P\frac{\partial F}{\partial x} + Q\frac{\partial F}{\partial y} = 0 \quad \text{and} \quad \xi\frac{\partial F}{\partial x} + \eta\frac{\partial F}{\partial y} = \Phi(F). \tag{2.4}$$

Eliminating the singular case $Q\xi = P\eta$, equations (2.4) imply the following two equations,

$$\frac{\partial F}{\partial x} + \frac{Q\Phi}{\xi Q - \eta P}, \quad \frac{\partial F}{\partial y} = -\frac{P\Phi}{\xi Q - \eta P}, \tag{2.5}$$

https://doi.org/10.1515/9783111387499-002

or we can write (2.5) as

$$\frac{Qdx - Pdy}{\xi Q - \eta P} = \frac{\partial F}{\Phi(F)}. \tag{2.6}$$

Since the expression $dF/\Phi(F)$ in (2.6) is an exact differential, equation (2.1) admits the group with the given operator X if

$$\mu = \frac{1}{\xi Q - \eta P} \tag{2.7}$$

is an integrating factor (also known as the Lie integrating factor) of equation (2.1). It has been shown for the first time by Sophus Lie in Ref. [49].

Let us illustrate it by some examples.

Example 2.1. Solve the following nonlinear ODE (Riccati equation):

$$\frac{dy}{dx} + y^2 - \frac{2}{x^2} = 0. \tag{2.8}$$

Solution. We can start with looking the simplest transformation $\overline{x} = kx$ and $\overline{y} = ly$. Then, in the new variables equation (2.8) is written as

$$\frac{d\overline{y}}{d\overline{x}} + \overline{y}^2 - \frac{2}{\overline{x}^2} = \frac{l}{k}y' + l^2y^2 - \frac{1}{k^2}\frac{2}{x^2}. \tag{2.9}$$

The invariance condition requires that the form of a differential equation remains the same. So, we must have

$$\frac{d\overline{y}}{d\overline{x}} + \overline{y}^2 - \frac{2}{\overline{x}^2} = \lambda\left(\frac{dy}{dx} + y^2 - \frac{2}{x^2}\right), \tag{2.10}$$

where

$$\lambda = \frac{l}{k} = l^2 = \frac{1}{k^2}. \tag{2.11}$$

The relations (2.11) imply that

$$l = \frac{1}{k}, \quad k > 0. \tag{2.12}$$

We can set

$$k = e^a,$$

so that

$$\overline{x} = xe^a, \quad \overline{y} = ye^{-a}. \tag{2.13}$$

The vector field of the transformation (2.13) is given by (see also Example 2.1)

$$\xi = \frac{\partial \overline{x}}{\partial a}\bigg|_{a=0} = x, \quad \eta = \frac{\partial \overline{y}}{\partial a}\bigg|_{a=0} = -y. \tag{2.14}$$

Thus, the generator of the group is

$$X = x\frac{\partial}{\partial x} - y\frac{\partial}{\partial y}. \tag{2.15}$$

We say that the group G generated by the operator (2.15) is admitted by the differential equation (2.8) or that the differential equation (2.8) is invariant under the transformation (2.15). Now the question is: How can we solve equation (2.8) using the symmetry (2.15)?

Let us write the differential equation (2.8) in differential form as

$$dy = \left(y^2 - \frac{2}{x^2}\right)dx = 0. \tag{2.16}$$

Then we have

$$\xi = x, \quad \eta = -y, \quad Q = y^2 - \frac{2}{x^2}.$$

As follows from formula (2.7), the Lie integrating factor is

$$\mu = \frac{1}{\xi Q - \eta P} = \frac{1}{xy^2 - \frac{2}{x} - y} = \frac{x}{x^2y^2 - xy - 2}. \tag{2.17}$$

Let us multiply equation (2.16) by its integrating factor μ. We have

$$\frac{xdy}{x^2y^2 - xy - 2} + \frac{1}{x^2y^2 - xy - 2}\frac{x^2y^2 - 2}{x}dx = 0, \tag{2.18}$$

which is an exact equation and equivalent to the form $d\Phi = 0$, so that the function Φ can easily be determined. □

Example 2.2. Determine the function Φ from Example 2.1.

Solution. We note that

$$\frac{x^2y^2 - 2}{x} = y + \frac{x^2y^2 - xy - 2}{x},$$

and thus we can rewrite equation (2.18) as

$$\frac{xdy}{x^2y^2 - xy - 2} + \left(\frac{1}{x^2y^2 - xy - 2}\left[y + \frac{x^2y^2 - xy - 2}{x}\right]\right)dx = 0. \tag{2.19}$$

The latter equation (2.19) can be simplified to the form

$$\frac{xdy + ydx}{x^2y^2 - xy - 2} + \frac{dx}{x} = d\left(\ln x + \frac{1}{3}\ln\frac{xy - 2}{xy + 1}\right) = 0. \tag{2.20}$$

Integrating equation (2.20), we have

$$\frac{xy - 2}{xy + 1} = \frac{C}{x^3}. \tag{2.21}$$

Solving equation (2.20) for y, we obtain

$$y = \frac{2x^3 + C}{x(x^3 - C)}. \tag{2.22}$$

□

Example 2.3. Solve the following linear nonhomogeneous ODE:

$$\frac{dy}{dx} + R(x)y = Q(x). \tag{2.23}$$

Solution. Since equation (2.23) is linear, the admitted group can be found from the linear *superposition principle*: if $y_0(x)$ is a solution of a homogeneous equation

$$\frac{dy_0}{dx} + R(x)y_0 = 0,$$

then equation (2.23) admits the group $\bar{y} = y + ay_0(x)$ generated by the operator

$$X = y_0(x)\frac{\partial}{\partial y}, \tag{2.24}$$

where

$$y_0 = e^{-\int R(x)dx}.$$

Let us write the given equation (2.23) in the form (2.1), i. e.,

$$(Q(x) - R(x)y)dx - dy = 0.$$

Then, according to (2.7) we have

$$\mu = \frac{1}{\xi Q - \eta P} = \frac{1}{0 \cdot Q - y_0 P} = -\frac{1}{y_0}.$$

Thus, we can write the original equation in exact form as

$$\frac{dy}{y_0(x)} + \frac{R(x)y - Q(x)dx}{y_0(x)} = dF.$$

We have

$$\frac{\partial F}{\partial y} = e^{\int R(x)dx}, \quad \frac{\partial F}{\partial x} = R(x)ye^{\int R(x)dx} - Q(x)e^{\int R(x)dx}. \tag{2.25}$$

From the first equation of (2.25) we have

$$F = ye^{\int R(x)dx} + \phi(x),$$

and substituting in the second equation of (2.25) we obtain

$$\phi(x) = -\int Q(x)e^{\int R(x)dx}dx.$$

Now, since $dF = 0$, we have $F = C$, i. e.,

$$ye^{\int R(x)dx} - \int Q(x)e^{\int R(x)dx}dx = C.$$

Thus, the general solution of equation (2.23) is

$$y(x) = e^{-\int R(x)dx}\left(\int Q(x)e^{\int R(x)dx}dx + C\right).$$

□

Example 2.4. Solve the following nonlinear ODE:

$$\frac{dy}{dx}(2x + y^3) = y. \tag{2.26}$$

Solution. Let us write the equation in differential form,

$$ydx - (2x + y^3)dy = 0. \tag{2.27}$$

Since x and dx appear linearly in (2.27), the equation itself will be linear if we consider x as an unknown function and y will be regarded as an independent variable. So, let us rewrite equation (2.27) as

$$\frac{dx}{dy} - \frac{2}{y}x = y^2. \tag{2.28}$$

Let x_0 be a solution of the associated homogeneous equation

$$\frac{dx_0}{dy} - \frac{2}{y}x_0 = 0. \tag{2.29}$$

Then equation (2.29) admits the group $\overline{x} = x + ax_0(y)$ generated by the operator

$$X = x_0(y)\frac{\partial}{\partial y}, \tag{2.30}$$

where

$$y_0 = y^2.$$

Let us now write equation (2.28) as

$$\left(y^2 + \frac{2}{y}\right)dy - dx = 0. \tag{2.31}$$

Then

$$\mu = \frac{1}{\xi Q - \eta P} = -\frac{1}{x_0},$$

so equation (2.31) can be written in exact form as

$$\frac{dx}{x_0} - \left(\frac{2}{y}\frac{x}{x_0} + \frac{y^2}{x_0}\right)dy = dF$$

and

$$\frac{\partial F}{\partial x} = y^2, \quad \frac{\partial F}{\partial y} = \frac{2x}{y^3} + 1. \tag{2.32}$$

From the first equation of (2.32) we have $F = y^2 x + \phi(y)$, and substituting it into the second equation of (2.32) we obtain

$$\phi'(y) = \frac{2x}{y^3} + 1 - 2xy,$$

so that

$$\phi(y) = -\frac{x}{y^2} + y - xy^2.$$

Now, since $dF = 0$, we have $F = C$, i. e.,

$$F = y^2 x - \frac{x}{y^2} + y - xy^2 = y - \frac{x}{y^2} = C.$$

Thus, the general solution of equation (2.28) is

$$x = y^2(y - C).$$

□

2.1.2 Integration using canonical variables

Consider the first-order ODE in the form

$$y' = f(x,y). \tag{2.33}$$

Suppose we know that equation (2.33) is invariant under the group generated by the operator

$$X = \xi(x,y)\frac{\partial}{\partial x} + \eta(x,y)\frac{\partial}{\partial y}. \tag{2.34}$$

Here is the three-step algorithm to solve equation (2.33) with the known symmetry (2.34).

Step 1. Find the canonical variables (t, u) by solving

$$X(t) = \xi(x,y)\frac{\partial t}{\partial x} + \eta(x,y)\frac{\partial t}{\partial y} = 1, \tag{2.35}$$

$$X(u) = \xi(x,y)\frac{\partial u}{\partial x} + \eta(x,y)\frac{\partial u}{\partial y} = 0. \tag{2.36}$$

Step 2. Rewrite the original equation (2.33) in the canonical variables t and u by letting $u = u(t)$ and express y' in terms of t, u and the derivative u'. Then the equation is written in the integrable form

$$\frac{du}{dt} = g(u). \tag{2.37}$$

Step 3. Integrate equation (2.37) to get the solution in the form $u = \phi(t, c)$. Then use back substitution $t = t(x,y)$ and $u = u(x,y)$.

Example 2.5. Solve the following nonlinear ODE:

$$\frac{dy}{dx} + y^2 - \frac{2}{x^2} = 0. \tag{2.38}$$

Solution. We already know from Example 2.1 that the symmetry of equation (2.38) is given by

$$X = x\frac{\partial}{\partial x} - y\frac{\partial}{\partial y}. \tag{2.39}$$

Step 1. Find the canonical variables.

The characteristic equation associated with the generator X is

$$\frac{dx}{x} = -\frac{dy}{y} \quad \Rightarrow \quad \ln x + \ln y = C,$$

and thus the first canonical variable is

$$u = xy.$$

Indeed,

$$X(u) = x\frac{\partial(xy)}{\partial x} - y\frac{\partial(xy)}{\partial y} = x(y) - y(x) = 0.$$

Next we find the second canonical variable t via $X(t) = 1$:

$$x\frac{\partial t}{\partial x} - y\frac{\partial t}{\partial y} = 1 \quad \Rightarrow \quad \frac{dx}{x} = \frac{dt}{1} \quad \Rightarrow \quad t = \ln|x|.$$

Step 2. Write the original equation (2.38) in the canonical variables. First, we express the derivative in terms of the canonical variables as follows:

$$\frac{dy}{dx} = \frac{d}{dx}\left(\frac{u}{x}\right) = \frac{\frac{du}{dx}x - u}{x^2} = -\frac{u}{x^2} + \frac{1}{x}\frac{du}{dt}\frac{dt}{dx} = \frac{u}{x^2} + \frac{1}{x}\frac{du}{dt},$$

since $dt/dx = 1$. Thus, the original equation is written as

$$\frac{dy}{dx} + y^2 - \frac{2}{x^2} = \frac{1}{x^2}\left(\frac{du}{dt} + u^2 - u - 2\right) = 0,$$

which means that

$$\frac{du}{dt} = -(u^2 - u - 2). \tag{2.40}$$

Step 3. Finally, we integrate (2.40):

$$\frac{du}{u^2 - u - 2} = -dt. \tag{2.41}$$

Using partial fractions, we have

$$\frac{1}{u^2 - u - 2} = \frac{1}{3}\left(\frac{1}{u-2} - \frac{1}{u+1}\right),$$

so we can integrate (2.41) to obtain

$$\ln\left(\frac{u-2}{u+1}\right) = -3t + \ln C,$$

which gives us the solution for u,

$$u = \frac{C + 2e^{3t}}{e^{3t} - C}, \tag{2.42}$$

where $t = \ln|x|$ and $u = xy$. Writing the solution (2.42) in original variables, we have

$$y = \frac{2x^3 + C}{x(x^3 + C)}. \tag{2.43}$$

□

2.1.3 Integration of ordinary differential equations admitting groups

First, as a straightforward illustrative example of Theorem 1.4, let us find a general form of the first-order ODEs for $y = y(x)$ that are invariant under nonuniform dilation

$$X = x\frac{\partial}{\partial x} + 2y\frac{\partial}{\partial y}.$$

To this end, we need to extend the action of the operator of the form

$$X = \xi(x,y)\frac{\partial}{\partial x} + \eta(x,y)\frac{\partial}{\partial y}$$

on the first derivative by using the prolongation formula (1.148),

$$\underset{1}{X} = X + \zeta_i^\alpha \frac{\partial}{\partial u_i^\alpha},$$

where (see also formula (1.149))

$$\zeta_i = D_i(\eta) - u_j D_i(\xi^j). \tag{2.44}$$

In our case, we have

$$\underset{1}{X} = X + \zeta_1 \frac{\partial}{\partial y_x}.$$

Using the order of variables in the extended space (x, y, y_x), from the first prolongation formula (2.44) we have

$$\zeta_1 = \zeta_x = D_x(\eta) - y_x D_x(\xi), \quad D_x = \frac{\partial}{\partial x} + y_x\frac{\partial}{\partial y}.$$

Thus,

$$\underset{1}{X} = \xi(x,y)\frac{\partial}{\partial x} + \eta(x,y)\frac{\partial}{\partial y} + [\eta_x + (\eta_y - \xi_x)\dot{y} - \dot{y}^2\xi_y]\frac{\partial}{\partial y_x}, \tag{2.45}$$

i. e.,

$$\underset{1}{X} = x\frac{\partial}{\partial x} + 2y\frac{\partial}{\partial y} + y_x\frac{\partial}{\partial y_x}.$$

The characteristic system is deduced from the operator X as

$$\frac{dx}{x} = \frac{dy}{2y} = \frac{dy_x}{y_x},$$

from which we find the basis of invariants

$$J_1 = \frac{y}{x^2}, \quad J_2 = \frac{y_x}{x}.$$

According to Theorem 1.4, any invariant first-order equation can be written in the form $J_2 = F(J_1)$, or in our case,

$$\frac{dy}{dx} = xF\left(\frac{y}{x^2}\right). \tag{2.46}$$

Equation (2.46) represents a general form of the first-order ODE admitting the nonuniform dilation transformation $\overline{x} = e^a x, \overline{y} = e^{2a} y$.

Example 2.6. Let us find a general form of the first-order ODEs for $y = y(x)$ that are invariant under the group of transformations determined by the generator

$$X = \phi(x)\frac{\partial}{\partial z}, \tag{2.47}$$

where $\phi(x)$ is a solution of the following linear homogeneous equation:

$$\phi'(x) + A(x)\phi(x) = 0. \tag{2.48}$$

Solution. Since $\phi(x)$ is a solution of (2.48), we have

$$\phi(x) = e^{-\int A(x)dx}. \tag{2.49}$$

We can use the following simple change of variable:

$$z = y^m, \quad m \neq 0, \quad m \neq 1. \tag{2.50}$$

According to formula (1.119), we have

$$\overline{X} = X(\overline{z}^i)\frac{\partial}{\partial \overline{z}^i},$$

i. e., the change of variable (2.50) transforms the generator (2.47) to the form

$$X = \phi(x)y^{1-m}\frac{\partial}{\partial y}. \tag{2.51}$$

Now, using equation (2.48) let us write the first prolongation of the operator (2.51). We have

$$X = \phi(x)\frac{y}{y^m}\frac{\partial}{\partial y} \quad \Longrightarrow \quad \underset{1}{X} = X + \zeta_1\frac{\partial}{\partial y_x},$$

where

$$\zeta_1 = D_x(\eta)$$

and

$$\eta = \phi(x)\frac{y}{y^m}, \quad D_x = \frac{\partial}{\partial x} + y_x\frac{\partial}{\partial y}.$$

Due to equation (2.48), we have $\phi'(x) = -A(x)\phi(x)$, so

$$\zeta_1 = \eta_x + y_x\eta_y = -\frac{\phi(x)}{y^m}Ay + \frac{\phi(x)}{y^m}y_x - \frac{\phi(x)}{y^m}my_x.$$

Thus,

$$\underset{1}{X} = \frac{\phi(x)}{y^m}\left[y\frac{\partial}{\partial y} - (Ay + (m-1)y_x)\frac{\partial}{\partial y_x}\right]. \tag{2.52}$$

We can see immediately that one of the invariants of the generator (2.52) is x. The second invariant can be found from the characteristic system

$$\frac{dy}{y} = -\frac{dy_x}{(m-1)y_x A_y},$$

which is more convenient to rewrite as

$$\frac{dy_x}{dy} + \frac{(m-1)}{y}y_x + A = 0. \tag{2.53}$$

Since x is the invariant, we regard $A = A(x)$ as a constant value. Hence, equation (2.53) is a linear nonhomogeneous ODE of the form (2.23) considered in Example 2.1. So, changing the variables $x = y$ and $y_x = y$, we can use the result of Example 2.1, and we can write (2.53) as the following exact equation:

$$y_x e^{\int \frac{m-1}{y}dy} + A(x)\int e^{\int \frac{m-1}{y}dy}dy = y_x y^{m-1} + \frac{A(x)}{m}y^m = C. \tag{2.54}$$

Hence, as the second invariant of the prolonged generator $\underset{1}{X}$ we can use

$$J = y_x y^{m-1} + \frac{A(x)}{m}y^m. \tag{2.55}$$

The corresponding G/H factor system $J = B(x)$ is

$$\frac{dy}{dx} + \frac{A(x)}{1-n}y = B(x)y^n, \tag{2.56}$$

where we denote $n = 1 - m$. The resulting equation (which is a Bernoulli equation) represents the first-order ODE for $y = y(x)$ that is invariant under the group of transformations determined by the generator (2.47). □

Example 2.7. Let us solve the Bernoulli equation

$$\frac{dy}{dx} + \frac{A(x)}{1-n}y = B(x)y^n \tag{2.57}$$

by using the generator (2.51),

$$X = \phi(x)y^{1-m}\frac{\partial}{\partial y}, \tag{2.58}$$

admitted by equation (2.57).

Solution. As we already know from Example 2.6, x is the first invariant. The second invariant u can be found by equation (1.120), $X(u) = 1$, i. e.,

$$\phi(x)y^n\frac{\partial u}{\partial y} = 1. \tag{2.59}$$

As follows from (2.59),

$$u = \frac{y^{1-n}}{(1-n)\phi(x)}. \tag{2.60}$$

So the Bernoulli equation can be integrated by the change of variable (2.60), i. e., in terms of u, the original equation (2.57) is written as

$$\phi(x)\frac{du}{dx} = B(x). \tag{2.61}$$

Integrating (2.61) we have

$$u = \int \frac{B(x)}{\phi(x)}dx + C. \tag{2.62}$$

Substituting this expression for u and $\phi(x)$ given by (2.61) into (2.60) we arrive at the general form of the integral for the Bernoulli equation:

$$y^{1-n} = (1-n)e^{-\int A(x)dx}\left[\int B(x)e^{-\int A(x)dx}dx + C\right]. \tag{2.63}$$

□

Example 2.8. Solve the Bernoulli equation

$$\frac{dy}{dx} - \frac{4}{x}y = x\sqrt{y} \tag{2.64}$$

by using the general formula (2.63).

Solution. We set $n = 1/2$, $A = -2/x$, and $B = x$ in formula (2.63) to obtain

$$\sqrt{y} = e^{\int \frac{2}{x}dx}\left[\frac{1}{2}\int xe^{-\int \frac{2}{x}dx}dx + C\right],$$

so

$$y = x^4(\ln\sqrt{x} + C)^2. \qquad \square$$

Example 2.9. Integrate for the Bernoulli equation

$$\frac{dy}{dx} + \frac{A(x)}{1-n}y = B(x)y^n \tag{2.65}$$

by using the general formula (2.63).

Solution. We already know from Example 2.7 that equation (2.65) admits the generator

$$X = \phi(x)y^n\frac{\partial}{\partial y}. \tag{2.66}$$

First, let us transform the original equation (2.65) by dividing it by y^n, multiplying by $(-n+1)$, and introducing a new function z via $z = y^{-n+1}$. The resulting equation for z is

$$\frac{dz}{dx} + A(x)z = (1-n)B(x). \tag{2.67}$$

Now, generator X in terms of the new function z is transformed to

$$X = \phi(x)(1-n)\frac{\partial}{\partial z}. \tag{2.68}$$

We can write equation (2.67) in differential form as

$$dz = [(1-n)B(x) - A(x)z]dx. \tag{2.69}$$

So, according to formula (2.7), the integrating factor of equation (2.69) is

$$\mu = \frac{1}{\xi Q - \eta P} = \frac{1}{\phi(x)(1-n)}. \tag{2.70}$$

Let us quickly check that the function μ given by formula (2.70) is the actual integrating factor. Indeed, multiplying equation (2.69) by μ we have

$$\frac{dz}{\phi(x)(1-n)} + \frac{A(x)z - (1-n)B(x)}{\phi(x)(1-n)}dx = 0.$$

The function μ is the integrating factor if the following equation is true:

$$\frac{\partial \mu Q}{\partial x} = \frac{\partial \mu P}{\partial z}, \tag{2.71}$$

where

$$\frac{\partial \mu Q}{\partial x} = -\frac{\phi'(x)}{\phi^2(x)(1-n)}, \quad \frac{\partial \mu P}{\partial z} = \frac{A(x)}{\phi(x)(1-n)}. \tag{2.72}$$

But since $\phi(x)$ is a solution of a linear homogeneous equation (see (2.48))

$$\phi'(x) + A(x)\phi(x) = 0, \tag{2.73}$$

we have

$$A(x) = -\frac{\phi'(x)}{\phi(x)},$$

so

$$\frac{\partial \mu P}{\partial z} = -\frac{\phi'(x)}{\phi(x)(1-n)},$$

and thus equation (2.71) is true. □

Example 2.10. Find the general form of differential equations for $y = y(x)$ admitting the group with the generator

$$X = \frac{\partial}{\partial x}. \tag{2.74}$$

Solution. We start with the first prolongation

$$\underset{1}{X} = \frac{\partial}{\partial x} + \zeta_1 \frac{\partial}{\partial y_x}, \quad \zeta_1 = \zeta_x = D_x(\eta) - y_x D_x(\xi), \quad D_x = \frac{\partial}{\partial x} + y_x \frac{\partial}{\partial y},$$

where $\eta = 0$ and $\xi = 1$. Thus,

$$\zeta_1 = \eta_x + y_x \eta_y - y_x \xi_x - y_x^2 \xi_y = 0,$$

so

$$\underset{1}{X} = \frac{\partial}{\partial x}.$$

The first invariant is $J_1 = y$ and the second invariant is found from the characteristic equation associated with the generator $\underset{1}{X}$:

$$\frac{dx}{1} = \frac{dy_x}{0} \quad \Rightarrow \quad J_2 = y_x.$$

The corresponding G/H factor system is $J_2 = F(J_1)$, so the general form of the first-order ODEs will be $y_x = F(y)$. □

Example 2.11. Find the general form of differential equations for $y = y(x)$ admitting the group with the generator

$$X = \frac{\partial}{\partial y}. \tag{2.75}$$

Solution. We start with the first prolongation

$$\underset{1}{X} = \frac{\partial}{\partial y} + \zeta_1 \frac{\partial}{\partial y_x}, \quad \zeta_1 = \eta_x + y_x \eta_y - y_x \xi_x - y_x^2 \xi_y = 0,$$

where $\eta = 1$ and $\xi = 0$. Thus,

$$\underset{1}{X} = \frac{\partial}{\partial y}.$$

The first invariant is $J_1 = x$ and the second invariant is $J_2 = y_x$. The corresponding G/H factor system is $J_2 = F(J_1)$, so the general form of the first-order ODEs will be $y_x = F(x)$. □

Example 2.12. Find the general form of differential equations for $y = y(x)$ admitting the group with the generator

$$X = l\frac{\partial}{\partial x} - k\frac{\partial}{\partial y}. \tag{2.76}$$

Solution. We start with the first prolongation

$$\underset{1}{X} = l\frac{\partial}{\partial x} - k\frac{\partial}{\partial y} + \zeta_1 \frac{\partial}{\partial y_x}, \quad \zeta_1 = D_x(\eta) - y_x D_x(\xi), \quad D_x = \frac{\partial}{\partial x} + y_x \frac{\partial}{\partial y},$$

where $\eta = -k$ and $\xi = l$. Thus,

$$\zeta_1 = \eta_x + y_x \eta_y - y_x \xi_x - y_x^2 \xi_y = 0,$$

so

$$\underset{1}{X} = X = l\frac{\partial}{\partial x} - k\frac{\partial}{\partial y}.$$

The characteristic equations are

$$\frac{dx}{l} = -\frac{dy}{k} = \frac{dy_x}{0}.$$

Thus,

$$\begin{aligned} \frac{dx}{l} = -\frac{dy}{k} \quad &\Rightarrow \quad \frac{1}{l}x + \frac{1}{k}y = C \quad \Rightarrow \quad J_1 = kx + ly, \\ \frac{dx}{l} = \frac{dy_x}{0} \quad &\Rightarrow \quad J_2 = y_x. \end{aligned}$$

The corresponding G/H factor system is $J_2 = F(J_1)$, so the general form of the first-order ODEs will be $y_x = F(kx + ly)$. □

Example 2.13. Find the general form of differential equations for $y = y(x)$ admitting the group with the generator

$$X = x\frac{\partial}{\partial x} + y\frac{\partial}{\partial y}. \tag{2.77}$$

Solution. In this case, we have $\eta = y$ and $\xi = x$. Thus,

$$\zeta_1 = \eta_x + y_x\eta_y - y_x\xi_x - y_x^2\xi_y = y_x - y_x = 0,$$

so

$$\underset{1}{X} = X = x\frac{\partial}{\partial x} + y\frac{\partial}{\partial y}.$$

The characteristic equations are

$$\frac{dx}{x} = \frac{dy}{y} = \frac{dy_x}{0}.$$

Thus,

$$\begin{aligned} \frac{dx}{x} = \frac{dy}{y} \quad &\Rightarrow \quad \ln x - \ln y = \ln C \quad \Rightarrow \quad J_1 = \frac{y}{x}, \\ \frac{dx}{x} = \frac{dy_x}{0} \quad &\Rightarrow \quad J_2 = y_x. \end{aligned}$$

The corresponding G/H factor system is $J_2 = F(J_1)$, so the general form of the first-order ODEs will be $y_x = F(\frac{y}{x})$. □

Example 2.14. Find the general form of differential equations for $y = y(x)$ admitting the group with the generator

$$X = x\frac{\partial}{\partial x} + ky\frac{\partial}{\partial y}. \tag{2.78}$$

Solution. In this case, we have $\eta = ky$ and $\xi = x$. Thus,

$$\zeta_1 = ky_x - y_x = y_x(k-1),$$

so the first prolongation of X is

$$\underset{1}{X} = x\frac{\partial}{\partial x} + ky\frac{\partial}{\partial y} + y_x(k-1)\frac{\partial}{\partial y_x}.$$

The characteristic equations are

$$\frac{dx}{x} = \frac{dy}{ky} = \frac{dy_x}{(k-1)y_x}.$$

Thus,

$$\begin{aligned} \frac{dx}{x} = \frac{dy}{ky} \quad &\Rightarrow \quad k\ln x - \ln y = \ln C \quad \Rightarrow \quad J_1 = \frac{y}{x^k}, \\ \frac{dx}{x} = \frac{dy_x}{(k-1)y_x} \quad &\Rightarrow \quad \ln x^{k-1} - \ln y_x = \ln C \quad \Rightarrow J_2 = \frac{y_x}{x^{k-1}}. \end{aligned}$$

The corresponding G/H factor system is $J_2 = F(J_1)$, so the general form of the first-order ODEs will be

$$y_x = x^{k-1}F\left(\frac{y}{x^k}\right).$$

□

Example 2.15. Find the general form of differential equations for $y = y(x)$ admitting the group with the generator

$$X = e^{-\int P(x)dx}\frac{\partial}{\partial y}. \tag{2.79}$$

Solution. In this case, we have

$$\eta = e^{-\int P(x)dx}, \quad \xi = 0.$$

Thus,

$$\zeta_1 = -P(x)e^{-\int P(x)dx},$$

so the first prolongation of X is

$$\underset{1}{X} = e^{-\int P(x)dx}\frac{\partial}{\partial y} - P(x)e^{-\int P(x)dx}\frac{\partial}{\partial y_x}.$$

The characteristic equations are

$$\frac{dx}{x} = \frac{dy}{ky} = \frac{dy_x}{(k-1)y_x}.$$

The first invariant is $J_1 = x$ and the second invariant is found from the characteristic equations

$$\frac{dy}{e^{-\int P(x)dx}} = -\frac{dy_x}{P(x)e^{-\int P(x)dx}} \quad \Rightarrow \quad P(x)dy + y_x = C.$$

The corresponding G/H factor system is $J_2 = F(J_1)$, so the general form of the first-order ODEs will be

$$y_x + P(x)y = F(x).$$

$$\frac{dx}{x} = \frac{dy}{ky} = \frac{dy_x}{(k-1)y_x}. \qquad \square$$

Example 2.16. Find the general form of differential equations for $y = y(x)$ admitting the group with the generator

$$X = x\frac{\partial}{\partial x} + \frac{y}{x}\frac{\partial}{\partial y}. \tag{2.80}$$

Solution. In this case, we have $\eta = y/x$ and $\xi = 1$. Thus,

$$\zeta_1 = \eta_x + y_x\eta_y - y_x\xi_x - y_x^2\xi_y = -\frac{y}{x^2} + \frac{y_x}{x},$$

so the first prolongation of X is

$$\underset{1}{X} = x\frac{\partial}{\partial x} + \frac{y}{x}\frac{\partial}{\partial y} + \left(\frac{y_x}{x} - \frac{y}{x^2}\right)\frac{\partial}{\partial y_x}.$$

The characteristic equations are

$$\frac{dx}{1} = \frac{x}{y}dy = \frac{dy_x}{-\frac{y}{x^2} + \frac{y_x}{x}}.$$

Thus,

$$\frac{dx}{x} = \frac{dy}{y} \quad \Rightarrow \quad J_1 = \frac{y}{x}, \tag{2.81}$$

$$\frac{dx}{x} = \frac{dy_x}{y_x - J_1} \quad \Rightarrow \quad \frac{dy_x}{dx} = \frac{y_x}{x} - \frac{J_1}{x}. \tag{2.82}$$

Equation (2.82) can be written as a linear nonhomogeneous equation

$$\frac{dy_x}{dx} - \frac{y_x}{x} - \frac{J_1}{x} \tag{2.83}$$

and its solution is

$$y_x = c_1 J_1 x + J_1.$$

The corresponding G/H factor system is $c_1 = F(J_1)$, i. e.,

$$y_x = F(J_1)J_1 + J_1 x = F\left(\frac{y}{x}\right)\frac{y}{x} + \frac{y}{x}x,$$

so the general form of the first-order ODEs will be

$$y_x = \frac{y}{x} + yF\left(\frac{y}{x}\right).$$

□

Example 2.17. Find out what type of equations with a known admissible generator (2.74), (2.75), (2.76), (2.77), (2.78), or (2.80) the equation

$$x\frac{dy}{dx} - y = x^m \tag{2.84}$$

belongs to and find its integrating factor.

Solution. Let us do the change of variables

$$y = z^p. \tag{2.85}$$

Then we can rewrite equation (2.84) in terms of z as

$$xmz^{m-1}\frac{dz}{dx} - z^m = x^m. \tag{2.86}$$

Since the resulting equation (2.86) is homogeneous, we can look for the scaling transformations

$$\overline{z} = az, \quad \overline{x} = bx. \tag{2.87}$$

Substituting the transformations (2.87) into equation (2.86), we see that $a = b$ and hence equation (2.86) admits the group with the generator (2.77), i. e.,

$$X = x\frac{\partial}{\partial x} + z\frac{\partial}{\partial z}.$$

In this case, since $\xi = x$ and $\eta = z$, we have

$$\zeta_1 = \eta_x + z_x\eta_z - z_x\xi_x - z_x^2\xi_z = z_x - z_x = 0,$$

so

$$\underset{1}{X} = x\frac{\partial}{\partial x} + z\frac{\partial}{\partial z}.$$

The characteristic equations are

$$\frac{dx}{x} = \frac{dz}{z} = \frac{dz_x}{0}.$$

Thus,

$$\frac{dx}{x} = \frac{dz}{z} \quad \Rightarrow \quad \ln x - \ln z = \ln C \quad \Rightarrow \quad J_1 = \frac{z}{x},$$
$$\frac{dx}{x} = \frac{dz_x}{0} \quad \Rightarrow \quad J_2 = z_x.$$

The corresponding G/H factor system is $J_2 = F(J_1)$, so the general form of the first-order ODEs will be $z_x = F(\frac{z}{x})$. In order to find the integrating factor, we will write the differential form

$$Qdx + Pdz = 0,$$

i. e.,

$$(z^m + x^m)dx - (xmz^{m-1})dz = 0.$$

Thus, the integrating factor is found by formula (2.7) as

$$\mu = \frac{1}{xz^m + x^{m+1} - xmz^m}. \tag{2.88}$$

□

2.2 Second-order equations

Just as when considering first-order equations for $y = y(x)$ and using the first prolongation of the generator X, in accordance with Theorem 1.4, second-order equations admitting the generator X can be written in terms of second-order differential invariants that are functions of four *independent* variables y, y, y_x, y_{xx}.

Let us start with writing the general form of the generator X in the form

$$X = \xi(x,y)\frac{\partial}{\partial x} + \eta(x,y)\frac{\partial}{\partial y}. \tag{2.89}$$

The first prolongation of the generator X is

$$\underset{1}{X} = X + \zeta_1\frac{\partial}{\partial y_x},$$

where

$$\zeta_1 = D_x(\eta) - y_x D_x(\xi), \quad D_x = \frac{\partial}{\partial x} + y_x \frac{\partial}{\partial y},$$

so

$$\zeta_1 = \eta_x + y_x \eta_y - y_x \xi_x - y_x^2 \xi_y = 0.$$

Hence,

$$\underset{1}{X} = \xi(x,y)\frac{\partial}{\partial x} + \eta(x,y)\frac{\partial}{\partial y} + [\eta_x + (\eta_y - \xi_x)y_x - y_x^2\xi_y]\frac{\partial}{\partial y_x}. \tag{2.90}$$

Similarly, the second prolongation of the generator X is

$$\underset{2}{X} = \xi\frac{\partial}{\partial x} + \eta\frac{\partial}{\partial y} + \zeta_1\frac{\partial}{\partial y_x} + \zeta_{11}\frac{\partial}{\partial y_{xx}}, \tag{2.91}$$

where we are going to use the second prolongation formula (1.151),

$$\zeta_{ij}^{\alpha} = D_j(\zeta_i^{\alpha}) - u_{ik}^{\alpha} D_j(\xi^k),$$

i. e.,

$$\begin{aligned}
\zeta_{11} = \zeta_{xx} &= D_x(\zeta_1) - y_{xx}D_x(\xi) \\
&= \zeta_{1x} + y_x\zeta_{1y} + y_{xx}\zeta_{1y_x} - y_{xx}(\xi_x + y_x)\xi_y \\
&= \eta_{xx} + y_x(\eta_{yx} - \xi_{xx}) - y_x^2\xi_{yx} + y_x[\eta_{xy} + y_x(\eta_{yy} - \xi_{xy}) - y_x^2\xi_{yy}] \\
&\quad + y_{xx}[(\eta_y - \xi_x) - 2y_x\xi_y] - y_{xx}(\xi_x + y_x\xi_y) \\
&= \eta_{xx} + (2\eta_{xy} - \xi_{xx})y_x + (\eta_{yy} - 2\xi_{xy})y_x^2 - y_x^3\xi_{yy} + (\eta_y - 2\xi_x - 3y_x\xi_y)y_{xx}.
\end{aligned}$$

Thus,

$$\begin{aligned}
\underset{2}{X} &= \xi\frac{\partial}{\partial x} + \eta\frac{\partial}{\partial y} + [\eta_x + (\eta_y - \xi_x)y_x - y_x^2\xi_y]\frac{\partial}{\partial y_x} \\
&\quad + [\eta_{xx} + (2\eta_{xy} - \xi_{xx})y_x + (\eta_{yy} - 2\xi_{xy})y_x^2 - y_x^3\xi_{yy} + (\eta_y - 2\xi_x - 3y_x\xi_y)y_{xx}]\frac{\partial}{\partial y_{xx}}.
\end{aligned} \tag{2.92}$$

There is a way to obtain second-order differential invariants from first-order and zero-order invariants using differentiation. The algorithm can be described as follows.

2.2.1 Algorithm

As we can see from the generator $\underset{1}{X}$ given by (2.90), there is one zero-order independent invariant and one second-order independent invariant. Let us denote them by $u(x,y)$ and $v(x,y,y_x)$. Let k and l be arbitrary constants. The first-order differential equation

$$v(x,y,y_x) = ku(x,y) - l = 0 \tag{2.93}$$

admits the generator X, since the left side of (2.93) is a differential invariant. If k is fixed and l is varied, then we obtain an infinite family of invariant equations. In view of this invariance, the set of all integral curves of the resulting family of equations will be invariant under transformations with the operator X. However, the specified set of curves coincides with the set of integral curves of the second-order differential equation obtained from (2.93) by excluding the parameter l by differentiation. Consequently, after the transformation of the group, each solution of the second-order equation $dv - kdu = 0$ or $dv/du = k$ is transformed into some solution of the same equation. Let us denote $w = dv/du$. Then we can see that the equation $w - k = 0$ admits the generator X, so the expression

$$\underset{2}{X}(w-k) \equiv \underset{2}{X}w$$

vanishes on the solution curves $w - k = 0$. However, this is true for any values of k and the expression $\underset{2}{X}w$ is independent of k. Thus, $\underset{2}{X}w = 0$ is an invariant.

This algorithmic description can be summarized as the following statement.

Theorem 2.1. *Let the invariant $u(x,y)$ and the differential invariant $v(x,y,y_x)$ be known for a given operator*

$$X = \xi(x,y)\frac{\partial}{\partial x} + \eta(x,y)\frac{\partial}{\partial y}.$$

Then, by differentiation, we obtain a second-order differential invariant w given by

$$w = \frac{dv}{du} = \frac{\frac{dv}{dx}}{\frac{du}{dx}} = \frac{v_x + y_x v_y + y_{xx} v_{yx}}{u_x + y_x u_y} = \frac{Dv}{Du}. \tag{2.94}$$

Any other differential invariant up to the second order is a function of u, v, and w.

According to Theorem 2.1, any second-order ODE admitting the group with the generator X can be written in terms of differential invariants u, v, and w of the generator X. Therefore, solving this invariant presentation with respect to w and using (2.94), we can write the second-order ODE as the first-order ODE

$$\frac{dv}{du} = F(u,v). \tag{2.95}$$

If we can find the first integral

$$\Phi(u,v,C) = 0 \tag{2.96}$$

of equation (2.95), then the solution technique of solving the given second-order equation is reduced to a simple integration. Indeed, substitution of the known functions

$u(x,y)$ and $v(x,y,y_x)$ into (2.96) leads us to a first-order equation admitting X due to invariance of u and v.

Let us illustrate this by an example.

Example 2.18. Reduce the second-order ODE

$$\frac{d^2y}{dx^2} + P(x)y = 0 \tag{2.97}$$

to a first-order ODE.

Solution. The admitted group can be found from the superposition principle: if y is a solution of equation (2.97), then this equation admits the group $\bar{y} = y + ay$, whose generator is

$$Y = y(x)\frac{\partial}{\partial x}.$$

Then

$$\underset{1}{Y} = \xi\frac{\partial}{\partial x} + \eta\frac{\partial}{\partial y} + \zeta_1\frac{\partial}{\partial y_x},$$

where

$$\eta = y, \quad \xi = 0, \quad \zeta_1 = D_x(\eta) - y_x D_x(\xi), \quad D_x = \frac{\partial}{\partial x} + y_x\frac{\partial}{\partial y},$$

so

$$\zeta_1 = \eta_x + y_x\eta_y - y_x\xi_x - y_x^2\xi_y = y_x,$$

i. e.,

$$\underset{1}{Y} = y\frac{\partial}{\partial y} + y_x\frac{\partial}{\partial y_x}.$$

The characteristic equations are

$$\frac{dx}{0} = \frac{dy}{y} = \frac{dy_x}{y_x}.$$

Thus,

$$J_1 = x, \quad J_2 = \frac{y_x}{y}.$$

So we choose $u = x$ and $v = y_x/y$. Then according to (2.94) we have

$$\frac{dv}{du} = \frac{(\frac{y_x}{y})_x + y_x(\frac{y_x}{y})_y + y_{xx}(\frac{y_x}{y})_{yx}}{x_x + y_x x_y} = \frac{y_{xx}}{y} - v^2. \tag{2.98}$$

We can rewrite (2.98) as

$$\frac{y_{xx}}{y} = \frac{dv}{du} + v^2,$$

and thus the original equation (2.97) can be written as a first-order ODE as follows:

$$\frac{dv}{du} + v^2 + P(x) = 0. \qquad \square$$

2.2.2 Exact second-order equations

Sometimes it is possible to solve second-order differential equations by quadratures like exact first-order equations. If we are given a second-order nonlinear differential equation written in the form

$$\phi(x, y, y') = 0, \tag{2.99}$$

we call it *exact* if

$$D_x(\phi(x, y, y')) = 0, \tag{2.100}$$

where

$$D_x = \frac{\partial}{\partial x} + y'\frac{\partial}{\partial y} + y''\frac{\partial}{\partial y'} + \cdots + y^{(s+1)}\frac{\partial}{\partial y^s} \tag{2.101}$$

is a differential operator with respect to the independent variable x and the notation $y^{(s+1)}$ means the $(s+1)$-st derivative of y with respect to x. In fact, the relation (2.100) holds if

$$\frac{\delta\phi}{\delta y} = 0, \tag{2.102}$$

where

$$\frac{\delta F}{\delta y} = \frac{\partial F}{\partial y} - D_x\left(\frac{\partial F}{\partial y'}\right) + D_x^2\left(\frac{\partial F}{\partial y''}\right) \tag{2.103}$$

is a *variational derivative.*

Let us illustrate this by an example.

Example 2.19. Let us solve the nonlinear differential equation

$$\phi(x,y,y') = xyy'' + x(y')^2 + 3yy' = 0. \tag{2.104}$$

Solution. First, let us check if it is exact. In our case, from the definition (2.103) we have

$$\begin{aligned}\frac{\delta\phi}{\delta y} &= xy'' + 3y' - \left(\frac{\partial}{\partial x} + y'\frac{\partial}{\partial y} + y''\frac{\partial}{\partial y'}\right)(2xy' + 3y) + D_x\left[\left(\frac{\partial}{\partial x} + y'\frac{\partial}{\partial y} + y''\frac{\partial}{\partial y'}\right)(xy)\right]\\ &= xy'' + 3y' - 2y' - 3y' - 2xy'' + (y' + y' + xy'') = 0,\end{aligned}$$

so equation (2.102) holds and thus equation (2.104) is exact. Now, given that the equation is exact, how can we solve it? We note that equation (2.100) is true, i. e.,

$$xyy'' + x(y')^2 + 3yy' = D_x(\phi(x,y,y')) = \phi_x + y'\phi_y + y''\phi_{y'}. \tag{2.105}$$

We combine the terms at y'' in equations (2.105):

$$\frac{\partial\phi}{\partial y'} = xy.$$

From this, it follows that

$$\phi = xyy' + \alpha(x,y), \tag{2.106}$$

where α is an arbitrary function. Then, from equation (2.105) we have

$$x(y')^2 + 3yy' = y'(xy' + \alpha_y) + yy' + \alpha_x,$$

i. e.,

$$2yy' = y'\alpha_y + \alpha_x. \tag{2.107}$$

Similarly, let us combine the terms at y' in equation (2.107):

$$2y = \alpha_y,$$

from which we have

$$\alpha = y^2 + \beta(x),$$

where β is an arbitrary function of integration. Thus,

$$\phi(x,y,y') = xyy' + y^2 + C, \tag{2.108}$$

where C is an arbitrary constant. Since the resulting equation (2.108) is the Bernoulli equation, it can easily be solved to find $y(x)$. □

2.3 Exercises

Exercise 2.1. Show how to obtain equation (2.20) from equation (2.19).

Exercise 2.2. Check that y given by (2.22) is a solution of equation (2.8).

Exercise 2.3. Check that the group $\bar{y} = y + ay_0(x)$ is generated by the operator (2.24) in Example 2.3.

Exercise 2.4. Find the nonuniform dilation transformation generated by the operator (2.78).

Exercise 2.5. Solve the Bernoulli equation

$$\frac{dy}{dx} - 4x^2y = xy^{\frac{2}{3}}$$

by using the general formula (2.63).

Exercise 2.6. Check that μ given by formula (2.88) is the integrating factor of equation (2.86).

Exercise 2.7. Solve equation (2.108) and check that it is a solution of the original equation (2.104).

Exercise 2.8. Check if the equation $yy'' + (y')^2 + \frac{3yy'}{x} = 0$ is exact. Remark: this equation is obtained from equation (2.104) by dividing it by x.

3 Illustration: invariant solutions as internal singularities of nonlinear differential equations

If a model is given by an ODE, one can integrate it by quadratures using Lie group methods, provided that the equation in question has enough symmetries. If the equation is nonlinear, its general solution is given, as a rule, by an implicit formula involving quadratures. Often, it is difficult to investigate the behavior of such solutions, and hence to use them for understanding a related physical process, even in cases where the quadratures can be worked out in terms of elementary functions. Application of Lie's integration method reveals that nonlinear equations may have, along with the *generic solution* (involving n parameters for n-th-order equations), certain *singular solutions* involving less than n parameters. We call them *internal singularities*, in contrast to *external singularities* manifested by the form of the nonlinear equation.

We devote the present example to an investigation of a possibility provided by internal singularities for augmenting our knowledge about a behavior of generic solutions given implicitly or numerically. We will illustrate the situation by considering the following nonlinear second-order ODE:

$$y'' = \frac{y'}{y^2} - \frac{1}{xy}. \tag{3.1}$$

The external singularities of this equation are the axes $x = 0$ and $y = 0$. We cannot see from equation (3.1) the internal singularities, but they are found by Lie group analysis. Namely, equation (3.1) has the two-parameter family of solutions

$$C_1 y + C_2 x + x \ln\left|C_1 \frac{y}{x} - 1\right| + C_1^2 = 0, \quad C_1, C_2 = \text{const.}, C_1 \neq 0, \tag{3.2}$$

and the following one-parameter families of solutions:

$$y = \pm\sqrt{2x + Cx^2}, \quad C = \text{const.} \tag{3.3}$$

$$y = Kx, \quad K = \text{const.} \tag{3.4}$$

The solutions (3.3) and (3.4) are unstable under small perturbations of initial conditions, but attract the solutions of perturbed initial value problems. This fact simplifies investigations of the generic solution (3.2), e. g., near the external singularity $x = 0$. The statement will be illustrated by several examples.

https://doi.org/10.1515/9783111387499-003

3.1 Invariant solutions

3.1.1 Derivation of invariant solutions

The symmetry algebra of equation (3.1) is the two-dimensional Lie algebra L_2 spanned by the operators

$$X_1 = x^2\frac{\partial}{\partial x} + xy\frac{\partial}{\partial y}, \quad X_2 = 2x\frac{\partial}{\partial x} + y\frac{\partial}{\partial y}. \tag{3.5}$$

Let us find the invariant solutions of equation (3.1) with respect to any operator $X \in L_2$, i. e., with respect to

$$X = \alpha X_1 + \beta X_2, \quad \alpha, \beta = \text{const.}$$

If $\beta = 0$, the operator X is proportional to X_1 and the corresponding solution is the invariant solution provided by the operator X_1. To find this invariant solution, we obtain the invariant $J(x,y)$ by solving the equation $X_1(J) = 0$, i. e., a first integral of the characteristic equation

$$\frac{dx}{x^2} = \frac{dy}{xy} \quad \text{or} \quad \frac{dx}{x} = \frac{dy}{y}.$$

Integrating the latter equation, we obtain the invariant

$$J = \frac{y}{x}.$$

The invariant solution is given by $J = K$, or $y = Kx$, $K = \text{const}$. One can readily verify that $y = Kx$ solves equation (3.1) for any constant K. Thus, we have obtained the one-parameter family of solutions given by equation (3.4):

$$y = Kx, \quad K = \text{const.}$$

If $\beta \neq 0$, one can take X in the form

$$X = X_2 + CX_1 = (2x + Cx^2)\frac{\partial}{\partial x} + (y + Cxy)\frac{\partial}{\partial y}, \quad C = \text{const.}$$

We proceed as above. Namely, let us integrate the characteristic equation

$$\frac{dx}{2x + Cx^2} = \frac{dy}{(1 + Cx)y}.$$

Rewriting it in the form

$$\frac{dy}{y} = \frac{(1 + Cx)dx}{(2 + Cx)x} = \frac{dx}{2x} + \frac{Cdx}{2(2 + Cx)}$$

and integrating, we obtain

$$y = A\sqrt{2x + Cx^2}, \quad A = \text{const.}$$

Substitution of equation (3.1) yields $A = \pm 1$, and hence we arrive at the one-parameter family of solutions (3.3):

$$y = \pm\sqrt{2x + Cx^2}, \quad C = \text{const.}$$

The invariant solutions (3.3) and (3.4) play an exceptional part in investigating qualitative properties of solutions of equation (3.1). Namely, they are given in simple explicit forms compared to the implicit solution (3.2) and describe perfectly the behavior of neighboring solutions, as demonstrated below by way of examples. This kind of utilization of invariant solutions can be useful for qualitative analysis of numerical solutions as well. It is also interesting to compare our approach with the treatment of invariant solutions as *separatrices* and *envelopes* for first-order ODEs (see Section 3.10.1 in *Bluman & Anco* [9]).

3.1.2 Behavior of invariant solutions

Case 1. Figure 3.1 shows the solutions (3.3) and (3.4) in the interval $0 \leqslant x \leqslant 2$, passing through the points

$$x_0 = 1, \quad y_0 = \pm 1,$$

i. e., the solution (3.3) with $C = -1$ and $\pm$ sign,

$$y = \pm\sqrt{2x - x^2}, \tag{3.6}$$

and the solution (3.4) with $K = \pm 1$,

$$y = \pm x. \tag{3.7}$$

Case 2. Additionally, Figure 3.1 shows the solutions (3.3) and (3.4) in the interval $0 \leqslant x \leqslant 2$, passing through the points

$$x_0 = \frac{1}{2}, \quad y_0 = \pm\frac{\sqrt{17}}{4},$$

i. e., the solution (3.3) with $C = 1/4$ and $\pm$ sign,

$$y = \pm\sqrt{2x + \frac{1}{4}x^2}, \tag{3.8}$$

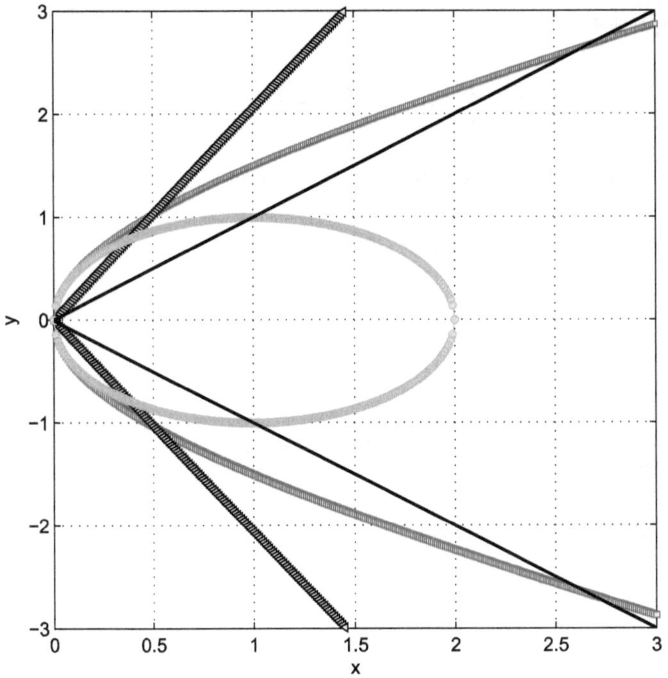

Figure 3.1: Singular solutions.

and the solution (3.4) with $K = \pm\sqrt{17}/2$,

$$y = \pm\frac{\sqrt{17}}{2}x. \tag{3.9}$$

3.2 Perturbation of singular solutions

3.2.1 Application to the first case

Let us consider the solution (3.7) with the plus sign, $y = x$. It solves the initial value problem at $x = 1$ with the initial conditions

$$y_0 = 1, \quad y_1 = 1.$$

We will take the following perturbation of the above initial conditions:

$$y_0 = 1, \quad y_1 = 1 + \varepsilon. \tag{3.10}$$

The solution with these initial conditions is given by equation (3.2). The corresponding values of the constants C_1 and C_2 are found by the following general procedure.

We differentiate equation (3.2),

$$C_1y' + C_2 + \ln\left|C_1\frac{y}{x} - 1\right| + C_1\frac{xy' - y}{C_1y - x} = 0, \tag{3.11}$$

and write equations (3.2) and (3.11) using given initial conditions

$$x = x_0, \quad y = y_0, \quad y' = y_1$$

to obtain

$$C_1y_0 + C_2x_0 + x_0\ln\left|C_1\frac{y_0}{x_0} - 1\right| + C_1^2 = 0, \tag{3.12}$$

$$C_1y_1 + C_2 + \ln\left|C_1\frac{y_0}{x_0} - 1\right| + C_1\frac{x_0y_1 - y_0}{C_1y_0 - x_0} = 0. \tag{3.13}$$

If we substitute in equation (3.13) the expression

$$C_2 + \ln\left|C_1\frac{y_0}{x_0} - 1\right| = -\frac{C_1}{x_0}(y_0 + C_1) \tag{3.14}$$

obtained from equation (3.12) and divide the result by C_1, we get

$$y_1 - \frac{1}{x_0}(y_0 + C_1) + \frac{x_0y_1 - y_0}{C_1y_0 - x_0} = 0.$$

Whence, upon multiplying by $C_1y_0 - x_0$ and dividing again by C_1, we easily find

$$C_1 = \frac{x_0}{y_0} - y_0 + x_0y_1. \tag{3.15}$$

Substituting this into equation (3.14), we finally obtain the second constant:

$$C_2 = 1 + y_0y_1 - \frac{x_0}{y_0^2} - x_0y_1^2 - 2\frac{x_0y_1}{y_0} - \ln\left|y_0y_1 - \frac{y_0^2}{x_0}\right|. \tag{3.16}$$

Applying equations (3.15) and (3.16) to the initial conditions (3.10), i. e., letting

$$x_0 = 1, \quad y_0 = 1, \quad y_1 = 1 + \varepsilon,$$

we obtain

$$C_1 = 1 + \varepsilon, \quad C_2 = -(2 + 3\varepsilon + \varepsilon^2 + \ln|\varepsilon|). \tag{3.17}$$

Thus, the solution with the initial conditions (3.10) is given by equation (3.2) with C_1, C_2 given by equations (3.17).

Consider now the solution (3.7) with the minus sign, $y = -x$. It solves the initial value problem at $x = 1$ with the initial conditions

$$y_0 = -1, \quad y_1 = -1.$$

Noting that equation (3.1) is invariant under the reflection $y \to -y$, we take the following modification of the perturbation (3.10):

$$y_0 = -1, \quad y_1 = -(1 + \varepsilon). \tag{3.18}$$

Proceeding as in the case of the initial conditions (3.10), we find that the solution with the initial conditions (3.18) is given by equation (3.2) with

$$C_1 = -(1 + \varepsilon), \quad C_2 = -(2 + 3\varepsilon + \varepsilon^2 + \ln|\varepsilon|). \tag{3.19}$$

The expression (3.19) can also be obtained from (3.17) by reflection $y \to -y$.

Note that the function (3.6),

$$y = \pm\sqrt{2x - x^2},$$

solves the initial value problem at $x = 1$ with the initial conditions

$$y_0 = \pm 1, \quad y_1 = 0.$$

Therefore, in order to see the behavior of solution between the singular solutions (3.6) and (3.7) as well as near these solutions, Figures 3.2 and 3.3 show the solution (3.2) with C_1, C_2 given by (3.17) and (3.19) for the values of the parameter ε as given in the following table:

Figure 3.2, panel (a)	**Figure 3.2, panel (b)**
$\varepsilon = -1.2$	$\varepsilon = -0.2$
$\varepsilon = -0.8$	$\varepsilon = +0.2$
$\varepsilon = -0.6$	$\varepsilon = +0.5$

(3.20)

The behavior of solutions that are distant from the singular solutions (3.6) and (3.7) is presented in Figures 3.4 and 3.5, showing the solution (3.2) with C_1, C_2 given by (3.17) and (3.19) for the following values of the parameter ε:

$$\varepsilon = 1, \quad \varepsilon = 5.0, \quad \text{and} \quad \varepsilon = 10.0. \tag{3.21}$$

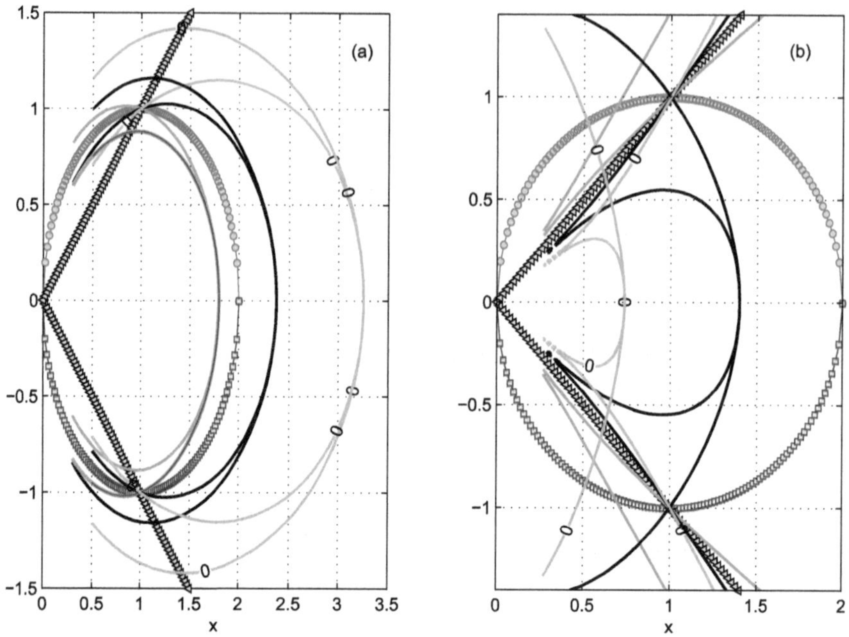

Figure 3.2: Comparison of the singular solutions (3.6) and (3.7) from case 1 with the perturbed solution (3.2) for small positive and negative values of the parameter ε in which the constants C_1, C_2 are given by (3.17) and (3.19). Panel (a) shows the perturbed singular solutions for $\varepsilon = -1.2$, $\varepsilon = -0.8$, and $\varepsilon = -0.6$. Panel (b) shows the perturbed singular solutions for $\varepsilon = -0.2$, $\varepsilon = +0.2$, and $\varepsilon = +0.5$.

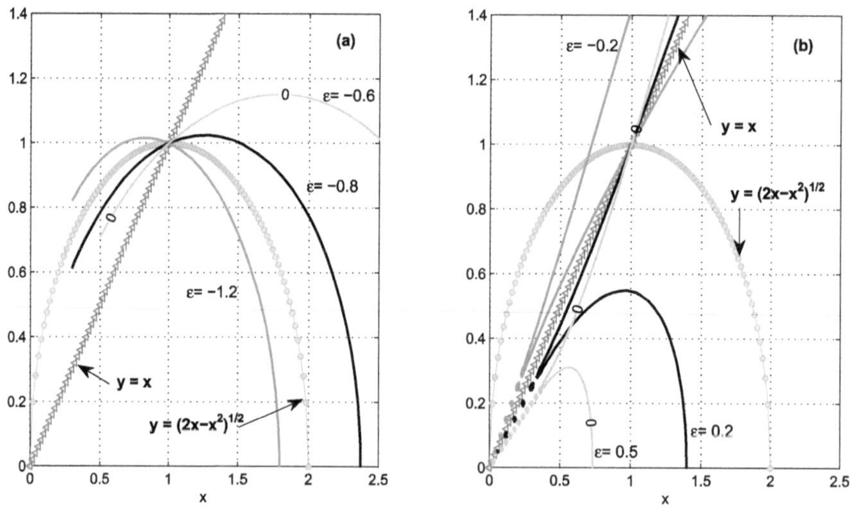

Figure 3.3: More comprehensive illustration for the same ε as in Figure 3.2 in which C_1, C_2 are given by (3.17) only. Panel (a) compares the singular solutions (3.6) and (3.7) with the perturbed singular solutions for negative values of the parameter: $\varepsilon = -1.2$, $\varepsilon = -0.8$, $\varepsilon = -0.6$. Panel (b) compares the singular solutions (3.6) and (3.7) with the perturbed singular solutions for positive values of the parameter: $\varepsilon = -0.2$, $\varepsilon = +0.2$, $\varepsilon = +0.5$.

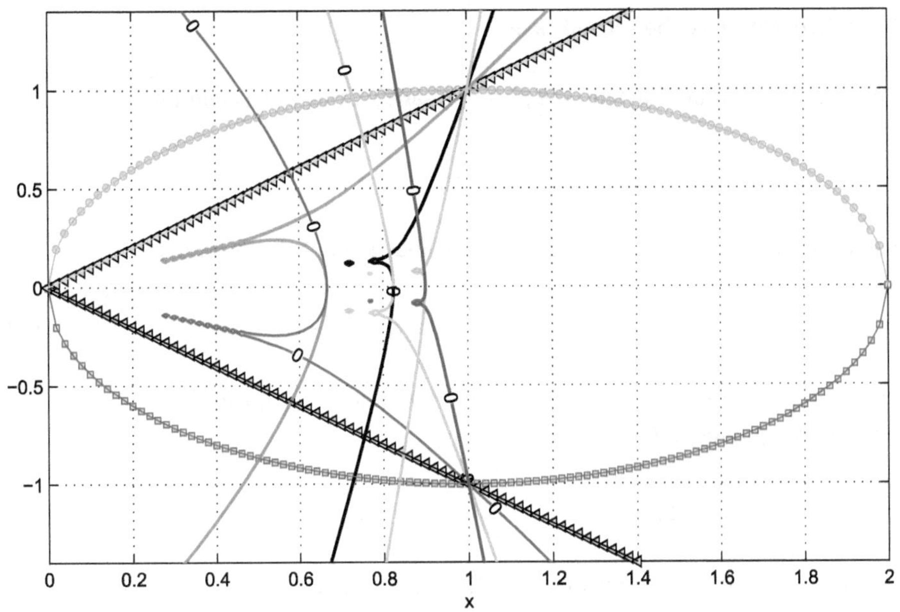

Figure 3.4: Additional illustration showing the behavior of perturbed invariant solutions that are distant from the singular solutions (3.6) and (3.7). Here the simulations were done for the constants C_1, C_2 given by (3.17) and (3.19) and for the following larger values of the parameter: $\varepsilon = +1.0$, $\varepsilon = +5.0$, and $\varepsilon = +10.0$.

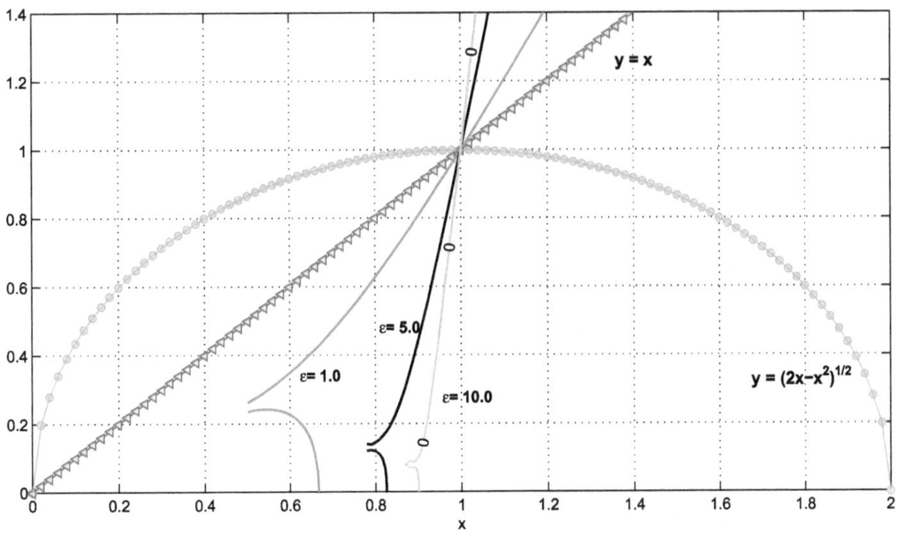

Figure 3.5: More comprehensive illustration for the same positive values of ε as in Figure 3.4 but for values of C_1, C_2 given by (3.17) only.

3.2.2 Application to the second case

The singular solution (3.8) with the plus sign satisfies the initial conditions

$$x_0 = \frac{1}{2}, \quad y_0 = \frac{\sqrt{17}}{4}, \quad y_1 = \frac{9}{2\sqrt{17}}.$$

Let us take the following perturbation of the above initial conditions:

$$x_0 = \frac{1}{2}, \quad y_0 = \frac{\sqrt{17}}{4}, \quad y_1 = \frac{9}{2\sqrt{17}} + \varepsilon. \tag{3.22}$$

Proceeding as before, one can see that the solution satisfying the initial conditions (3.22) is given by equation (3.2) with

$$C_1 = \frac{\varepsilon}{2}, \quad C_2 = -\frac{\sqrt{17}}{4}\varepsilon - \frac{\varepsilon^2}{2} - \ln\left|-1 + \frac{\sqrt{17}}{4}\varepsilon\right|. \tag{3.23}$$

In the case of the solution (3.8) with the minus sign we have the initial conditions

$$x_0 = \frac{1}{2}, \quad y_0 = -\frac{\sqrt{17}}{4}, \quad y_1 = -\frac{9}{2\sqrt{17}}$$

and we take the following modification of the perturbed initial conditions (3.22):

$$x_0 = \frac{1}{2}, \quad y_0 = -\frac{\sqrt{17}}{4}, \quad y_1 = -\left(\frac{9}{2\sqrt{17}} + \varepsilon\right). \tag{3.24}$$

Then the solution is given by equation (3.2) with

$$C_1 = -\frac{\varepsilon}{2}, \quad C_2 = -\frac{\sqrt{17}}{4}\varepsilon - \frac{\varepsilon^2}{2} - \ln\left|-1 + \frac{\sqrt{17}}{4}\varepsilon\right|. \tag{3.25}$$

Noting that the singular solution (3.9) has the initial conditions

$$x_0 = \frac{1}{2}, \quad y_0 = \pm\frac{\sqrt{17}}{4}, \quad y_1 = \pm\frac{\sqrt{17}}{2},$$

Figures 3.6 and 3.7 show the solution (3.2) with C_1, C_2 given by (3.25) for the values of the parameter ε as given in the following table:

Panel (a) in Figures 3.6 and 3.7	Panel (b) in Figures 3.6 and 3.7
$\varepsilon = -1.5$	$\varepsilon = +1.5$
$\varepsilon = -0.5$	$\varepsilon = +0.5$
$\varepsilon = -0.1$	$\varepsilon = +0.1$

(3.26)

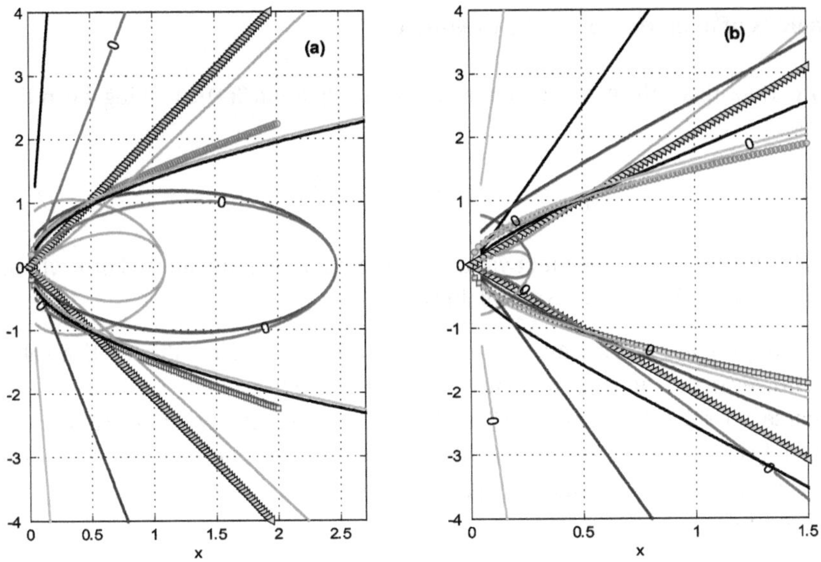

Figure 3.6: Comparison of the singular solutions (3.8) and (3.9) from case 2 with the perturbed solution (3.2) for positive and negative values of the parameter ε in which the constants C_1, C_2 are given by (3.23) and (3.25). Panel (a) shows the perturbed singular solutions for the following negative values of the parameter: $\varepsilon = -1.5$, $\varepsilon = -0.5$, and $\varepsilon = -0.1$. Panel (b) shows the perturbed singular solutions for the following positive values of the parameter: $\varepsilon = +1.5$, $\varepsilon = +0.5$, and $\varepsilon = +0.1$.

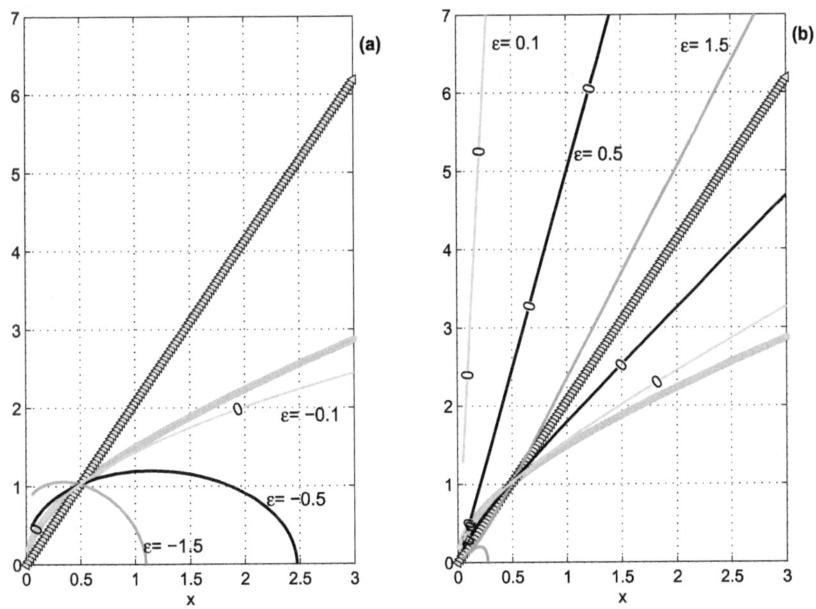

Figure 3.7: More comprehensive illustration for the same values of positive of ε as in Figure 3.6 but for the constants C_1, C_2 given by (3.23) only. In panel (a) the singular solutions (3.8) and (3.9) are compared with the perturbed singular solutions with the following negative values of the parameter: $\varepsilon = -1.5$, $\varepsilon = -0.5$, $\varepsilon = -0.1$. In panel (b), the singular solutions (3.8) and (3.9) are compared with the perturbed singular solutions with the following positive values of the parameter: $\varepsilon = 1.5$, $\varepsilon = 0.5$, $\varepsilon = 0.1$.

3.2.3 Analysis of numerical or implicit solutions

Problem 1. Investigate the behavior of the solution of equation (3.1) satisfying the initial conditions

$$x_0 = 3, \quad y_0 = \frac{\sqrt{15}}{2}, \quad y_1 = \frac{1}{8}. \tag{3.27}$$

In particular, estimate the maximum value of the solution and the interval of x where the solution is defined and has positive values.

Solution. The problem can be investigated numerically or by using the implicit solution (3.2). However, the internal singularities provide a simpler way.

Of course, the solution formulae (3.3) and (3.4) do not furnish the explicit solution to the problem because they are not compatible with the initial conditions (3.27). Indeed, the solutions (3.3) and (3.4) passing through the point (x_0, y_0) with the coordinates given in (3.27) have the forms

$$y = \sqrt{2x - \frac{1}{4}x^2} \tag{3.28}$$

and

$$y = \frac{\sqrt{15}}{6}x, \tag{3.29}$$

respectively. Differentiating (3.28) and (3.29) at $x = x_0$, we obtain the initial values

$$x_0 = 3, \quad y_0 = \frac{\sqrt{15}}{2}, \quad y_1 = \frac{1}{2\sqrt{15}} \tag{3.30}$$

and

$$x_0 = 3, \quad y_0 = \frac{\sqrt{15}}{2}, \quad y_1 = \frac{\sqrt{15}}{6}, \tag{3.31}$$

respectively. In both of them, y_1 is different from that in (3.27), and hence functions (3.28) and (3.29) do not provide the explicit solution to our initial value problem.

However, the initial conditions (3.30), unlike (3.31), are very close to (3.27). Therefore, we conclude that function (3.28) gives a good approximation to the solution of equation (3.1) with the initial conditions (3.27). In particular, using (3.28), we estimate the maximum value of the solution to be $y_{max} \approx 2$ and the interval of x where the solution is defined and has positive values to be approximately $0 < x \leqslant 8$. In order to verify our expectations, Figure 3.8 (panel (a)) shows the function (3.28) and the implicit solution to the problem given by equation (3.2) with

$$C_1 = \frac{3}{2}\left(\frac{1}{4} - \frac{1}{\sqrt{15}}\right), \quad C_2 = \frac{1}{16}\left(\frac{49}{20} - \frac{9}{\sqrt{15}}\right) - \ln\left(\frac{5}{4} - \frac{\sqrt{15}}{16}\right). \tag{3.32}$$

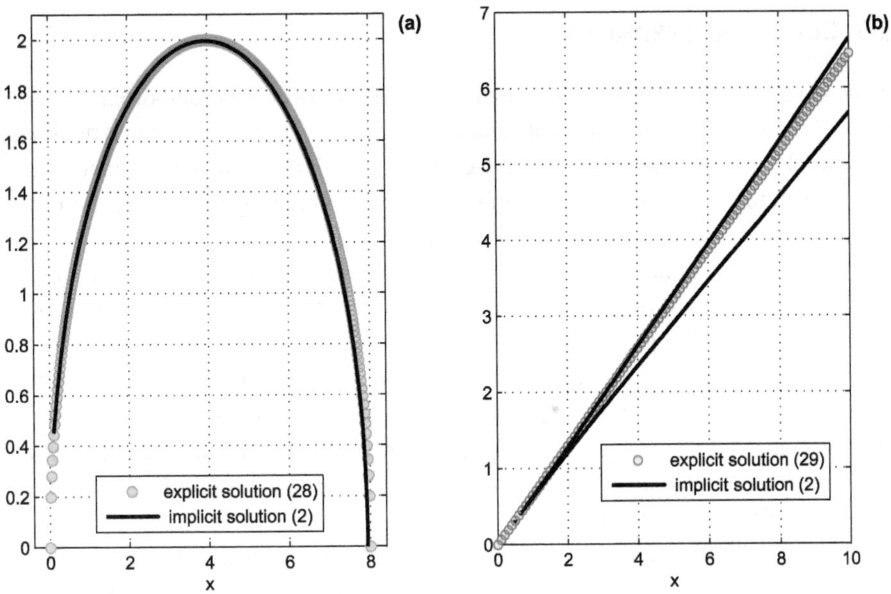

Figure 3.8: Comparison of explicit solutions (3.28) and (3.29) with implicit solution (3.2). Panel (a) shows the implicit solution (3.2) in which the constants C_1, C_2 are given by formulae (3.32). Panel (b) shows the implicit solution (3.2) in which the constants C_1, C_2 are given by formulae (3.34). The lower branch of the implicit solution (3.2) does not satisfy $y_0 = \sqrt{15}/2$ at $x = 3$, so it should be regarded as an odd branch of the solution.

The constants C_1, C_2 in (3.32) are obtained by substituting (3.27) in equations (3.15) and (3.16). □

Problem 2. Investigate the behavior of the solution of equation (3.1) satisfying the initial conditions

$$x_0 = 3, \quad y_0 = \frac{\sqrt{15}}{2}, \quad y_1 = \frac{2}{3}. \tag{3.33}$$

Solution. Reasoning as in Problem 1, we conclude that function (3.29) gives a good approximation to the solution of equation (3.1) with the initial conditions (3.33). In order to verify our expectations, Figure 3.8 (panel (b)) shows function (3.29) and the implicit solution to the problem given by equation (3.2) with the following values of the constants C_1, C_2:

$$C_1 = \frac{21}{\sqrt{15}}, \quad C_2 = -\left(\frac{17}{15} + \frac{3}{\sqrt{15}} + \ln\left|\frac{\sqrt{15}}{3} - \frac{5}{4}\right|\right), \tag{3.34}$$

obtained by substituting (3.33) in equations (3.15) and (3.16). □

3.3 Concluding remarks

Using Lie group methods one can integrate by quadratures any second-order ODE $y'' = f(x, y, y')$ admitting a two-dimensional Lie algebra. However, if the equation is nonlinear, its general solution is given prevalently by an implicit formula involving quadratures. It is difficult to employ that type of implicit solutions to investigate the behavior of particular solutions satisfying given initial conditions.

We have shown that nonlinear equations have internal singularities. They are defined by invariant solutions and provide patterns for qualitative investigations of implicit and numerical solutions.

As a model example, we have investigated the internal singularities of the nonlinear equations (3.1) provided by the invariant solutions (3.3)–(3.4). In this example, the totality of all solutions to equation (3.1) is given by the one-parameter families of singular solutions (3.3)–(3.4) and the two-parameter family of implicit solutions (3.2) written in terms of elementary functions.

Each singular solution (3.3) and (3.4) passes through every point (x_0, y_0) with $x_0 \neq 0$, $y_0 \neq 0$. One of the remarkable features of these solutions is that, using continuous perturbations of initial conditions, one can gradually pass from one branch of the singular solution, defined in one interval, to another branch of the singular solution, defined in a different interval. For example, Figure 3.3 illustrates a passage from the singular solution (3.3), defined in the bounded interval $(0, 2)$, to the singular solution (3.4), defined in the unbounded interval $(0, \infty)$.

We have demonstrated using several examples that internal singularities make it possible to find good approximations of solutions to the Cauchy problem by virtue of perturbations of initial data. The procedure is illustrated by Problems 1 and 2.

One can see, even from this possibly narrow modeling scenario, that internal singularities provide a useful tool for finding or analyzing exact and approximate solutions to the Cauchy problem for nonlinear differential equations. They can be utilized in investigating nonlinear mathematical models appearing in various physical, engineering, and other real-life problems.

3.4 Exercises

Exercise 3.1. Check that (3.2) and (3.4) are solutions of equation (3.1).

Exercise 3.2. Check that the symmetries X_1 and X_2 given by (3.5) are admitted by equation (3.1).

Exercise 3.3. Find the invariant of the operator $X = x^2 \frac{\partial}{\partial x} + xy \frac{\partial}{\partial y}$.

Exercise 3.4. Find the invariant of the operator $X = (2x + Cx^2)\frac{\partial}{\partial x} + (y + Cxy)\frac{\partial}{\partial y}$, where C is a constant.

Exercise 3.5. Check that a general solution of the ODE

$$\frac{dy}{y} = \frac{dx}{2x} + \frac{Cdx}{2(2 + Cx)}$$

is given by

$$y = A\sqrt{2x + Cx^2}, \quad A = \text{const.}$$

4 Modeling scenario 1: blood flow of variable density

Mathematical modeling and numerical simulations have become important tools to better understand the human cardiovascular system in recent years. One of the main goals in investigating blood flow in the aortic system is to understand arteriosclerosis and the related phenomena as well as their dependence on the blood flow structure. In particular, the aorta and arteries have low resistance to blood flow compared with the arterioles and capillaries. When the ventricle contracts, a volume of blood is rapidly ejected into the arterial vessels. Since the outflow to the arterioles is relatively slow because of their high resistance, the arteries are inflated to accommodate the extra blood volume. During diastole, the elastic recoil of the arteries forces the blood out into the arterioles. Thus, the elastic properties of the arteries help to convert the pulsatile flow of blood from the heart into a more continuous flow through the rest of the circulatory system. Hemodynamics is a term used to describe the mechanisms that affect the dynamics of blood circulation [22, 23, 66]. In reality, an accurate model of blood flow in the arteries would include the following realistic features: (i) the flow is pulsatile, with a time history containing major frequency components up to the eighth harmonic of the heart period; (ii) the arteries are elastic and tapered tubes; (iii) the geometry of the arteries is complex and includes tapered, curved, and branching tubes; and (iv) in small arteries, the viscosity depends upon vessel radius and shear rate [13]. Such a complex model has never been accomplished. But each of the features above has been "isolated," and qualitative if not quantitative models have been derived. As is so often the case in the engineering analysis of a complex system, the model derived is a function of the major phenomena one wishes to illustrate.

In this modeling example we will model and examine the general trend of possible solutions associated with the governing equations describing a simple one-dimensional blood flow that would depict blood progressing within a thin and elastic pulsating artery. In reality, for many flow situations, the changes of density due to changes in pressure associated with the flow are very small but not zero. In our simulations, the density is assumed variable for the following reason: we treat blood not as not a homogeneous fluid but as a suspension of particles (red blood cells [RBCs], white blood cells, platelets) in a fluid called plasma. Blood particles must be taken into account in the rheological model in smaller arterioles and capillaries since their size becomes comparable to that of the vessel [33, 30, 53, 21]. In particular, as has been discussed in Ref. [43], RBCs exhibit a unique deformability, which enables them to change shape reversibly in response to an external force. Human RBCs have the ability to undergo large deformations when subjected to external stresses, which allows them to pass through capillaries that are narrower than the diameter of a resting RBC. In fact, RBCs are more deformable than any other biomaterial. RBCs are biconcave discs, typically 6–8 μm in diameter and 2 μm thick, and their deformation can involve a change in cell curvature, a uniaxial deformation, or an area expansion. In mammals, RBCs are nonnucleated and consist of a concentrated hemoglobin solution enveloped by a highly flexible membrane. The deformability

https://doi.org/10.1515/9783111387499-004

of RBCs plays an important role in their main function, the transport of gases (O_2 and CO_2) via blood circulation [14, 43].

The purpose of this modeling scenario is to present the method of solution for blood flow from two different perspectives. One method is an approximate perturbed method to obtain approximate solutions and the second method is the invariance method presented in the first part of the book to find exact solutions. The latter method is extended in consecutive chapters on geophysical flows.

4.1 Mathematical modeling

To put the deformability of blood due to pressure in perspective, we consider a multi-component system of total volume V, with

$$V = \sum_i V_i, \tag{4.1}$$

where V_i is the subvolume of component i in the system. The (isothermal) compressibility of the system is

$$\kappa = -\frac{1}{V}\left(\frac{\partial V}{\partial P}\right)_T = -\frac{1}{V}\sum_i \left(\frac{\partial V_i}{\partial P}\right)_T. \tag{4.2}$$

But the compressibility of each component is

$$\kappa_i = -\frac{1}{V_i}\left(\frac{\partial Vi}{\partial P}\right)_T. \tag{4.3}$$

Therefore, (4.2) reduces to

$$\kappa = \frac{1}{V}\sum_i V_i \kappa_i. \tag{4.4}$$

Finally, denoting the volume fraction of each component of the system by α_i, we have

$$\kappa = \sum_i \alpha_i \kappa_i. \tag{4.5}$$

In our case, the main contribution to the compressibility and deformability of blood comes from RBCs.

The method of obtaining a general solution of the governing equations for the given model is considered from two points of view. The first approach represents a singular perturbation theory providing explicit approximate solutions to the model and the second one is based on group theoretical modeling that provides the exact solutions written in an implicit form. The main goal is to compare these two approaches and determine the pros and cons of each proposed approach.

The range of developed models or models being developed extends from lumped models to complicated three-dimensional fluid structure models [64, 62]. In this modeling scenario we consider a simple one-dimensional model of blood flow in a vessel. This model and generally all one-dimensional models are not suitable for describing blood flow in complicated morphological regions such as regions with stenosis or bifurcations. However, these situations can also be covered to a certain extent. On the one hand, these models can be used as an alternative to the more complex three-dimensional fluid structure models or in conjunction with them in a geometrical multi-scale fashion, as explained in Ref. [25]. On the other hand, the computational complexity of one-dimensional models is several orders of magnitude lower than that of multi-dimensional models. A few decades ago, the multi-scale approach attracted more interest. In the multi-scale approach, one-dimensional models may be coupled on the one hand with lumped-parameter models [62] based on a system of ODEs [64, 79] and on the other hand with three-dimensional fluid structure models, as discussed in Refs. [26] and [27]. In the latter case they may also provide a way of implementing more realistic boundary conditions for 3D calculations, or they can be used for the numerical acceleration of a three-dimensional Navier–Stokes solver in a multi-level multi-scale scheme. Additionally, one-dimensional models give a good description of the propagation of pressure waves in arteries [71, 60]; hence, they can be successfully used to investigate the effects on pulse waves of geometrical and mechanical arterial modifications, due, e. g., to the presence of stenosis or to the placement of stents or prostheses [25].

In order to describe a problem in mathematical terms, one must make use of the basic laws that govern the elements of the problem. Within the frame of the present modeling approach, we start with conservation laws for mass and momentum and consider a perfect compressible fluid propagating along a tube with longitudinal coordinate x and slowly varying cross-section $a(x,t)$. Because of the pressure gradient in the blood, the artery wall must deform. The elastic restoring force in the wall makes it possible for waves to propagate. In terms of one-dimensional modeling, we assume that the artery radius $r(x,t)$ varies from the constant mean r_0 in time and along the artery (in x). We denote the local cross-sectional area as $a = \pi r^2$ and the averaged velocity as $v(x,t)$. Consider a fixed geometrical volume between x and $x + dx$, through which fluid moves in and out. Conservation of mass requires

$$\frac{\partial a}{\partial t} + \frac{\partial (va)}{\partial t} = 0. \tag{4.6}$$

We next assume that the time rate of momentum change in the volume is balanced by the net influx of momentum through the two ends and the pressure force acting on all sides. The rate of momentum change M is given by

$$M = \frac{\partial (\rho va)}{\partial t}, \tag{4.7}$$

where $\rho(x,t)$ is the density of fluid associated with the density of the blood consisting of blood plasma and RBCs. In reality, the RBC fraction may include large viscosity variations, stressing the importance of accounting for the non-Newtonian effects (see, e. g., [78], where the Quemada viscosity model [61] is used to account for the non-Newtonian viscosity behavior).

The net rate of momentum influx is

$$-\frac{\partial(\rho v^2 a)}{\partial x}dx = -\rho v\frac{\partial(va)}{\partial x} - \rho va\frac{\partial v}{\partial x}. \tag{4.8}$$

The net pressure force at the two ends is given by $-\partial(pa)/\partial x$, while that on the sloping wall is

$$2\pi rp\frac{\partial r}{\partial x} = p\frac{\partial a}{\partial x}. \tag{4.9}$$

The sum of all pressure forces P is given by

$$P = -a\frac{\partial p}{\partial x}. \tag{4.10}$$

Balancing the momentum by equating M given by (4.7) to the sum of (4.8) and P given by (4.10), we get, after making use of mass conservation (4.6),

$$\frac{\partial v}{\partial t} + v\frac{\partial v}{\partial x} = -\frac{1}{\rho}\frac{\partial p}{\partial x}. \tag{4.11}$$

Arterial pulse propagation varies along the circulatory system as a result of the complex geometry and nonuniform structure of the arteries. In order to study the basic arterial pulse characteristics, we assume an idealized case of an infinitely long circular elastic tube that contains slightly compressible blood, which is a suspension of particles in what is basically water. As such, its compressibility will be mainly due to the RBCs, as explained above. We can think of it as a two-phase homogeneous nonviscous fluid flow of water and gas bubbles. If we apply pressure to the water/gas mixture, the overall density will decrease as the gas is compressed, leading to the mixture continuity equation that, under the assumption of zero relative velocity, reduces to the equivalent single-phase flow of density ρ, i. e.,

$$\frac{\partial \rho}{\partial t} + \frac{\partial}{\partial x}(\rho v) = 0. \tag{4.12}$$

In addition, empirical constitutive laws are needed to relate pressure and density such as equations of state

$$p = p(\rho, S), \tag{4.13}$$

where S denotes the entropy. Since in our modeling no temperature gradient is imposed externally and the gradient of the flow is not too large, we ignore thermal diffusion. The fluid motion is adiabatic; entropy $S = S_0$ is constant. As a result, $p = p(\rho, S_0)$ depends only on density. As is well known from thermodynamics,

$$\left(\frac{\partial p}{\partial \rho}\right)_S = \gamma \left(\frac{\partial p}{\partial \rho}\right)_T, \tag{4.14}$$

where T is the temperature and γ is the ratio of specific heats at constant pressure and constant volume. Furthermore, since pressure is a function of density only, we can write $p = p(\rho)$. Expanding this function in a Taylor series about the equilibrium density ρ_0, we have

$$p = p_0 + \frac{p'(\rho_0)}{1!}(\rho - \rho_0) + \frac{p''(\rho_0)}{2!}(\rho - \rho_0)^2 + \cdots, \tag{4.15}$$

where p_0 is the equilibrium pressure at which $\rho = \rho_0$. Since $\rho - \rho_0$ is small, we can neglect the second- and higher-order terms and write

$$p = p_0 + \lambda(\rho - \rho_0), \tag{4.16}$$

where λ is a constant. From this equation it follows that

$$\frac{\partial p}{\partial x} = \lambda \frac{\partial \rho}{\partial x} \quad \text{and} \quad \frac{\partial p}{\partial t} = \lambda \frac{\partial \rho}{\partial t}. \tag{4.17}$$

Since the force due to gravity is neglected, combining (4.11) and (4.12), we arrive at the governing equations of motion for unknowns velocity $v(x,t)$, pressure $p(x,t)$, and density $\rho(x,t)$ that are written as follows:

$$\frac{\partial v}{\partial t} + v\frac{\partial v}{\partial x} = -\frac{1}{\rho}\frac{\partial p}{\partial x}, \tag{4.18}$$

$$\frac{\partial \rho}{\partial t} + \frac{\partial}{\partial x}(\rho v) = 0, \tag{4.19}$$

$$p = \lambda \rho, \tag{4.20}$$

where t is time, x is a spatial variable, and λ is a constant.

We eliminate the pressure from these equations by differentiating equation (4.18) with respect to t and using the equation of state (4.20) to get

$$\frac{\partial^2 v}{\partial t^2} + \frac{\partial v}{\partial t}\frac{\partial v}{\partial x} + v\frac{\partial^2 v}{\partial x \partial t} = -\frac{\lambda}{\rho}\left(\frac{\partial^2 \rho}{\partial x \partial t} - \frac{1}{\rho}\frac{\partial \rho}{\partial t}\frac{\partial \rho}{\partial x}\right). \tag{4.21}$$

Using equation (4.19), we can rewrite (4.21) as

$$\frac{\partial^2 v}{\partial t^2} + \frac{\partial}{\partial x}\left(v\frac{\partial v}{\partial t}\right) = \lambda\frac{\partial^2 v}{\partial x^2} + \frac{\partial}{\partial x}\left(\frac{\lambda}{\rho}v\frac{\partial \rho}{\partial x}\right). \tag{4.22}$$

As follows from equations (4.18) and (4.20), we have

$$\frac{\lambda v}{\rho}\frac{\partial \rho}{\partial x} = -v\frac{\partial v}{\partial t} - v^2\frac{\partial v}{\partial x}. \tag{4.23}$$

Substituting this result into (4.22), we arrive at the following single equation for $v(x,t)$:

$$\frac{\partial^2 v}{\partial t^2} + \frac{\partial}{\partial x}\left(2v\frac{\partial v}{\partial t} + v^2\frac{\partial v}{\partial x}\right) = \lambda\frac{\partial^2 v}{\partial x^2}. \tag{4.24}$$

4.2 First approach: approximate analysis

In order to identify the resonant input in the model, we start with an approximate solution in the form of a naive expansion,

$$v(x,t) = v_0 + \sum_{i=1}^{\infty} \varepsilon^i v_i(x,t), \tag{4.25}$$

where ε is a small parameter and v_0 = const. is an exact trivial solution of equation (4.24).

4.2.1 Failure of the direct approach

We substitute the expansion (4.25) into (4.24) and collect powers of ε.

Problem $0(\varepsilon^1)$ gives the following equation:

$$\frac{\partial^2 v_1}{\partial t^2} - \lambda\frac{\partial^2 v_1}{\partial x^2} + \frac{\partial}{\partial x}\left(2v_0\frac{\partial v_1}{\partial t} + v_0^2\frac{\partial v_1}{\partial x}\right) = 0. \tag{4.26}$$

We look for a solution of equation (4.26) in the form

$$v_1(x,t) = A_1 \sin\theta_1 + B_1\cos\theta_1, \tag{4.27}$$

where A_1, B_1 are constants and $\theta_1 = k_1 x - \omega_1 t$, in which k_1 and ω_1 are wave number and frequency of the primary wave, respectively. Solution (4.27) is valid provided the dispersion relation is satisfied:

$$\omega_1^2 - 2v_0 k\omega_1 - k_1^2(\lambda - v_0^2) = 0. \tag{4.28}$$

So the $0(\varepsilon^1)$ problems represent a hyperbolic model with two wave modes,

$$\omega_1 = k_1(v_0 \pm \sqrt{\lambda}). \tag{4.29}$$

Problem $0(\varepsilon^2)$ gives the following equation:

$$\frac{\partial^2 v_2}{\partial t^2} - \lambda\frac{\partial^2 v_2}{\partial x^2} + \frac{\partial}{\partial x}\left(2v_0\frac{\partial v_1}{\partial t} + v_0^2\frac{\partial v_2}{\partial x}\right) = -\frac{\partial}{\partial x}\left(2v_1\frac{\partial v_1}{\partial t} + 2v_0v_1\frac{\partial v_1}{\partial x}\right). \tag{4.30}$$

In view of presentation (4.27), the right-hand side in equation (4.30) is written as

$$-\frac{\partial}{\partial x}\left(2v_1\frac{\partial v_1}{\partial t} + 2v_0v_1\frac{\partial v_1}{\partial x}\right) = \delta(A_1^2 - B_1^2)\cos(2\theta_1) - 2\delta A_1B_1\sin(2\theta_1), \tag{4.31}$$

where we denote

$$\delta = 2k_1(\omega_1 - v_0k_1). \tag{4.32}$$

We look for a particular solution in the form

$$v_2^{(p)} = H\cos(2\theta_1) + R\sin(2\theta_1). \tag{4.33}$$

Substituting this solution into (4.31), we obtain the following expressions for H and R:

$$(H, R) = \frac{\delta(B_1^2 - A_1^2,\ 2A_1B_1)}{4[\omega_1^2 - 2v_0k\omega_1 - k_1^2(\lambda - v_0^2)]}. \tag{4.34}$$

As expected, because of the dispersion relation (4.28), the right-hand side of equation (4.30) corresponds to resonance and yields the secular terms.

If we look for a particular solution of the form

$$v_2^{(p)} = t\widetilde{H}\cos(2\theta_1) + t\widetilde{R}\sin(2\theta_1), \tag{4.35}$$

the resonance input would be removed since, in this case, the constants $\widetilde{H}$ and $\widetilde{R}$ would be

$$(\widetilde{H}, \widetilde{R}) = -k_1\left(A_1B_1, \frac{A_1^2 - B_1^2}{2}\right), \tag{4.36}$$

so the particular solution would have the form

$$v_2^{(p)} = -ktA_1B_1\cos(2\theta_1) - \frac{k_1}{2}t(A_1^2 - B_1^2)\sin(2\theta_1). \tag{4.37}$$

Since $v_2^{(p)}$ grows linearly in time, the term $\varepsilon^2 v_2$ would become comparable to εv_1 for large values of time t (e. g., when time is of order $\frac{1}{\varepsilon}$), as shown in Figure 4.1.

In particular, Figure 4.1 shows the qualitative behavior of the time series of the second-order approximation of solution with secular terms (4.37) for the values $k = 1$ and 2 and $\lambda = 1$ and 2. The following values of parameters have been chosen: $\varepsilon = 0.1$, $v_0 = 1$, $x = 2$, $A_1 = 0.18$, $B_1 = 0.19$, and ω_1 is determined by the dispersion relation (4.29).

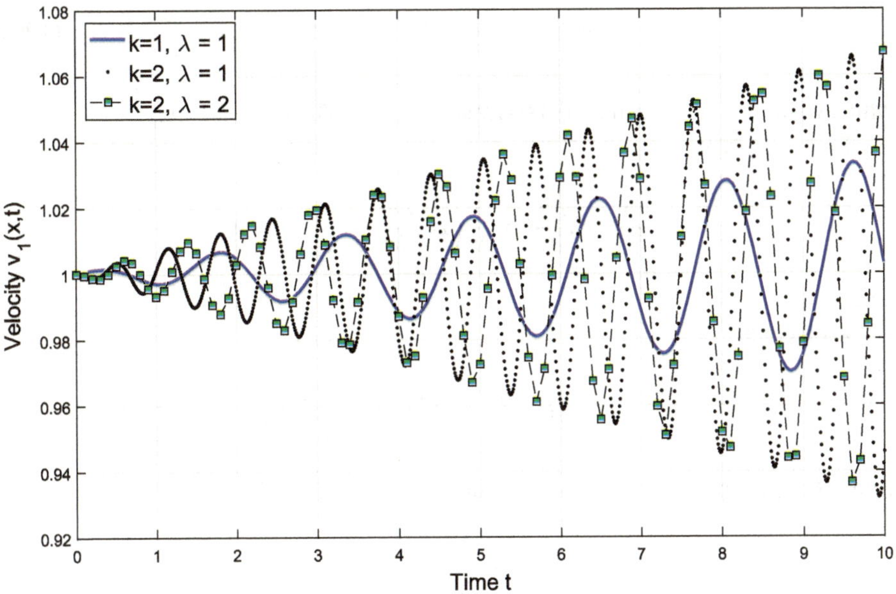

Figure 4.1: Visualization of the approximate solution $v(x,t)$ with secular terms in v_2^p.

Since we are interested only in general solutions, the choice of constants is arbitrary and we are focused on qualitative analysis.

As seen from (4.37), the first terms of the expansion (4.25) provide a local (small t) approximation, at most. The shortcoming of (4.25) is related to the breakdown of the straightforward approach to nonlinear perturbation analysis of equation (4.24), but it is more transparent to explanations. The nonlinear terms in (4.24) will slowly, but accumulatively, absorb energy and damp the motion. Hence, even though the term $\varepsilon v_2^{(p)}$ itself is small, the long-term effects are crucial and the solution cannot be described as being periodic plus a small correction. The consequence for the naive expansion (4.25) is that the ordering requirement $v_0 > \varepsilon v_1 > \cdots$ is violated. However, it may be instructive to try and fail in order to understand the nature of the resonance phenomena.

4.2.2 Multi-scale approach

We introduce the multi-scale approach based on the new variable

$$\tau = \varepsilon t. \tag{4.38}$$

We now consider the fast scale t and the slow scale τ as independent variables. We rewrite equation (4.24) in terms of the new variable (4.38) and modify the series expansion (4.25) to the form

$$v(x,t,\tau) = v_0 + \sum_{i=1}^{\infty} \varepsilon^i v_i(x,t,\tau), \tag{4.39a}$$

which yields the perturbation hierarchy similar to (4.26) and (4.30), i. e.,
Problem $0(\varepsilon^1)$:

$$\frac{\partial^2 v_1}{\partial t^2} - \lambda\frac{\partial^2 v_1}{\partial x^2} + \frac{\partial}{\partial x}\left(2v_0\frac{\partial v_1}{\partial t} + v_0^2\frac{\partial v_1}{\partial x}\right) = 0$$

and
Problem $0(\varepsilon^2)$:

$$\frac{\partial^2 v_2}{\partial t^2} - \lambda\frac{\partial^2 v_2}{\partial x^2} + \frac{\partial}{\partial x}\left(2v_0\frac{\partial v_1}{\partial t} + v_0^2\frac{\partial v_2}{\partial x}\right) = F + Q, \tag{4.40}$$

where F represents the right-hand side of equation (4.30) and

$$Q = -2\left[\frac{\partial^2 v_1}{\partial t\partial\tau} - v_0\frac{\partial^2 v_1}{\partial x\partial t}\right] \tag{4.41}$$

appears because of scaling (4.38). Since the derivative with respect to the fast variable appears only at $0(\varepsilon^2)$, the problem at $0(\varepsilon^1)$ is identical to equation (4.26). The slow time variable τ is implicit in the constant of integration and the most general real-valued solution of the $0(\varepsilon^1)$ problem can be written as

$$v_1(x,t,\tau) = A_1(\tau)e^{i\theta_1} + A_1^* e^{-\theta_1}. \tag{4.42}$$

With this solution in hand, equation (4.40) reads

$$F + Q = 2(\omega_1 - v_0 k_1)\left[i\left(\frac{dA_1}{d\tau}e^{i\theta_1} - \frac{dA_1^*}{d\tau}e^{-i\theta_1}\right) - 2k_1\left(A_1^2 e^{2i\theta_1} + A_1^{*2}e^{-2i\theta_1}\right)\right]. \tag{4.43}$$

To avoid secular terms we require

$$\left(\frac{dA_1}{d\tau} + \frac{dA_1^*}{d\tau}\right)\sin\theta_1 = -2k_1(A_1^2 + A_1^{*2})\cos(2\theta_1), \tag{4.44}$$

$$\left(\frac{dA_1}{d\tau} - \frac{dA_1^*}{d\tau}\right)\cos\theta_1 = 2k_1(A_1^{*2} - A_1^{*2})\sin(2\theta_1). \tag{4.45}$$

The natural choice is to set $A_1^* = 0$. Then, squaring and adding the resulting equations for A_1, we arrive at a single equation:

$$\frac{dA_1}{d\tau} = 2k_1 A_1^2. \tag{4.46}$$

Solving equation (4.46), we write the solution in the form

$$v(x,t) = v_0 + \varepsilon \frac{\cos k_1 x - \omega_1 t}{2k_1 \varepsilon t + c_1} + 0(\varepsilon^2), \tag{4.47}$$

where c_1 is a constant of integration and ω_1 is related to k_1 and λ by the dispersion relation (4.28), i. e.,

$$\omega_1 = k_1(v_0 \pm \sqrt{\lambda}).$$

Figure 4.2 shows the qualitative behavior of the time series of the solution (4.47) for the values of $k_1 = 1$ and $k_2 = 4$ when $c_1 = 1.5$ and for the same values of parameters ε, v_0, x, and A_1 as we used above to visualize the time series of $v_2^{(p)}$ shown in Figure 4.1.

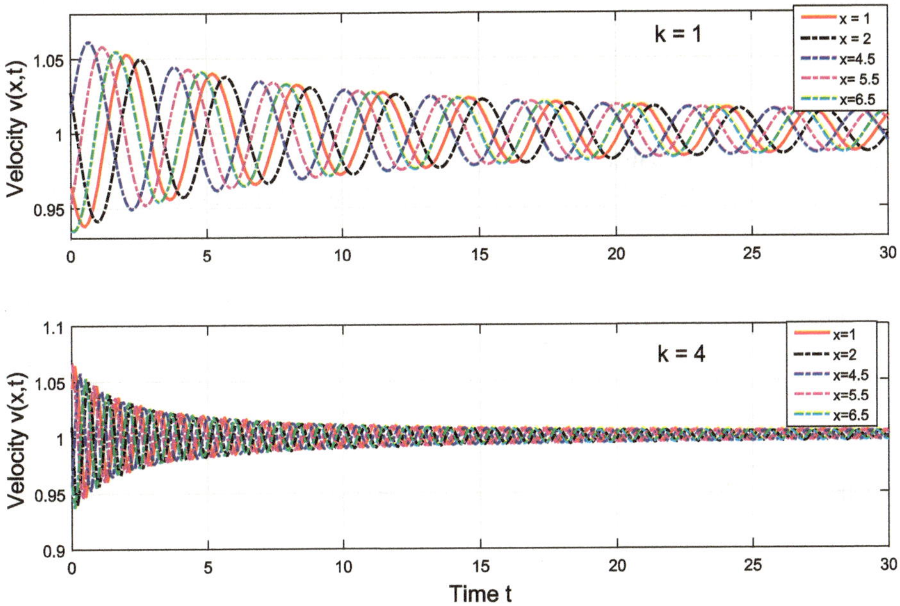

Figure 4.2: Visualization of the approximate solution $v(x,t)$.

A question of particular interest is the investigation of asymptotic stability of the trivial solution v_0. This will be done in the next section.

4.2.3 Stability of perturbed steady flows

We note that the stationary solution

$$v = U(x) \tag{4.48}$$

solves equation (4.24) in the stationary case, i. e.,

$$\frac{\partial}{\partial x}\left(v^2\frac{\partial v}{\partial x}\right) = \lambda\frac{\partial^2 v}{\partial x^2}, \tag{4.49}$$

which can be integrated to give the exact solution of the form

$$U^3(x) - 3\lambda U(x) = M_1 x + M_2, \tag{4.50}$$

where M_1 and M_2 are constants. We denote

$$\sigma = 4M_1 x + 4M_2 + 4\sqrt{(M_1 x)^2 + 2M_1 M_2 x - 4\lambda^3 + M_2^2}. \tag{4.51}$$

Then the only real branch of the solution for $U(x)$ of equation (4.49) can be written explicitly as

$$U(x) = \frac{1}{2}\frac{\sigma^{\frac{2}{3}} + 4\lambda}{\sigma^{\frac{1}{3}}}. \tag{4.52}$$

Figure 4.3 is used to visualize the stationary flow $U(x)$ given by (4.52) for different values of the constant $\lambda = \lambda_1, \lambda = \lambda_2$ and the following fixed values of parameters: $\lambda_1 = 5, \lambda_2 = 1$, $M_1 = 5$, and $M_2 = 1$.

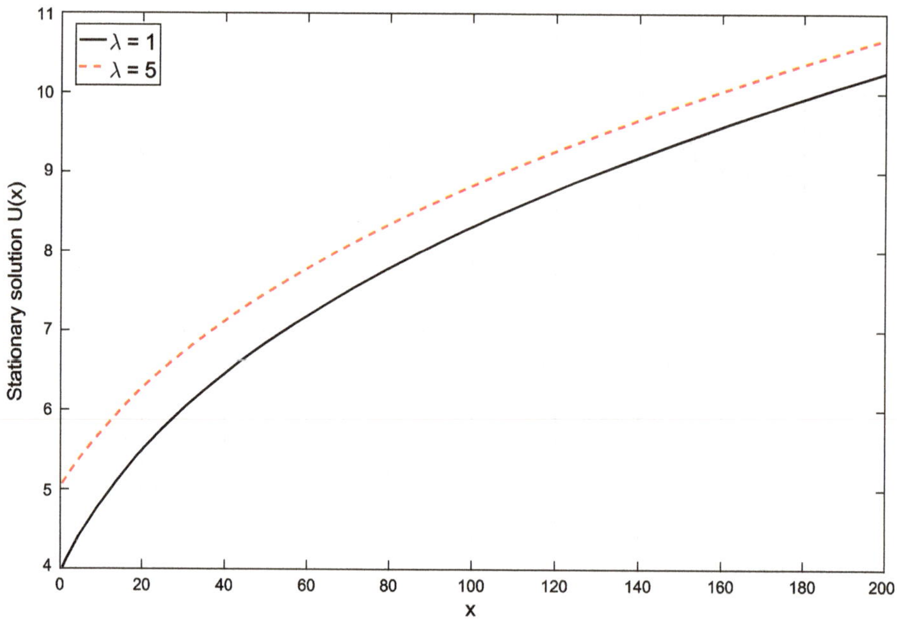

Figure 4.3: Visualization of the stationary solution $U(x)$.

Since, as Figure 4.2 shows, $U(x)$ is growing linearly in x, we classify $U(x)$ as a nonphysical solution.

Let us now look for a nonstationary solution of equation (4.24) that is close to $U(x)$ in the form

$$v(x,t) = U(x) + \varepsilon\tilde{v}(x,t), \tag{4.53}$$

where ε is a small parameter and $\tilde{v}$ denotes the perturbation. This procedure is largely formal. Mathematics ideally requires proof that the solution of the complete equations in question for $\varepsilon \to 0$ has a solution of the approximate equations at zeroth order of ε (at least asymptotically). In fact, this ideal is achieved in very rare cases; researchers are usually limited to the formal construction of an approximate model. The justification is based on physical intuition which opens a wide scope. It is clear that, at the same time, the role of the criterion of practice is greatly increased.

We assume a perturbation of the form of a plane harmonic wave propagating in the positive x direction,

$$\tilde{v}(x,t) = Ae^{i(kx-\omega t)}, \tag{4.54}$$

where A is a constant amplitude, k is a wave number, and ω is the angular frequency of the oscillations. Substituting the presentations (4.53) and (4.54) into equation (4.24) and collecting the terms of order ε^0, we get the nonlinear equation for the mean flow

$$\frac{\partial}{\partial x}\left(U^2\frac{\partial U}{\partial x}\right) - \lambda\frac{\partial^2 U}{\partial x^2} = 0, \tag{4.55}$$

which coincides with (4.49) and thus $U(x)$ has the form (4.52).

Similarly, collecting and separating the real and imaginary parts of the terms of order ε, we get the equations

$$\omega^2 - 2U\omega k + k^2(U-\lambda) - 2\frac{\partial}{\partial x}\left(U\frac{\partial U}{\partial x}\right) = 0 \tag{4.56}$$

and

$$\frac{\omega}{k}\frac{\partial}{\partial x}[U(1-U)] = 0. \tag{4.57}$$

For progressing wave-like solutions, equations (4.56) and (4.57) imply that there is another particular exact solution of equation (4.49) (and, correspondingly, of equation (4.24)),

$$v(x,t) = v_0 = \text{const.}, \tag{4.58}$$

provided that ω and k satisfy the dispersion relation (4.28), i. e.,

$$\omega^2 - 2v_0\omega k + (v_0 - \lambda)k^2, \tag{4.59}$$

with two known wave modes given by (4.29). As one can expect, since the flow is away from frictional boundaries, the dispersion relation (4.28) confirms asymptotic stability of the mean constant flow (4.58).

Figure 4.4 shows snapshots of the perturbed flow $v(x,t) = v_0 + \varepsilon\tilde{v}(x,t)$ at initial time $t = 0$ (red line) and later times of $t = 0.5, 1.0, 1.5$, and 1.8 units (plotted in different colors) for different values of the wave number k and the following fixed values of parameters: $v_0 = 1$ (for $k = 1$), $v_0 = 2$ (for $k = 2$), $v_0 = 3$ (for $k = 10$), and $v_0 = 4$ (for $k = 25$), $a = 1$, and $\varepsilon = 0.1$.

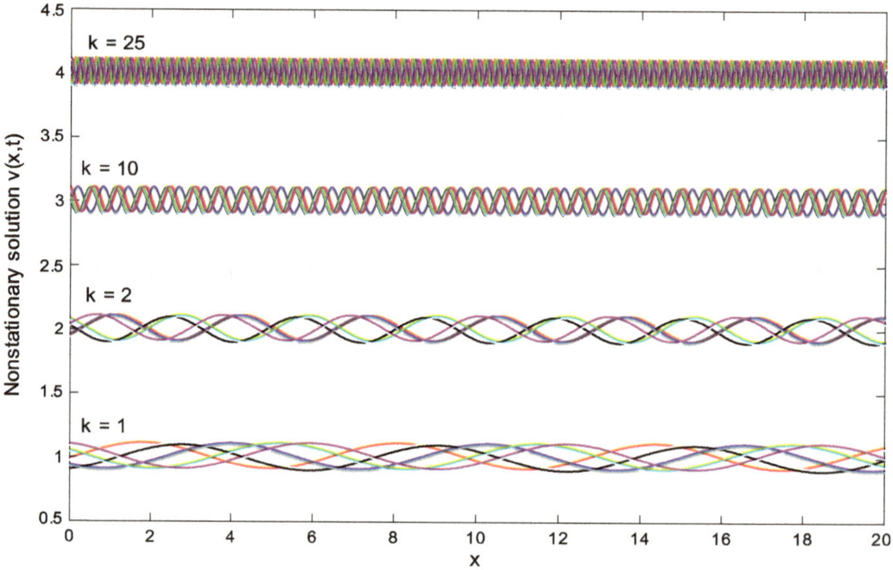

Figure 4.4: Visualization of the perturbed constant flow v_0 for different values of wave number k.

4.3 Second approach: group theoretical point of view

A simple inspection shows that equation (4.24) admits the one-parameter translation group

$$\bar{t} = t + a_1, \quad \bar{x} = x + a_2$$

of t and x and the one-parameter group of uniform scaling transformations in the (t,x)-plane

$$\bar{t} = te^{a_3}, \quad \bar{x} = xe^{a_3}.$$

The above transformation groups have the following generators (called also infinitesimal symmetries):

$$X_1 = \frac{\partial}{\partial t}, \quad X_2 = \frac{\partial}{\partial x}, \quad X_3 = t\frac{\partial}{\partial t} + x\frac{\partial}{\partial x}. \tag{4.60}$$

Various linear combinations of the generators (4.60) can serve for constructing group-invariant solutions of equation (4.60).

4.3.1 Traveling waves

Traveling waves are invariant solutions with respect to any linear combination of the translation generators X_1 and X_2. We take the linear combination in the form

$$X_1 + mX_2 = \frac{\partial}{\partial t} + m\frac{\partial}{\partial x}, \qquad m = \text{const}. \tag{4.61}$$

The operator (4.61) has two independent invariants, v and $z = x - mt$. According to the theory of invariant solutions, the invariant solution has the representation

$$v = f(z). \tag{4.62}$$

The representation (4.62) reduces equation (4.24) to the ODE

$$(m^2 - \lambda)f''^2 f' - 2mff')' = 0.$$

We integrate it once and obtain

$$(m^2 - \lambda)f'^2 f' - 2mff' = M_1, \quad M_1 = \text{const}.$$

The second integration gives the cubic equation

$$f^3 - 3mf^2 + 3(m^2 - \lambda)f = M_1\mu + M_2 \tag{4.63}$$

for determining $f(\mu)$. Thus, the traveling wave solution (4.62) is determined by the cubic equation (4.63) and involves three arbitrary parameters m, M_1, M_2. Cardano's solution for the cubic equation allows to express the traveling waves in radicals.

Remark. In the special case $m = 0$ in (4.61), equations (4.62) and (4.63) give the stationary solution $v = U(x)$ given explicitly by the cubic equation (4.50).

4.3.2 Similarity solution

The invariant solution with respect to the generator

$$X_3 = t\frac{\partial}{\partial t} + x\frac{\partial}{\partial x} \tag{4.64}$$

of the uniform scaling transformation group is known as a similarity solution. The operator (4.64) has two independent invariants, v and

$$\xi = \frac{x}{t}. \tag{4.65}$$

The invariant solution has the representation

$$v = g(\xi). \tag{4.66}$$

Calculations show that the representation (4.66) reduces equation (4.24) to the second-order ODE

$$[(g-\mu)^2 - \lambda]g'' + 2(g-\mu)g'^2 - 2(g-\mu)g' = 0.$$

Rewriting this equation in the form

$$[(g-\mu)^2 g']' - \lambda g'' = 0 \tag{4.67}$$

and integrating once, we arrive at the first-order equation

$$[(g-\mu)^2 - \lambda]g' = C_1, \quad C_1 = \text{const.} \tag{4.68}$$

We transform equation (4.68) into separable form

$$(w^2 - \lambda)\frac{dw}{d\mu} = -w^2 + \lambda + C_1 \tag{4.69}$$

by introducing the new dependent variable

$$w = g - \mu. \tag{4.70}$$

Separating the variables,

$$\frac{w^2 - \lambda}{\lambda + C_1 - w^2} dw = d\mu,$$

we write equation (4.69) as

$$-dw + \frac{C_1}{\lambda + C_1 - w^2} dw = d\mu,$$

whence upon integration

$$C_1 \int \frac{dw}{\lambda + C_1 - w^2} = w + \mu. \tag{4.71}$$

Evaluating the integral in (4.71) we obtain the following cases.

Case 1. $\mu + C_1 = 0$. Then the integration in equation (4.71) gives

$$-\frac{\lambda}{w} = w + \mu + C_2,$$

whence

$$w = \frac{1}{2}[-(\mu + C_2) \pm \sqrt{(\mu + C_2)^2 - 4\lambda}]. \tag{4.72}$$

Case 2. $\mu + C_1 < 0$. Then we evaluate the integral by writing

$$C_1 \int \frac{dw}{\lambda + C_1 - w^2} = -C_1 \int \frac{dw}{-(\lambda + C_1) + w^2}$$

and write equation (4.71) as

$$-\frac{C_1}{\sqrt{-(\lambda + C_1)}} \arctan \frac{w}{\sqrt{-(\lambda + C_1)}} = w + \mu + C_2. \tag{4.73}$$

Case 3. $\mu + C_1 > 0$. In this case equation (4.71) becomes

$$\frac{C_1}{2\sqrt{\lambda + C_1}} \ln \frac{\sqrt{\lambda + C_1} + w}{\sqrt{\lambda + C_1} - w} = w + \mu + C_2 \tag{4.74}$$

when $w^2 < \mu + C_1$ and

$$\frac{C_1}{2\sqrt{\lambda + C_1}} \ln \frac{w + \sqrt{k + C_1}}{w - \sqrt{\lambda + C_1}} = w + \mu + C_2 \tag{4.75}$$

when $w^2 > \mu + C_1$.

Based on the above calculations we conclude that all similarity solutions based on the infinitesimal symmetry (4.64) are provided by equations (4.65), (4.66), (4.70), and (4.72)–(4.75).

4.4 Concluding remarks

The presented modeling example can be used as an approximation for a pulsating elastic artery. We have proposed two different points of view. The first approach represents a singular perturbation theory that formalizes the scale separation property by explicitly defining multiple scales that exist in the given nonlinear model with the goal of separating derivatives with respect to fast and slow scales into different orders of perturbation theory. The advantage of this approach is that it yields a solvable perturbative hierarchy of equations that provides useful perturbative information already at low orders in ε. However, the disadvantage of this approach is the need to identify the various scales *a priori* and, in the frame of the present modeling, the multi-scale approach cannot be brought beyond the leading order. Alternatively, the group theoretical approach provides *all possible* exact solutions of the nonlinear model (4.24) without any perturbations and, consequently, without introducing a small parameter ε, which is a significant advantage. In this modeling scenario we have provided the exact solutions that were obtained implicitly by solving the nonlinear ODEs, which have the deficiency of the latter approach. However, in terms of numerical simulations, the second approach seems more advantageous.

4.5 Exercises

Exercise 4.1. Show that equation (4.24) admits the one-parameter group of translations $\bar{t} = t + a_1, \bar{x} = x + a_2$ of t and x and the one-parameter group of uniform scaling transformations in the (t, x)-plane $\bar{t} = te^{a_3}, \bar{x} = xe^{a_3}$.

Exercise 4.2. Show that the transformations $\bar{t} = t + a_1, \bar{x} = x + a_2$ of t and x and the one-parameter uniform scaling transformations in the (t, x)-plane $\bar{t} = te^{a_3}, \bar{x} = xe^{a_3}$ form the group and check that these transformations correspond to the generators X_1, X_2, and X_3 given by (4.60).

Exercise 4.3. Find two independent invariants of the operator $X = \frac{\partial}{\partial t} + m\frac{\partial}{\partial x}$.

Exercise 4.4. Find two independent invariants of the operator $X = t\frac{\partial}{\partial t} + x\frac{\partial}{\partial x}$.

Exercise 4.5. Find the general presentation of a solution for the following ODE:

$$\frac{w^2 - \lambda}{\lambda + C_1 - w^2} dw = d\mu.$$

5 Modeling scenario 2: invariant solutions as dispersion relation

In fluid dynamics, physics, or electrical engineering, dispersion most often refers to frequency-dependent effects of wave propagation. For example, in the theory of surface waves, the dispersion relation generally refers to frequency dispersion, which means that waves of different wavelengths travel at different phase speeds. For example, oceanic surface gravity waves, moving under the forces of gravity, propagate faster at increasing wavelength. For a given wavelength, gravity waves in deeper water have a larger phase speed than those in shallow water. In contrast, capillary surface waves only undergoing surface tension forces propagate faster at shorter wavelengths. Besides frequency dispersion, water waves also exhibit amplitude dispersion. This is a nonlinear effect, by which waves of larger amplitude have a different phase speed from small-amplitude waves. In physical sciences, however, there are several other uses of the word "dispersion." In the presence of dispersion, wave velocity is no longer uniquely defined, giving rise to the distinction of phase velocity and group velocity. A well-known effect of phase velocity dispersion is the color dependence of light refraction that can be observed in prisms and rainbows. Dispersion occurs when pure plane waves of different wavelengths have different propagation velocities, so that a wave packet of mixed wavelengths tends to spread in space. The speed of a plane wave, u, is a function of the wave's wavelength L:

$$u = u(L).$$

The elementary sorts of waves being considered can be described by a *wave packet*, which is an expression of the wave in the form

$$u(x,t) = ae^{i(kx-\omega t)}, \tag{5.1}$$

where a is the *wave amplitude*, x is a position along the wave's direction of travel, t is the time at which the wave is described, k is the *wave number* defined by means of the *wavelength* L as

$$L = \frac{2\pi}{k}, \tag{5.2}$$

and ω is the *wave frequency* defined in terms of the *wave period* T as

$$T = \frac{2\pi}{\omega}. \tag{5.3}$$

The *phase* associated with the wave packet (5.1) is defined as $\theta = kx - \omega t$. Note that $d\theta = kdx - \omega dt$, which means that $d\theta = 0$ (i. e., the phase is constant) if $dx/dt = \omega/k$. The *phase* and *group velocity* vectors $\vec{c}_p$ and $\vec{c}_g$ are defined as

https://doi.org/10.1515/9783111387499-005

$$\vec{c}_p = \frac{\omega}{k}, \quad \vec{c}_g = \frac{d\omega}{dk}, \tag{5.4}$$

respectively. According to the definition (5.4), the phase denotes the rate at which the phase propagates through the medium. Also, using the definition (5.4), we can write the wave packet (5.1) in the form

$$u(x,t) = ae^{ik(x-ct)}. \tag{5.5}$$

5.1 Preliminaries

We start by considering a polynomial equation

$$\sum_{s=0}^{m}\sum_{l=0}^{n} b_{sl}\left(\frac{\partial}{\partial x}\right)^s\left(\frac{\partial}{\partial t}\right)^l u(x,t) = 0, \tag{5.6}$$

in which b_{sl} are constants and Einstein summation is assumed. For the sake of brevity,

$$\sum_{i=1}^{n} c_i x^i = c_1 x^1 + c_2 x^2 + \cdots + c_n x^n$$

is expressed as $c_i x^i$. Whenever the subscript and superscript are the same, it is customary to use the Einstein summation convention to make expressions simple. However, the superscript of x does not mean exponent; it generally means the i-th component. It may also denote the index of coordinates.

Substitution of wave packet (5.1) into (5.4) yields the following constraint for the wave parameters ω and k:

$$F(\omega,k) = b_{sl}(ik)^s(-i\omega)^l = 0, \tag{5.7}$$

which is known as the *dispersion relation* of the wave solution of the form (5.1).

Example 5.1 (One-dimensional wave motion). In the case when $n = m = 2$, equation (5.6) is written as the following second-order PDE:

$$b_{11}\frac{\partial}{\partial x}\frac{\partial u}{\partial t} + b_{12}\frac{\partial}{\partial x}\frac{\partial^2 u}{\partial t^2} + b_{21}\frac{\partial^2}{\partial x^2}\frac{\partial u}{\partial t} + b_{22}\frac{\partial^2}{\partial x^2}\frac{\partial^2 u}{\partial t^2} = 0. \tag{5.8}$$

Let us substitute the wave packet (5.1) into the PDE (5.8). Since differentiation with respect to x produces the term ik and differentiation with respect to t produces the term $-i\omega$, i. e.,

$$\frac{\partial}{\partial x} \to ik, \quad \frac{\partial}{\partial t} \to -i\omega,$$

we obtain the dispersion relation ω and k in the form

$$F(\omega,k) = b_{11}(ik)(-i\omega) + b_{12}(ik)(-i\omega)^2 + b_{21}(ik)^2(-i\omega) + b_{22}(ik)^2(-i\omega)^2 = 0, \tag{5.9}$$

which is a quadratic equation for ω. □

Remark. We say that if $\omega''(k) = 0$ (i. e., if $\vec{c}_g = \text{const.}$), then the wave is *hyperbolic*, and if $\omega''(k) \neq 0$ (i. e., if $\vec{c}_g \neq \text{const.}$), then the wave is *dispersive*.

For example, let us find the wave modes and the phase speed for the wave equation

$$u_{tt} - a^2 u_{xx} = 0, \tag{5.10}$$

where a is a constant. We look for a solution of equation (5.10) in the form (5.1). The dispersion relation (5.7) is

$$\omega^2 - a^2k^2 = 0, \tag{5.11}$$

so the two wave modes are given by $\omega = \pm ak$ and the phase speed is $\omega/k = \pm a$. Also, since $\omega''(k) = 0$, the wave is hyperbolic.

Each solution (root) $\omega_p(k)$ ($p = 1, 2, \ldots$) of the dispersion relation (5.7) is called a *wave mode.*

In three-dimensional space, the wave packet (5.1) is

$$u(x,y,z,t) = ae^{i(kx+ly+mz-\omega t)} = ae^{i(\vec{k}\cdot\vec{x}-\omega t)},$$

where $\vec{k} = (k,l,m)$ is called a wave number vector and $|\vec{k}| = \sqrt{k^2+l^2+m^2}$.

Note that the phase velocity is the vector in three-dimensional space given by

$$\vec{c}_p = \left(\frac{\omega}{k}, \frac{\omega}{l}, \frac{\omega}{m}\right) = \frac{\omega}{\vec{k}}. \tag{5.12}$$

If we multiply (5.12) by $\vec{k}/\vec{k}$, we can write the phase velocity vector in the form $\vec{k}/|\vec{k}|$, where $|\vec{k}|$ represents the unit vector in the direction of wave number $\vec{k}$.

The following example illustrate the concept of the dispersion relation by means of *translations* and *uniform dilation.*

Example 5.2 (Two-dimensional wave motion). Consider the polynomial equation

$$P\left(\frac{\partial}{\partial x}, \frac{\partial}{\partial y}, \frac{\partial}{\partial t}\right)\phi(x,y,t) = \sum_{j=1}^{N}\left(\frac{\partial}{\partial x}\right)^{a_j}\left(\frac{\partial}{\partial y}\right)^{b_j}\left(\frac{\partial}{\partial t}\right)^{c_j}\phi = 0. \tag{5.13}$$

Consider the *uniform dilation*

$$\bar{x} = \varepsilon x, \quad \bar{y} = \varepsilon y, \quad \bar{t} = \varepsilon t, \tag{5.14}$$

which, for $\varepsilon \ll 1$, can be associated with *slow length and time scaling transformations.*

Show that, up to the order $0(\varepsilon)$, any wave of the form

$$\phi = A(\overline{x}, \overline{y}, \overline{t})e^{i(kx+ly-\omega t)} \tag{5.15}$$

propagates according to the law

$$A_\tau + \left(\frac{\partial \omega}{\partial k}, \frac{\partial \omega}{\partial l}\right)\nabla A = 0, \tag{5.16}$$

where

$$\nabla = \left(\frac{\partial}{\partial \overline{x}}, \frac{\partial \omega}{\partial \overline{y}}\right). \tag{5.17}$$

Solution. We look for a solution of equation (5.13) in the form of the wave packet (5.1), i. e.,

$$\phi = A(\overline{x}, \overline{y}, \overline{t})e^{i(kx+ly-\omega t)}. \tag{5.18}$$

Let us denote $\theta = kx + ly - \omega t$ and calculate the partial derivatives of ϕ, e. g.,

$$\phi_x = (\varepsilon A_{\overline{x}} + ikA)e^{i\theta} = \left(\frac{\partial}{\partial \overline{x}} + ik\right)(Ae^{i\theta}), \tag{5.19}$$

$$\phi_{xx} = \left(\varepsilon\frac{\partial}{\partial \overline{x}} + ik\right)^2 (Ae^{i\theta}) = (\varepsilon^2 A_{\overline{xx}} + 2ik\varepsilon A_{\overline{x}} - k^2 A)e^{i\theta}, \tag{5.20}$$

$$\phi_t = (\varepsilon A_{\overline{t}} - i\omega A)e^{i\theta}, \tag{5.21}$$

$$\phi_{tt} = \left(\varepsilon\frac{\partial}{\partial \overline{t}} - i\omega\right)^2 (Ae^{i\theta}) = (\varepsilon^2 A_{\overline{tt}} - 2i\omega A_{\overline{t}} - \omega^2 A)e^{i\theta}, \tag{5.22}$$

and so forth.

Thus, substituting the wave packet (5.18) into equation (5.13) and keeping the terms of $0(\varepsilon)$, we obtain

$$\sum_{j=1}^{N}\left[(ik)^{a_j} + a_j(ik)^{a_j-1}\varepsilon\frac{\partial}{\partial \overline{x}}\right]\left[(il)^{b_j} + b_j(il)^{b_j-1}\varepsilon\frac{\partial}{\partial \overline{y}}\right]\left[(-i\omega)^{c_j} + c_j(-i\omega)^{c_j-1}\varepsilon\frac{\partial}{\partial \overline{t}}\right] = 0. \tag{5.23}$$

At the leading order, we have

$$\sum_{j=1}^{N}[(ik)^{a_j}(il)^{b_j}(-i\omega)^{c_j}] = 0, \tag{5.24}$$

which is the dispersion relation.

At the order $0(\varepsilon)$, we have

$$\sum_{j=1}^{N}[a_j(ik)^{a_j-1}(il)^{b_j}(-i\omega)^{c_j}A_{\overline{x}}+b_j(ik)^{a_j}(il)^{b_j-1}(-i\omega)^{c_j}A_{\overline{y}}+c_j(ik)^{a_j}(il)^{b_j}(-i\omega)^{c_j-1}A_{\overline{t}}]=0. \quad (5.25)$$

We can write equation (5.25) in the form

$$\sum_{j=1}^{N}\{c_j(ik)^{a_j}(il)^{b_j}(-i\omega)^{c_j}\}\left[A_{\overline{t}}+\frac{a_j(ik)^{a_j-1}(il)^{b_j}(-i\omega)^{c_j}}{c_j(ik)^{a_j}(il)^{b_j}(-i\omega)^{c_j-1}}A_{\overline{x}}+\frac{b_j(ik)^{a_j}(il)^{b_j-1}(-i\omega)^{c_j}}{c_j(ik)^{a_j}(il)^{b_j}(-i\omega)^{c_j-1}}A_{\overline{y}}\right]=0. \quad (5.26)$$

Let us differentiate the dispersion relation (5.24) with respect to k. We obtain

$$\sum_{j=1}^{N}[a_j(ik)^{a_j-1}(il)^{b_j}(-i\omega)^{c_j}+c_j(ik)^{a_j}(il)^{b_j}(-i\omega)^{c_j-1}\omega_k]=0. \quad (5.27)$$

Similarly, differentiation of the dispersion relation (5.24) with respect to k yields

$$\sum_{j=1}^{N}[b_j(ik)^{a_j}(il)^{b_j-1}(-i\omega)^{c_j}+c_j(ik)^{a_j}(il)^{b_j}(-i\omega)^{c_j-1}\omega_l]=0. \quad (5.28)$$

Thus, we can express ω_k from (5.27) as

$$\omega_k=\frac{a_j(ik)^{a_j-1}(il)^{b_j}(-i\omega)^{c_j}}{c_j(ik)^{a_j}(il)^{b_j}(-i\omega)^{c_j-1}},$$

and similarly we can express ω_l from (5.28) as

$$\omega_k=\frac{b_j(ik)^{a_j}(il)^{b_j-1}(-i\omega)^{c_j}}{c_j(ik)^{a_j}(il)^{b_j}(-i\omega)^{c_j-1}}.$$

Comparing with the expressions in the brackets of (5.26), we can write (5.26) in the form

$$A_{\overline{t}}+\omega_k A_{\overline{x}}+\omega_l A_{\overline{y}}=0, \quad (5.29)$$

which is the required form (5.16). □

5.2 Group theoretical derivation of dispersion relations

5.2.1 Homogeneous linear equations

We will illustrate the essence of our approach by considering the wave equation (5.10),

$$u_{tt}-a^2u_{xx}=0.$$

We note first that the representation of a particular solution of equation (5.10) in the form (5.1) (we take for the sake of simplicity $A = 1$),

$$u(x,t) = e^{i(kx-\omega t)},$$

is not invariant under a change of variables. Indeed, let us replace the dependent variable u by a new dependent variable v defined as follows:

$$v = \ln u. \tag{5.30}$$

Then equation (5.10) becomes

$$v_{tt} + v_t^2 - a^2 v_{xx} - a^2 v_x^2 = 0. \tag{5.31}$$

This nonlinear equation does not have a particular solution of the wave packet form (5.1). But it has the particular solution

$$v = kx - \omega t. \tag{5.32}$$

It is obtained from the solution (5.1) by the substitution (5.30). Substituting (5.32) in equation (5.31) we arrive again at the dispersion relation (5.11):

$$\omega^2 - a^2 k^2 = 0.$$

However, in deriving the solution (5.32) and the dispersion relation (5.11) we do not have the physical reasoning as in the case of the solution (5.1). Hence, the physical concept of dispersion relations is not invariant under changes of variables. This fact makes it difficult to extend the concept of dispersion relations to nonlinear equations. Therefore we will give an invariant, group theoretic, derivation of the dispersion relation (5.11).

We will use the invariance of the wave equation (5.10) under the two-parameter group consisting of the time and space translations with the generators

$$X_1 = \frac{\partial}{\partial t}, \quad X_2 = \frac{\partial}{\partial x} \tag{5.33}$$

and look for the invariant solutions with respect to the linear combination

$$X = kX_1 + \omega X_2 \tag{5.34}$$

of the generators (5.33). Using the invariants u and $kx - \omega t$ for the operator (5.34), we obtain the following general form of the invariant solutions:

$$u = \phi(kx - \omega t). \tag{5.35}$$

Substitution of (5.35) in equation (5.10) gives

$$(\omega^2 - a^2k^2)\phi'' = 0, \tag{5.36}$$

where ϕ'' is the second derivative of the function $\phi(kx-\omega t)$ with respect to its argument. Equation (5.36) yields the dispersion relation (5.11) unless $\phi(kx-\omega t)$ is the linear function.

A similar result holds for the homogeneous linear equation (5.6) of the order N, i. e., when $s + l = N$. In this case equation (5.36) is replaced by

$$F_N(\omega, k)\phi^{(N)} = 0, \tag{5.37}$$

where

$$F_N(\omega, k) = \sum_{s+l=N} b_{sl}k^s(-\omega)^l.$$

Thus, we have the following result for *homogeneous* linear equations.

Proposition 5.1. *Function* (5.35) *is an invariant solution of the homogeneous linear equation* (5.6),

$$\sum_{s+l=N} b_{sl}\left(\frac{\partial}{\partial x}\right)^s\left(\frac{\partial}{\partial t}\right)^l u(x,t) = 0, \tag{5.38}$$

if either

$$\phi^{(N)}(kx - \omega t) = 0,$$

i. e., if $\phi(kx-\omega t)$ is a polynomial of degree $N-1$, or the parameters ω, k solve the equation

$$F_N(\omega, k) = 0. \tag{5.39}$$

Equation (5.39) guarantees that (5.35) with arbitrary ϕ is an invariant solution for equation (5.38). The algebraic equation (5.39) is called a dispersion relation *for the differential equation (5.38).*

Remark 5.1. Our approach is applicable to *appropriate* nonlinear equations as well. Let us apply it, e. g., to equation (5.31). This equation also admits the translation generators (5.33). These generators lead again to the general form (5.35) of invariant solutions, namely,

$$v = \psi(kx - \omega t). \tag{5.40}$$

Substitution of (5.40) in equation (5.31) gives

$$(\omega^2 - a^2k^2)(\psi'' + \psi'^2) = 0. \tag{5.41}$$

Whence, using Proposition 5.1, we obtain the dispersion relation (5.11). Equation (5.33) shows that if the dispersion relation (5.11) is not satisfied, then $\psi'' + \psi'^2 = 0$, i. e., one obtains the simple invariant solution (5.40):

$$v = \ln|C_1 + C_2(kx - \omega t)|.$$

5.2.2 Nonhomogeneous polynomial linear equations

In order to understand the situation with general nonhomogeneous linear equations, we will consider equation (5.8):

$$b_{00}u + b_{01}\frac{\partial u}{\partial t} + b_{02}\frac{\partial^2 u}{\partial t^2} + b_{10}\frac{\partial u}{\partial x} + b_{11}\frac{\partial}{\partial x}\frac{\partial u}{\partial t} + b_{12}\frac{\partial}{\partial x}\frac{\partial^2 u}{\partial t^2}$$
$$+ b_{20}\frac{\partial^2 u}{\partial x^2} + b_{21}\frac{\partial^2}{\partial x^2}\frac{\partial u}{\partial t} + b_{22}\frac{\partial^2}{\partial x^2}\frac{\partial^2 u}{\partial t^2} = 0.$$

We substitute function (5.35) in equation (5.8) and obtain

$$\begin{aligned} &b_{00}\phi + [b_{10}k - b_{01}\omega]\phi' + [b_{02}\omega^2 - b_{11}k\omega + b_{20}k^2]\phi'' \\ &+ [b_{12}k\omega^2 - b_{21}k^2\omega]\phi''' + b_{22}k^2\omega^2\,\phi^{(iv)} = 0. \end{aligned} \tag{5.42}$$

One can separate the variables in equation (5.41) if and only if ϕ, ϕ', ϕ'', ϕ''', and $\phi^{(iv)}$ are proportional with constant coefficients. For this purpose, it suffices to let

$$\phi' = \lambda\phi, \quad \lambda = \text{const.}$$

The above equation yields that (5.35) is the exponential function

$$u = Ae^{\lambda(kx - \omega t)} \tag{5.43}$$

with arbitrary constants A and λ. Taking $\lambda = i$ one obtains the wave packet (5.1). Ignoring the trivial case $A = 0$ we obtain from (5.41) the dispersion relation

$$\begin{aligned} &b_{00} + [b_{10}k - b_{01}\omega]\lambda + [b_{02}\omega^2 - b_{11}k\omega + b_{20}k^2]\lambda^2 \\ &+ [b_{12}k\omega^2 - b_{21}k^2\omega]\lambda^3 + b_{22}k^2\omega^2\,\lambda^4 = 0. \end{aligned} \tag{5.44}$$

Equation (5.44) coincides with the dispersion relation (5.9) upon setting $\lambda = i$.

Thus, in the case of nonhomogeneous linear equations (5.6) Proposition 5.1 is replaced by the following statement.

Proposition 5.2. *The dispersion relation for an arbitrary nonhomogeneous linear equation* (5.6) *is the algebraic equation*

$$F(\omega, k) \equiv \sum_{s=0}^{m} \sum_{l=0}^{n} b_{sl} k^s (-\omega)^l \lambda^{s+l} = 0. \tag{5.45}$$

Equation (5.44) *guarantees that the exponential function* (5.43) *is an invariant solution for equation* (5.6).

5.3 Concluding remarks

The dispersion relation of wave motion, and especially of oceanic internal gravity waves, can be considered from a group analysis point of view. In this modeling scenario we have shown that a similar dispersion relation appears in nonlinear wave motion and it can be used to obtain the more general form of physically relevant solutions of nonlinear equations.

5.4 Exercises

Exercise 5.1. Check that the wave equation $u_{tt} - a^2 u_{xx} = 0$ admits the group consisting of the time and space translations with the generators

$$X_1 = \frac{\partial}{\partial t}, \quad X_2 = \frac{\partial}{\partial x}.$$

Exercise 5.2. Check that the dispersion relation of the wave equation $u_{tt} - a^2 u_{xx} = 0$ is given by formula (5.11).

Exercise 5.3. Show that the invariants of the operator $X = kX_1 + \omega X_2$ are given by u and $kx - \omega t$.

6 Modeling scenario 3: invariant solutions of nonlinear surface waves in the ocean and atmosphere

In terms of group theoretical modeling, the dynamics of a nonlinear model in question can often be described by means of the *symmetries* of Euler and Navier–Stokes equations (see, e. g., [39, 40, 37]). Although not discussed here, it is worth noting that the conservation laws can be deduced from Noether's theorem [56], which manifests a connection between symmetries and invariant solutions.

The ocean surface is in continual motion. Waves are the result of disturbance of the water surface; waves themselves represent a restoring force to calm the surface. Waves at the surface of oceans and lakes are orbital progressive waves. This type of wave forms at the boundary of two phases of different density, in this case air and water. The wave moves forward with a steady velocity, so it is called "progressive." The water itself moves very little: like the crowd in a football stadium doing "the wave," individual particles of water move up and then down, but do not follow the moving wave. The complete motion of the water particles is a circle, so that a small object floating on top of the wave actually describes a circle as the wave goes underneath it.

The wave period is the length of time it takes for a wave to pass a fixed point (crest to crest). The speed of a wave is equal to the wavelength divided by the wave period. Wave steepness is defined as the ratio of the wave height to the wavelength. When the wave builds and reaches a steepness greater than a ratio of 1:7, the wave breaks and spills forward. The wave has actually become too steep to support itself and gravity takes over. Breakers are normally associated with shorelines, where they are known as surf, but can occur anywhere in the ocean.

The passage of a wave only affects the water down to the wave base, which is half the wavelength. Below that depth there is negligible water movement. This is the part of the water column that submarines use for "clear sailing." Waves in water deeper than half their wavelength are known as deep water waves. Their speed in meters per second can be approximated as speed $= gT/2\pi$, where T is the wave period and g is the acceleration due to gravity ($9.8\,\mathrm{m/s^2}$).

Shallow water waves are those moving in water with a depth of less than one-twentieth of their wavelength. Waves approaching shallow water at a shoreline are in this category. In these waves, the orbits of water particles are flat ellipses rather than circles. Shallow water wave movements can be felt at the bottom, and their interaction with the bottom affects both wave and sea floor. Shallow water waves include both seismic sea waves (tsunamis) generated by earthquakes at sea and tide waves generated by the attractive forces of the Moon and the Sun on the ocean. Both of these wave types have such long wavelengths that average ocean depths are easily less than one-twentieth that value. The speed of shallow water waves decreases as the water depth decreases; it is equal to 3.1 times the square root of the depth. *Transitional waves* have

https://doi.org/10.1515/9783111387499-006

wavelengths between 2 and 20 times the water depth; their speed is controlled in part by water depth and in part by the wave period.

6.1 Preliminaries

Consider an irrotational flow of an incompressible inviscid fluid of uniform density confined between a free surface $y = \eta(x,z,t)$ and a variable stationary bottom $y = h(x,z)$. The equations and boundary conditions governing the fluid motion constitute the *Cauchy–Poisson free boundary problem*:

$$\Delta\varphi = 0 \quad \text{at } 0 < x < \infty, \quad -h(x,z) < y < \eta(x,z,t), \tag{6.1}$$

$$\frac{\partial\varphi}{\partial\vec{n}} = 0 \quad \text{at } y = -h(x,z), \tag{6.2}$$

$$\eta_t + \varphi_x\eta_x + \varphi_z\eta_z = \varphi_y \quad \text{at } y = \eta(x,z,t), \tag{6.3}$$

$$\varphi_t + \frac{1}{2}[\varphi_x^2 + \varphi_y^2 + \varphi_z^2] + g\eta = F(x,z,t) \quad \text{at } y = \eta(x,z,t), \tag{6.4}$$

where F in equation (6.4) represents the external forces, including the pressure. The first equation (6.1) represents the fact that the fluid motion is potential. The second boundary condition (6.2) stands for the no-leak condition through the solid bottom. The last two conditions (6.3) and (6.4) represent the kinematic and dynamic conditions at the free surface.

Additionally, the initial conditions are imposed as follows:

$$\eta(x,z,t)|_{t=0} = 0, \quad (\varphi_x = \varphi_y = \varphi_z)\big|_{t=0} = 0. \tag{6.5}$$

The Cauchy–Poisson problem consists in determining the motion of the fluid (we need to find φ) and the shape of the free boundary $\eta(x,z,t)$ for $t > 0$.

The main difficulty in solving the problem (6.1)–(6.5) is that the boundary conditions are nonlinear. Additionally, as the conditions (6.3) and (6.4) indicate, the unknown φ is given on the unknown boundary η.

Referring to Figure 6.1, consider next the following illustrative example of the Cauchy–Poisson problem which is simplified to the two-dimensional case with flat bottom.

Example 6.1. Consider a channel of average water depth H. Suppose that the average velocity in the domain of motion $-\infty < x < +\infty$, $0 < y < h(x,t)$ is given by

$$u(x,t) = \frac{1}{h}\int_0^h u_1(x,y,t)\,dy, \tag{6.6}$$

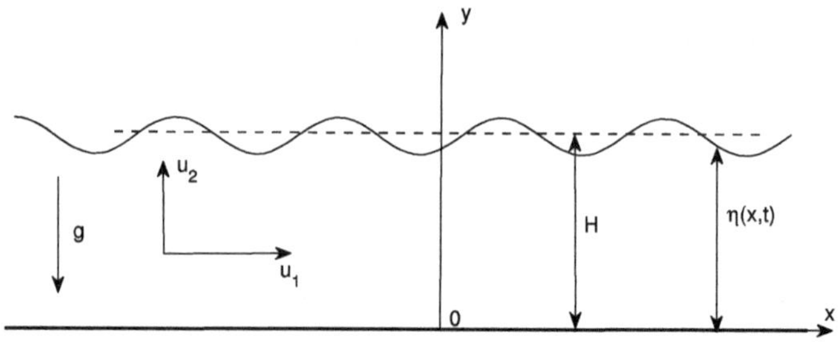

Figure 6.1: Two-dimensional free boundary problem with a flat bottom at $y = 0$.

where $h(x,t)$ is the variable water depth in the channel and u_1, u_2 are the horizontal and vertical components of the velocity vector $\vec{u}(x,y,t)$. Show that the kinematic condition

$$\frac{\partial h}{\partial t} + u_1\frac{\partial h}{\partial x} - u_2 = 0 \tag{6.7}$$

can be written in the conservative form

$$\frac{\partial A}{\partial t} + \frac{\partial B}{\partial x} = 0, \tag{6.8}$$

where A and B are some functions of x and t.

Solution. Let us introduce the stream function ψ by $u_1 = \psi_y$ and $u_2 = -\psi_x$. We assume a no-leak boundary condition at the bottom $y = 0$ and the dynamic boundary condition $p = p_{\text{atm}}$ at the free boundary $f(x,y,t) = y - h(x,t)$.

So, we have

$$p|_{y=h} = p_{\text{atm}} \quad \text{and} \quad \left.\frac{df}{dt}\right|_{y=h} = 0. \tag{6.9}$$

We also note that

$$u(x,t) = \frac{1}{h}\int_0^h \frac{\partial \psi}{\partial y}dy = \frac{1}{h}(\psi(x,h) - \psi(x,0)). \tag{6.10}$$

Since $\psi(x,0) = 0$, from (6.10) we have

$$\psi(x,h) = u(x,t)h. \tag{6.11}$$

Now, since

$$\int_0^h u_{1_x}dy = \int_0^h \frac{\partial}{\partial y}\left(\frac{\partial \psi}{\partial x}\right)dy = \psi_x|_0^h \tag{6.12}$$

and

$$-u_2(x,h) = \psi_x(x,h), \tag{6.13}$$

we can write the kinematic condition (6.7) as

$$h_t + u_1(x,h)h_x + \int_0^h u_{1_x}dy = 0. \tag{6.14}$$

Now, using the formula

$$\frac{\partial}{\partial y}\int_{\alpha(y)}^{\beta(y)} f(x,y)dy = \int_{\alpha(y)}^{\beta(y)} \frac{\partial f}{\partial y}dy + f[\beta_y - \alpha_y], \tag{6.15}$$

we write the kinematic condition (6.14) in the form

$$h_t + \frac{\partial}{\partial x}\int_0^h u_1(x,y,t)dy = 0, \tag{6.16}$$

or

$$h_t + \frac{\partial}{\partial x}\int_0^h \psi_y = 0. \tag{6.17}$$

□

6.2 Cauchy–Poisson free boundary problem

We consider an irrotational inviscid, nondiffusive, two-dimensional fluid flow vertically confined to lie between the rigid boundary S_0 given by $z = -Z(x,y,t)$ and the free boundary S given by $z = h(x,y,t)$. We study the motion by adopting a local Cartesian system of coordinates. We denote $\vec{x} = (x,y,z)$, where the coordinate x is chosen in the direction of wave propagation, which is transverse to the coordinate y, and the vertical coordinate z changes between the rigid bottom S_0 and the free surface S. Thus, the domain of the homogeneous fluid motion is given by

$$D_\xi = \{(x,y) \in \mathbb{R}^2, -Z(x,y,t) \leqslant z \leqslant \zeta(x,y,t)\},$$

where $Z(x,y,t)$ is a given function and $\zeta(x,y,t)$ is to be determined.

We rewrite the *Cauchy–Poisson free boundary problem* (6.1)–(6.4) in terms of the velocity potential as the following system of equations:

$$\Delta\varphi = 0 \quad (\vec{x} \in D_\xi), \tag{6.18}$$

$$Z_t + Z_x\varphi_x + Z_y\varphi_y + \varphi_z = 0 \quad (\vec{x} \in S_0), \tag{6.19}$$

$$\zeta_t + \varphi_x\zeta_x + \varphi_y\zeta_y = \varphi_z \quad (\vec{x} \in S), \tag{6.20}$$

$$\varphi_t + \frac{1}{2}|\nabla\varphi|^2 + g\zeta + \frac{p}{\rho} = 0 \quad (\vec{x} \in S). \tag{6.21}$$

The functions $\varphi(x, y, z, t)$ and $\zeta(x, y, t)$ are to be determined as solutions of equations (6.18)–(6.21) for $t > 0$ provided $\varphi(x, y, z, 0)$ and $\zeta(x, y, 0)$ are known functions.

Simplifying assumptions. First, we assume that the pressure on the free boundary is constant, i. e., $p(x, y, \zeta, t) = p_0 =$ const., where we set p_0 to be equal to the atmospheric pressure. Second, without loss of generality, we consider the case where the bottom is stationary and flat, i. e., we set $Z(x, y, z, t) = h_0 =$ const. > 0. Then the system (6.18)–(6.21) is simplified to the form

$$\Delta\varphi = 0 \quad (\vec{x} \in D_\xi), \tag{6.22}$$

$$\varphi_z = 0 \quad (z = -h_0), \tag{6.23}$$

$$\zeta_t + \varphi_x\zeta_x + \varphi_y\zeta_y = \varphi_z \quad (\vec{x} \in S), \tag{6.24}$$

$$\varphi_t + \frac{1}{2}|\nabla\varphi|^2 + g\zeta = 0 \quad (\vec{x} \in S). \tag{6.25}$$

6.3 Linear theory

The exact solution

$$\zeta_0 = 0, \quad \varphi_0 = u_0 x - \frac{1}{2}u_0^2 t \quad (u_0 = \text{const.}) \tag{6.26}$$

of system (6.22)–(6.25) corresponds to the flow with constant velocity u_0. That is to say, solution (6.26) represents a stationary fluid motion with $\vec{u} = (u_0, 0, 0)$. For small perturbations φ', ζ' defined via

$$\varphi = \varphi_0 + \varphi', \quad \zeta = \zeta', \tag{6.27}$$

the linearized problem (6.22)–(6.25) in the domain

$$D_0 = \{(x, y) \in \mathbb{R}^2, -h_0 \leqslant z \leqslant 0\} \tag{6.28}$$

has the following form:

$$\Delta\varphi' = 0 \quad (\vec{x} \in D_0), \tag{6.29}$$

$$\varphi'_z = 0 \quad (z = -h_0), \tag{6.30}$$

$$\zeta'_t + u_0\zeta'_x = \varphi'_z \quad (z = 0), \tag{6.31}$$

$$\varphi'_t + u_0\varphi'_x + g\zeta' = 0 \quad (z = 0). \tag{6.32}$$

We can look for a solution of (6.29)–(6.32) in the form of the elementary wave packets

$$\varphi' = \Phi(z)e^{i(kx+ly-\omega t)}, \quad \zeta' = He^{i(kx+ly-\omega t)}, \tag{6.33}$$

where $\vec{k} = (k, l)$ is a wave number vector and $H = \text{const}$. Substituting (6.33) into equations (6.29) and (6.30), we obtain the following two equations:

$$\Phi''(z) - m^2\Phi''(z) = 0 \tag{6.34}$$

and

$$\Phi'(-h_0) = 0, \tag{6.35}$$

where the prime means differentiation with respect to z and $m = (k^2 + l^2)^{\frac{1}{2}}$.

One can check by direct differentiation that the general solution of equation (6.34) is given by

$$\Phi(z) = A\cosh[m(z + h_0)] + B\sinh[m(z + h_0)], \tag{6.36}$$

where A and B are constants. It follows from the boundary condition (6.35) that $B = 0$. Additionally, substitution of the presentation (6.33) into the boundary conditions (6.31) and (6.32) yields the following system of linear equations for A and H:

$$i(\omega - ku_0)H + mA\sinh[mh_0] = 0, \tag{6.37}$$

$$gH - iA(\omega - ku_0)\cosh[mh_0] = 0. \tag{6.38}$$

A nontrivial solution of the system (6.37) and (6.38) exists if

$$\det\begin{vmatrix} i(\omega - ku_0) & m\sinh[mh_0] \\ g & -i(\omega - ku_0)\cosh[mh_0] \end{vmatrix} = 0. \tag{6.39}$$

The condition (6.39) yields the dispersion relation for surface gravity waves

$$(\omega - ku_0)^2 = gm\tanh[mh_0], \tag{6.40}$$

which shows that surface waves with a constant flow are dispersive waves with two wave modes, i. e.,

$$\omega = ku_0 \pm \sqrt{gm\tanh[mh_0]}. \tag{6.41}$$

6.3.1 Two-dimensional case

In the two-dimensional case, we set $l = 0$ and $m = |k|$. Then the dispersion relation (6.41) takes the form

$$\omega = ku_0 \pm \sqrt{gk\tanh[kh_0]}. \tag{6.42}$$

Since we can rewrite (6.42) in the form

$$\omega = u_0 k \pm \frac{\sqrt{gh_0}\sqrt{kh_0 \tanh[kh_0]}}{h_0},$$

the phase and the group velocities are given by

$$c = \frac{\omega}{k} = u_0 \pm \sqrt{gh_0}\sqrt{\frac{\tanh[kh_0]}{kh_0}}, \tag{6.43}$$

$$U = \frac{d\omega}{dk} = u_0 \pm \sqrt{gh_0} f(kh_0), \tag{6.44}$$

where f is defined by

$$f(\mu) = \frac{d}{d\mu}\sqrt{\mu \tanh[\mu]} = \frac{1}{2}\left(\sqrt{\frac{\tanh[\mu]}{\mu}} + \frac{1}{\cosh^2[\mu]}\sqrt{\frac{\mu}{\tanh[\mu]}}\right). \tag{6.45}$$

6.3.2 Three particular cases

Example 6.2 (Stationary waves). For stationary waves, $c = 0$. Thus, for the Froude number F_r defined by

$$F_r = \frac{|u_0|}{\sqrt{gh_0}}, \tag{6.46}$$

we obtain

$$F_r = \sqrt{\frac{\tanh[kh_0]}{kh_0}} \tag{6.47}$$

since

$$u_0 = \pm\sqrt{gh_0}\sqrt{\frac{\tanh[kh_0]}{kh_0}}.$$

Since $\tanh[kh_0] \leqslant kh_0$, we obtain that stationary waves may exist only if $|u_0| \leqslant \sqrt{gh_0}$. The latter condition represents subcritical flow. □

Example 6.3 (Infinitely deep fluid). If $h_0 \to \infty$, we have $\lim_{\mu\to\infty} \tanh[\mu] = 1$. So, in this limiting case,

$$c = u_0 \pm \sqrt{\frac{g}{k}}, \quad U = u_0 \pm \frac{1}{2}\sqrt{\frac{g}{k}}. \tag{6.48}$$

In particular, in the stationary case when u_0, we have $U = \frac{1}{2}c$. □

Example 6.4 (Long waves). The wavelength is given by $L = 2\pi/k$. Since

$$\lim_{\mu\to\infty} \frac{\tanh[\mu]}{\mu} = 1,$$

in the limiting case when $k \to \infty$, we have

$$c = u_0 \pm \sqrt{gh_0} = U,$$

i. e., the group and phase velocities are equal to each other, which means that in the long wave approximation, waves become hyperbolic. □

6.4 The Lagrange method for the long wave theory

In the system of coordinates where the bottom is given by $z = 0$ and water depth is h, equations (6.22)–(6.25) are written as

$$\Delta\varphi = 0 \quad (0 < z < h), \tag{6.49}$$

$$\varphi_z = 0 \quad (z = 0), \tag{6.50}$$

$$h_t + \varphi_x h_x + \varphi_y h_y = \varphi_z \quad (z = h(x,y,t)), \tag{6.51}$$

$$\varphi_t + \frac{1}{2}(\varphi_x^2 + \varphi_y^2 + \varphi_z^2) + gh = 0 \quad (z = h(x,y,t)). \tag{6.52}$$

The Lagrange method consists in the presentation of the velocity potential φ as the solution of the Cauchy problem for the Laplace equation (6.49) with the given data at $z = 0$. Due to equations (6.49)–(6.50), to find φ, it is enough to know $\varphi(x,y,0,t) = A(x,y,t)$. Indeed, let $A(x,y,t)$ be given. We look for the series solution

$$\varphi(x,y,z,t) = \sum_{n=0}^{\infty} z^{2n}\varphi_n(x,y,t) \tag{6.53}$$

with $\varphi_0(x,y,t) = A(x,y,t)$. Substituting the presentation (6.53) into the Laplace equation (6.49) we have

$$\sum_{n=0}^{\infty} z^{2n}\Delta_2\varphi_n + \sum_{n=1}^{\infty} z^{2n-2}(2n-1)2n\varphi_n = 0, \tag{6.54}$$

where

$$\Delta_2 = \partial_x^2 + \partial_y^2 \tag{6.55}$$

and Δ_2^n is the n-th degree of the horizontal Laplacian operator Δ_2. As follows from (6.54),

$$\Delta_2 \varphi_{n-1} + 2n(2n-1)\varphi_n = 0,$$

from which we obtain the following recurrence relation:

$$\varphi_n = -\frac{\Delta_2 \varphi_{n-1}}{2n(2n-1)}, \tag{6.56}$$

where $\varphi_0(x,y,t) = A(x,y,t)$ is a given function. Thus, we can rewrite (6.56) in terms of φ_0 as

$$\varphi_n = \frac{(-1)^n \Delta_2^n \varphi_0}{(2n)!}. \tag{6.57}$$

Correspondingly, the series solution (6.53) of equations (6.49)–(6.50) is given by

$$\varphi(x,y,z,t) = \sum_{n=0}^{\infty} (-1)^n \frac{z^{2n}}{(2n)!} \Delta_2^n A(x,y,t). \tag{6.58}$$

6.4.1 Nondimensionalization

It is useful to recast the model (6.49)–(6.52) in nondimensional form by taking L as the horizontal scale (wavelength) and H as the vertical scale (undisturbed water depth) and introduce nondimensional variables $(x', y', z', h', t', \varphi')$ via

$$(x,y) = L(x',y'), \quad (z,h) = H(z',h'), \quad \varphi = L\sqrt{gH}\varphi', \quad t = \frac{L}{\sqrt{gH}}t' \tag{6.59}$$

and define a small parameter $\varepsilon = H/L$. Then, omitting primes, system (6.49)–(6.52) can be written in nondimensional form as

$$\varepsilon^2(\varphi_{xx} + \varphi_{yy}) + \varphi_{zz} = 0 \quad (0 < z < h), \tag{6.60}$$

$$\varphi_z = 0 \quad (z = 0), \tag{6.61}$$

$$h_t + \varphi_x h_x + \varphi_y h_y = \varepsilon^{-2}\varphi_z \quad (z = h), \tag{6.62}$$

$$\varphi_t + \frac{1}{2}(\varphi_x^2 + \varphi_y^2) + \frac{1}{2}\varepsilon^{-2}\varphi_z^2 + h = 0 \quad (z = h). \tag{6.63}$$

Since in the nondimensional variables the Laplacian operator Δ_2 transforms as $\varepsilon^{-2}\Delta_2$, the Lagrangian presentation (6.58) for the velocity potential then becomes

$$\varphi(x,y,z,t) = \sum_{n=0}^{\infty}(-1)^n \varepsilon^{2n} \frac{z^{2n}}{(2n)!}\Delta_2^n A(x,y,t). \tag{6.64}$$

6.5 Shallow water equations

Since up to the order $0(\varepsilon^2)$ the presentation (6.64) is written as

$$\varphi(x,y,z,t) = A - \frac{1}{2}\varepsilon^2 z^2 \Delta_2 A + \frac{1}{24}\varepsilon^4 z^4 \Delta_2^2 A + 0(\varepsilon^6),$$

we compute

$$\begin{aligned} \varphi_x &= A_x + 0(\varepsilon^2), \quad \varphi_y = A_y + 0(\varepsilon^2), \\ \varphi_t &= A_t + 0(\varepsilon^2), \quad \varphi_z|_{z=h} = -\varepsilon^2 h \Delta_2 A + 0(\varepsilon^4). \end{aligned}$$

Thus, substituting the presentation (6.64) into the kinematic and dynamic conditions (6.62)–(6.63) and keeping the terms up to ε^2, we obtain the following system of equations:

$$h_t + (hA_x)_x + (hA_y)_y = 0, \tag{6.65}$$

$$A_t + \frac{1}{2}(A_x^2 + A_y^2) + h = 0. \tag{6.66}$$

The system (6.65)–(6.66) is known as the system of equations of *shallow water approximation.*

Introducing a vector $\vec{u} = (u,v) = (A_x, A_y)$, the latter system (6.65)–(6.66) can be rewritten as

$$h_t + \operatorname{div}(h\vec{u}) = 0, \tag{6.67}$$

$$u_t + \vec{u}\nabla\vec{u} + \nabla h = 0. \tag{6.68}$$

6.6 Nonlinear approximations

The Korteweg–de Vries (KdV) equation is a mathematical model of waves on shallow water surfaces. It is particularly notable as the prototypical example of an exactly solvable model, that is, a nonlinear PDE whose solutions can be exactly and precisely specified. The solutions in turn include prototypical examples of solitons. KdV can be solved by means of the inverse scattering transform. The mathematical theory behind the KdV equation is rich and interesting, and, in a broad sense, is a topic of active mathematical research. The equation is named after Diederik Korteweg and Gustav de Vries, who studied it (*Korteweg & de Vries* [44]), though the equation first appeared in (*Boussinesq* [12, p. 360]).

6.6.1 The second approximation

We consider the plane (y-independent) version of equations (6.60)–(6.63), i. e.,

$$\varepsilon^2\varphi_{xx} + \varphi_{zz} = 0 \quad (0 < z < h), \tag{6.69}$$

$$\varphi_z = 0 \quad (z = 0), \tag{6.70}$$

$$h_t + \varphi_x h_x = \varepsilon^{-2}\varphi_z \quad (z = h), \tag{6.71}$$

$$\varphi_t + \frac{1}{2}(\varphi_x^2 + \varepsilon^{-2}\varphi_z^2) + gh = 0 \quad (z = h). \tag{6.72}$$

In the second approximation, the Lagrangian presentation (6.64) yields

$$\varphi(x,z,t) = A - \frac{1}{2}\varepsilon^2 z^2 A_{xx} + \frac{1}{24}\varepsilon^4 z^4 A_{xxxx} + 0(\varepsilon^6). \tag{6.73}$$

From (6.73) we calculate the derivatives of φ with respect to x, z, and t on the free boundary at $z = h$ up to the order $0(\varepsilon^4)$ as

$$\varphi_t = A_t - \frac{1}{2}\varepsilon^2 h^2 A_{xxt}, \quad \varphi_x = A_x - \frac{1}{2}\varepsilon^2 h^2 A_{xxx}, \tag{6.74}$$

$$\varphi_z = -\varepsilon^2 h A_{xx} + \frac{1}{6}\varepsilon^4 h^3 A_{xxxx}. \tag{6.75}$$

Substitution of the latter expressions (6.74)–(6.75) into the kinematic and dynamic conditions (6.71)–(6.72) and keeping the terms up to the order $0(\varepsilon^4)$ gives the following *second-order shallow water approximation*:

$$h_t + \left(hA_x - \frac{1}{6}\varepsilon^2 h^3 A_{xxx}\right)_x = 0, \tag{6.76}$$

$$A_t + \frac{1}{2}A_x^2 + gh - \frac{1}{2}\varepsilon^2 h^2(A_{xxt} + A_x A_{xxx} - A_{xx}^2) = 0. \tag{6.77}$$

6.6.2 Linear analysis

We introduce small perturbations ζ and α via

$$h = h_0 + \zeta, \quad A = -gh_0 t + \alpha. \tag{6.78}$$

Linearization of (6.76)–(6.77) yields the system for small perturbations

$$\zeta_t + \left(h_0 a_x - \frac{\varepsilon^3 1}{6} h_0^3 a_{xxx}\right)_x = 0, \tag{6.79}$$

$$a_t + g\zeta - \frac{\varepsilon^2}{2} h_0^2 a_{xxt} = 0. \tag{6.80}$$

We now set $\varepsilon = 1$ (inverse scaling transformation) to get the following systems for the unknowns a and ζ:

$$\zeta_t + h_0 a_{xx} - \frac{1}{6} h_0^3 a_{xxxx} = 0, \tag{6.81}$$

$$a_t + g\zeta - \frac{1}{2} h_0^2 a_{xxt} = 0. \tag{6.82}$$

We look for solutions in the form

$$\zeta = F \exp[i(kx - \omega t)], \quad a = G \exp[i(kx - \omega t)]. \tag{6.83}$$

Substituting (6.83) into (6.81)–(6.82), we obtain the following linear system for F and G:

$$-i\omega F - k^2 h_0 G - \frac{1}{6} h_0^3 G k^4 = 0, \tag{6.84}$$

$$Fg - G\left(-i\omega - \frac{i\omega k^2}{2} h_0^2\right) = 0. \tag{6.85}$$

The nontrivial solution exists if

$$\det \begin{vmatrix} i\omega & h_0 k^2 + \frac{1}{6} h_0^3 k^4 \\ g & -i\omega(1 + \frac{1}{2} h_0^2 k^2) \end{vmatrix} = 0. \tag{6.86}$$

The latter equation (6.86) yields the dispersion relation

$$\omega^2 = g h_0 k^2 \frac{1 + \frac{1}{6} h_0^2 k^2}{1 + \frac{1}{2} h_0^2 k^2}. \tag{6.87}$$

6.6.3 Boussinesq approximation for surface gravity waves

We next derive the Boussinesq approximation from (6.76)–(6.77). To this end, we set

$$h = h_0 + \varepsilon^2 \zeta, \quad A = -g h_0 t + \varepsilon^2 a. \tag{6.88}$$

Substituting (6.88) into (6.76)–(6.77) and keeping the terms up to the order of ε^4, we obtain the following nonlinear system:

$$\zeta_t + h_0 a_{xx} + \varepsilon^2 \left(\zeta a_x - \frac{1}{6} h_0^3 a_{xxx}\right)_x = 0, \tag{6.89}$$

$$a_t + g\zeta + \frac{1}{2}\varepsilon^2(a_x^2 - h_0^2 a_{xxt}) = 0. \tag{6.90}$$

The system (6.89)–(6.90) is knows as the *Boussinesq approximation.*

6.6.4 KdV equation

We first eliminate ζ from (6.89)–(6.90). To this end we multiply equation (6.89) by $(-g)$ and add equation (6.90) differentiated by t to get

$$a_{tt} - gh_0 a_{xx} + \varepsilon^2\left\{a_x a_{xt} - \frac{h_0^2}{2}a_{xxtt} - g\zeta_x a_x - g\zeta a_{xx} + \frac{1}{6}gh_0^3 a_{xxxx}\right\} = 0. \tag{6.91}$$

We first note that

$$\frac{1}{2}(a_x^2 - h_0^2 a_{xxt})_t + (a_t a_x)_x = a_x a_{xt} - \frac{h_0^2}{2}a_{xxtt} + a_{tx}a_x + a_t a_{xx}.$$

Also, we note that

$$a_t a_x = -g\zeta a_x,$$

which follows from equation (6.90). Thus, equation (6.91) can be written as

$$a_{tt} - gh_0 a_{xx} + \varepsilon^2\left\{\frac{1}{2}(a_x^2 - h_0^2 a_{xxt})_t + (a_t a_x)_x + \frac{1}{6}gh_0^3 a_{xxxx}\right\} = 0. \tag{6.92}$$

Let us now assume that we are moving to the right at speed c_0 and consider the wave evolution in a slow time scale. The observed wave evolution is described then by the variables

$$\xi = x - c_0 t, \quad \tau = \varepsilon^2 t. \tag{6.93}$$

Since

$$\frac{\partial}{\partial t} \to c_0\frac{\partial}{\partial \xi} + \varepsilon^2\frac{\partial}{\partial \tau}, \tag{6.94}$$

equation (6.92) transforms to (up to the order $0(\varepsilon^2)$)

$$(c_0^2 - gh_0)a_{\xi\xi} - \varepsilon^2\left\{2c_0 a_{\xi\tau} + 3c_0 a_\xi a_{\xi\xi} + \left(\frac{1}{2}c_0^2 h_0^2 - \frac{1}{6}gh_0^3\right)a_{\xi\xi\xi}\right\} = 0. \tag{6.95}$$

This equation is satisfied if

$$c_0^2 = gh_0 \quad \text{(critical velocity)} \tag{6.96}$$

and

$$2a_{\xi\tau} + 3a_\xi a_{\xi\xi} + \frac{1}{3}c_0 h_0^2 a_{\xi\xi\xi\xi} = 0, \tag{6.97}$$

where equation (6.96) has been used.

We next rewrite equation (6.97) in terms of the nondimensional variables x, t, and u defined by

$$a_\xi = c_0 u, \quad \xi = \frac{1}{3}h_0 x, \quad \tau = \frac{2}{9}\frac{h_0}{c_0}t \tag{6.98}$$

to obtain the KdV equation

$$u_t + uu_x + u_{xxx} = 0. \tag{6.99}$$

6.6.5 Periodic and solitary solutions

Since the KdV equation (6.99) admits the Galileo transformation group, all of its solutions can be reduced to the stationary solutions which can be found by the double integration of (6.99) with respect to x. This leads to the first-order equation for $u(x)$,

$$3u'^2 = -u^3 + au + b, \tag{6.100}$$

where a and b are arbitrary constants. The solution of (6.100) is bounded on the whole real axis $-\infty < x < +\infty$, if the polynomial $-u^3 + au + b = 0$ has three real roots. If all three roots are distinct, then the solution is described by *periodic waves*. If one of the roots is of multiplicity two, then equation (6.100) has a *solitary wave*-like solution which has the form

$$u = u_0 + \frac{12k^2}{\cosh^2(kx)}. \tag{6.101}$$

Example 6.5. In this example, let us find stationary solutions of the stationary KdV equation (6.99) written in more general form as

$$\frac{\partial u}{\partial t} + \alpha u\frac{\partial u}{\partial x} + \beta\frac{\partial^3}{\partial x^3} = 0, \tag{6.102}$$

where α and β are constants.

Solution. We consider stationary solutions which depend only on $\xi = x - Vt$. Then equation (6.102) transforms to an ODE:

$$\frac{d}{dx}\left(\beta\frac{d^2u}{dx^2} + \frac{\alpha}{2}u^2 - Vu\right) = 0. \tag{6.103}$$

The latter equation (6.103) can be integrated once:

$$\beta\frac{d^2u}{dx^2}+\frac{\alpha}{2}u^2-Vu=0, \tag{6.104}$$

where the constant of integration is set to zero as we are looking for a solitary solution vanishing at infinity. Then equation (6.104) can be integrated once again with another zero constant of integration, i. e.,

$$\beta\left(\frac{du}{dx}\right)^2+\frac{\alpha}{3}u^2-Vu^2=0. \tag{6.105}$$

We then find from equation (6.105)

$$\frac{du}{dx}=\pm\left[\frac{u^2}{\beta}\left(V-\frac{\alpha}{3}u\right)\right]^{\frac{1}{2}}, \tag{6.106}$$

so that

$$\xi-\xi_0=\pm\int\left[\frac{u^2}{\beta}\left(V-\frac{\alpha}{3}u\right)\right]^{-\frac{1}{2}}du. \tag{6.107}$$

Let us make the change of variables

$$z=\sqrt{\frac{3V}{\alpha}-u},\quad u=\frac{3V}{\alpha}-z^2,\quad du=-2zdz. \tag{6.108}$$

We substitute (6.108) into (6.107) and obtain

$$\xi-\xi_0=\pm\sqrt{\frac{3\beta}{\alpha}}\int\frac{-2z}{\frac{3V}{\alpha}}dz=\pm\frac{\alpha}{3V}\sqrt{\frac{12\beta}{\alpha}}\int\frac{dz}{1-(z/\sqrt{3V/\alpha})}. \tag{6.109}$$

Evaluating the integral in (6.109), we obtain

$$\xi-\xi_0=\pm\sqrt{\frac{12\beta}{3V}}\tanh^{-1}\left(z\sqrt{\frac{\alpha}{3V}}\right). \tag{6.110}$$

Thus,

$$z=\pm\sqrt{\frac{3V}{\alpha}}\tanh\left[\sqrt{\frac{3V}{12\beta}}(\xi-\xi_0)\right]. \tag{6.111}$$

Finally,

$$u(\xi)=\frac{3V}{\alpha}\left\{1-\tanh^2\left[\sqrt{\frac{3V}{12\beta}}(\xi-\xi_0)\right]\right\}, \tag{6.112}$$

or, after simplifying,

$$u(\xi) = \frac{3V}{\alpha}\operatorname{sech}^2\left[\sqrt{\frac{3V}{12\beta}}(\xi - \xi_0)\right]. \tag{6.113}$$

If we denote $A = 3V/\alpha$, in which A represents the amplitude of a solitary wave, then we will rewrite (6.113) in the traditional form of the solution

$$u(\xi) = A\operatorname{sech}^2\left[\sqrt{\frac{\alpha A}{12\beta}}(\xi - \xi_0)\right], \tag{6.114}$$

in which the value

$$\Delta = \sqrt{\frac{12\beta}{\alpha A}}$$

represents the half-width of the soliton related to its amplitude. The velocity of the soliton is given by $V = \alpha A/3$. □

6.7 Modeling equatorial planetary atmospheric waves

Planetary waves are large-scale perturbations of the atmospheric dynamical structure that extend coherently over the Earth at a certain longitude. They are important because they have a significant influence on the wind speeds, temperature, ozone distribution, and other characteristics of the middle atmosphere structure. They also play an important role in global climate control and weather prediction ([73, 2]). In particular, it is well accepted that equatorial waves play an important role in the climate system as well as in weather and climate changes. Equatorial waves have been associated with large-scale perturbations of the atmospheric motion extending around the equator ([55, 7]). Also, it is believed that equatorial waves can be an important component of the long-term mean upwelling at the tropical tropopause, of widespread changes in the climate system, and in the absorption of harmful solar ultraviolet radiation ([35, 68]). Also, equatorial waves can be an important component in controlling the stratospheric temperature via atmospheric radiative heating ([5]).

In this modeling scenario, we aim to visualize atmospheric equatorial waves that are described by the particular class of the solution that is deduced as one of the invariant solutions of the nonlinear shallow water equations representing the free boundary of a two-dimensional model describing the nonstationary motion of an incompressible perfect fluid propagating around a solid circle of a sufficiently large radius with the gravity directed to its center. Surprisingly, here we will observe some correlation between two-dimensional atmospheric equatorial waves and the Fibonacci spiral. Probably, the best-known relationship that occurs in nature, ranging from the smallest to the largest

Figure 6.2: Fibonacci spirals observed in nature. These images were posted in Art/Science, Paul's Journal, Workshops. Available at: https://tumamocsketchbook.com/category/journal.

objects, is the Fibonacci sequence of numbers F_n defined by using the recursive relation $F_n = F_{n-1}+F_{n-2}$ with the assumed values $F_0 = F_1 = 1$. It can be observed in shells of snails, in the pattern found on the wings of dragonflies, and in galaxies, as shown in Figure 6.2. The symmetry and appearance of Fibonacci spiral patterns in nature attracted the attention from scientists, but a mathematical or physical explanation for their common occurrence in nature is still lacking.

We consider a two-dimensional motion of an incompressible perfect fluid which has a free boundary η and a solid bottom represented by a circle of radius R. The fluid is propagating within a curricular domain bounded below by a sufficiently large solid circle and above by a free boundary. It is assumed that the motion is irrotational and the pressure on a free boundary is constant. It is also assumed that the distance from the origin of the circle to the unperturbed free boundary is small compared to the radius of the circle, as shown schematically in Figure 6.3.

We introduce polar coordinates $x = r\cos\theta, y = r\sin\theta$ and use the following notation: R is the radius of the Earth, θ is a polar angle, r is the distance from the origin, and $h = h_0 + \eta(t,\theta)$, where h_0 is the level of the undisturbed atmosphere above the Earth and $\eta(t,\theta)$ is the level of disturbance of a free boundary, as shown schematically in Figure 6.3. It is supposed in what follows that $\theta \in [0, 2\pi]$ while $r \in [R, h(t,\theta)]$. The homogeneous gravity field $\vec{g}$ is assumed to be a constant and directed to the center of the Earth. The restriction $\theta \in [0, 2\pi]$ appears for the following reason: the velocity potential $\varphi(\varsigma)$ can be introduced by the analyticity of the complex potential $\vartheta(\varsigma) = \varphi + i\psi$, where $\varsigma = re^{i\theta}$ is the independent complex variable and $\psi(\varsigma)$ is the stream function. Correspondingly, the complex velocity $d\vartheta/d\varsigma$ is a single-valued analytic function of ς, although ϑ is not single-valued. In fact, when we turn around the bottom $r = R$ once, φ in-

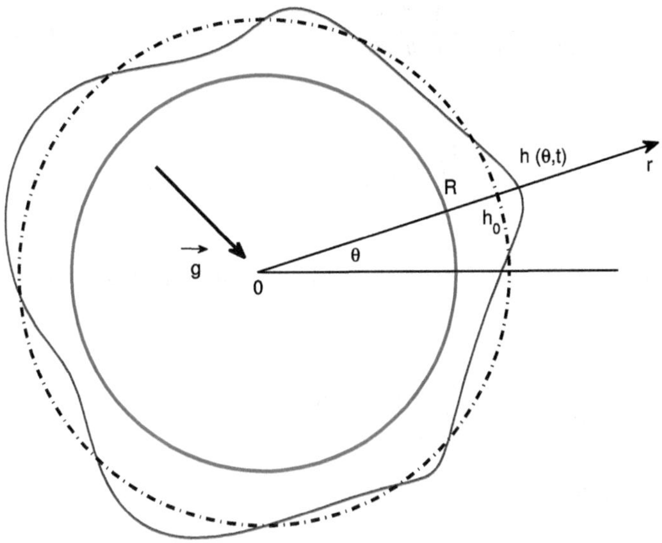

Figure 6.3: Schematic showing a longitudinal atmospheric motion circulating around the Earth.

creases by $-\int_0^{2\pi}\frac{\partial\psi}{\partial r}(R,\theta)d\theta$, which has a positive sign by the maximum principle (Hopf's lemma). Hence, if we remove the width of the annulus region $\theta = 0, r \in [R, R+h_0]$, then at every point (r,θ), the complex potential $\vartheta(\varsigma)$ is a single-valued analytic function.

We start with the usual assumption that the velocity field $\overrightarrow{v} = (v_r, v_\theta)$ satisfies the Euler equations and that the no-leak condition $v_r = 0$ on a solid bottom $r = R$ holds. We also assume the kinematic condition on the free boundary. Namely, the velocity on the free boundary $r = R + h(t,\theta)$ is tangential to the free boundary. We define the free boundary by the equation $f = r - h(\theta,t) = 0$ so that the kinematic condition is written as

$$\frac{df}{dt} = \frac{\partial f}{\partial t} + \overrightarrow{v}\nabla f = 0, \tag{6.115}$$

where

$$\nabla = \left(\frac{\partial}{\partial r}, \frac{1}{r}\frac{\partial}{\partial\theta}\right). \tag{6.116}$$

In what follows, it is assumed that the fluid motion is potential in the domain of the motion, which allows to introduce the stream function $\psi(t,r,\theta)$ via

$$v_r = -\frac{1}{r}\frac{\partial\psi}{\partial\theta}, \quad v_\theta = \frac{\partial\psi}{\partial r}, \tag{6.117}$$

so that the no-leak condition on the solid boundary can be written as $\psi(R,\theta,t) = 0$, whereas the kinematic condition (6.115) takes the form

$$\frac{\partial h}{\partial t} + \frac{1}{r}\frac{\partial \psi}{\partial \theta} + \frac{1}{r}\frac{\partial \psi}{\partial r}\frac{\partial h}{\partial \theta} = 0. \tag{6.118}$$

Since

$$v_r = -\frac{1}{r}\int_R^{R+h} \frac{\partial v_\theta}{\partial \theta} dr, \tag{6.119}$$

we can also write equation (6.118) at the free boundary $r = R + h$ as the mass balance equation

$$\frac{\partial h}{\partial t} + \frac{1}{R+h}\frac{\partial}{\partial \theta}\int_R^{R+h} v_\theta dr = 0. \tag{6.120}$$

We next define the average velocity $u(\theta, t)$ as

$$u(\theta, t) = \frac{1}{h}\int_R^{R+h} v_\theta(r, \theta, t) dr = \frac{1}{h}\psi(R+h, \theta, t). \tag{6.121}$$

In terms of the average velocity, the kinematic condition (6.118) is written as

$$\frac{\partial h}{\partial t} + \frac{1}{r}\int_R^{R+h} \frac{\partial v_\theta}{\partial \theta} dr + \frac{1}{r}\frac{\partial \psi}{\partial r}\frac{\partial h}{\partial \theta} = 0.$$

Finally, the dynamic condition is obtained from the requirement that the pressure p is constant at the free boundary $r = R + h(\theta, t)$. Thus, projection of the impulse equation

$$\frac{\partial \vec{v}}{\partial t} + \nabla\left(\frac{1}{2}|\vec{v}|^2\right) + \frac{1}{\rho}\nabla p = \vec{g} \tag{6.122}$$

on the tangential vector

$$\vec{\tau} = \left(\frac{1}{r}\frac{\partial h}{\partial \theta}, 1\right)$$

to the free boundary yields

$$\frac{1}{r}\frac{\partial h}{\partial \theta}\frac{\partial v_r}{\partial t} + \frac{\partial v_\theta}{\partial t} + \frac{1}{r}\frac{\partial h}{\partial \theta}\frac{\partial}{\partial r}\frac{1}{2}(v_r^2 + v_\theta^2) + \frac{1}{r}\frac{\partial}{\partial \theta}\frac{1}{2}(v_r^2 + v_\theta^2) + \frac{1}{\rho}\left(\frac{1}{r}\frac{\partial h}{\partial \theta}\frac{\partial p}{\partial r} + \frac{1}{r}\frac{\partial p}{\partial \theta}\right) = 0, \tag{6.123}$$

where ρ is a constant fluid density.

Thus, since $p|_{r=R+h} = \text{const.}$ and ψ is the harmonic function at the domain of the fluid motion, the model describing a longitudinal atmospheric motion around the Earth can be written as the following free boundary problem:

$$\frac{\partial^2\psi}{\partial\theta^2} + r^2\frac{\partial^2\psi}{\partial r^2} + r\frac{\partial\psi}{\partial r} = 0 \quad (R < r < R + h), \tag{6.124}$$

$$\psi(R, \theta, t) = 0, \tag{6.125}$$

$$\psi(R + h, \theta, t) = u(\theta, t)h, \tag{6.126}$$

$$\frac{\partial^2\psi}{\partial t\partial r} - \frac{1}{r^2}\frac{\partial h}{\partial\theta}\frac{\partial^2\psi}{\partial t\partial\theta} + \frac{1}{2r}\frac{\partial}{\partial\theta}\left[\frac{1}{r^2}\left(\frac{\partial\psi}{\partial\theta}\right)^2 + \left(\frac{\partial\psi}{\partial r}\right)^2\right] + \frac{g}{r}\frac{\partial h}{\partial\theta} = 0 \quad (r = R + h), \tag{6.127}$$

$$r\frac{\partial h}{\partial t} + \frac{\partial}{\partial\theta}(uh) = 0 \quad (r = R + h). \tag{6.128}$$

One can check by direct differentiation that there exists an exact stationary solution to the model (6.124)–(6.128) given by

$$h_0 = 0, \quad \psi_0 = -\frac{\Gamma}{2\pi}\log\left(\frac{r}{R}\right), \tag{6.129}$$

where Γ = const. is the intensity of the vortex (source) localized at the center of the Earth and is related with the rotation rate of the Earth (angular velocity $\Omega = 2\pi\text{rad/day} \approx 0.73 \times 10^{-4}\,\text{s}^{-1}$) by the equation $\Gamma = 2\pi\Omega R^2$.

The solution (6.129) corresponds to singular constant flow with an undisturbed free surface. Since the vortex is isolated (R is a solid boundary), this exact solution represents a flow whose streamlines can be visualized as concentric circles with the common center at the origin. We remark that, from a practical standpoint, it is useful to note that the fluid particles at the North and South Poles spin around themselves at a rate of $\Omega = 2\pi$ rad/day, whereas fluid particles in the polar (latitude) domain $\theta \in [\theta_0, \pi - \theta_0]$ do not spin around themselves but simply translate provided $\theta_0 \in (0, \frac{\pi}{2})$. Thus, the achievable meteorological flows rotating around the poles correspond to the flows that are being translated along the equatorial plane.

6.7.1 Shallow water approximation

It is useful to recast the model in nondimensional form by introducing the following dimensionless variables:

$$\theta = \hat{\theta}, \quad r = R + h_0\hat{r}, \quad h = h_0\hat{h}, \quad t = \frac{R\hat{t}}{\sqrt{gh_0}},$$
$$\psi = h_0\sqrt{gh_0}\hat{\psi}, \quad u = \sqrt{gh_0}\hat{u}. \tag{6.130}$$

We next introduce the parameter

$$\varepsilon = \frac{h_0}{R}. \tag{6.131}$$

Of course, water is shallow if the parameter ε is small. So, in the present model (6.124)–(6.128), the functions $\eta(\theta,t)$ and $\psi(r,\theta,t)$ are two unknown functions, whereas the parameter ε is given. Although shallow water theory is usually related to the case when the water depth is small relative to the wavelengths of the waves, we find it more appropriate to choose the radius of the Earth R as a natural physical scale since, in the frame of the present model, we consider waves with wavelengths of the order of the radius of the Earth.

The dynamic condition (6.127) is then nondimensionalized as follows:

$$\frac{\partial^2\psi}{\partial t\partial r}-\frac{\varepsilon^2}{(1+\varepsilon h)^2}\frac{\partial^2\psi}{\partial t\partial\theta}\frac{\partial h}{\partial\theta}+\frac{1}{2(1+\varepsilon h)}\frac{\partial}{\partial\theta}\left(\frac{\varepsilon^2}{(1+\varepsilon h)^2}\left(\frac{\partial\psi}{\partial\theta}\right)^2+\left(\frac{\partial\psi}{\partial r}\right)^2\right)+\frac{1}{(1+\varepsilon h)}\frac{\partial h}{\partial\theta}=0. \tag{6.132}$$

Following Lagrange's method, we represent the stream function ψ by the following series expansion: $\psi=\sum_n\varepsilon^n\psi^{(n)}$. Then the Laplace equation (6.124) takes the form

$$\begin{aligned}&\frac{\partial^2\psi^{(0)}}{\partial r^2}+\varepsilon\left(\frac{\partial^2\psi^{(1)}}{\partial r^2}+2r\frac{\partial^2\psi^{(0)}}{\partial r^2}+\frac{\partial\psi^{(0)}}{\partial r}\right)\\&+\varepsilon^2\left(\frac{\partial^2\psi^{(0)}}{\partial\theta^2}+\frac{\partial^2\psi^{(2)}}{\partial r^2}+2r\frac{\partial^2\psi^{(1)}}{\partial r^2}+r^2\frac{\partial^2\psi^{(0)}}{\partial r^2}+\frac{\partial\psi^{(1)}}{\partial r}+r\frac{\partial\psi^{(0)}}{\partial r}\right)+0(\varepsilon^3)=0.\end{aligned} \tag{6.133}$$

A comparison of the terms with the same order ε in equation (6.133) yields a recurrent system of differential equations for the determination of all functions $\psi^{(n)}$, i. e., the Lagrange method consists in presentation of ψ as the solution of the Cauchy problem with boundary conditions (6.125)–(6.126) for $\psi^{(0)}$ and zero boundary conditions for $\psi^{(1)}$ and $\psi^{(2)}$.

Thus, up to the order ε^2, the function ψ is determined as follows:

$$\psi=ur+\varepsilon\left(u\frac{r^2}{2}-uh\frac{r}{2}\right)+\varepsilon^2\left(uh\frac{r^2}{4}-u_{\theta\theta}\frac{r^3}{6}+u_{\theta\theta}h^2\frac{r}{6}-uh^2\frac{r}{4}\right). \tag{6.134}$$

Note that the unknowns u and h are related by the dynamic conditions (6.132) and the kinematic condition (6.128) which is written in nondimensional form as follows:

$$\frac{1}{2}\frac{\partial}{\partial t}(\varepsilon h^2+2h)+\frac{\partial}{\partial\theta}(uh)=0. \tag{6.135}$$

Using the Taylor series expansion

$$(1+\varepsilon h)^{-1}=1-\varepsilon h+\varepsilon^2h^2+0(\varepsilon^3) \tag{6.136}$$

and keeping the terms $0(\varepsilon^2)$, we write the dynamic condition (6.132) as

$$\frac{\partial^2\psi}{\partial t\partial r}+\frac{1}{2}\frac{\partial}{\partial\theta}\left(\varepsilon^2\left(\frac{\partial\psi}{\partial\theta}\right)^2+\left(\frac{\partial\psi}{\partial r}\right)^2\right)-\varepsilon^2\frac{\partial^2\psi}{\partial t\partial\theta}\frac{\partial h}{\partial\theta}$$

$$+\left(\frac{\varepsilon h}{2}\frac{\partial}{\partial\theta}\left(\frac{\partial\psi}{\partial r}\right)^2+\varepsilon h\frac{\partial h}{\partial\theta}\right)(\varepsilon h-1)+\frac{\partial h}{\partial\theta}=0. \tag{6.137}$$

Substituting ψ given by (6.134) into equation (6.137), we arrive at the following system of nonlinear shallow water equations (higher-order analog of the *Su–Gardner equations* [72]):

$$u_t+uu_\theta+h_\theta+\frac{\varepsilon}{2}(3hu_t-uh_t-u^2h_\theta+2hh_\theta)=0, \tag{6.138}$$

$$2h_t+uh_\theta+hu_\theta+\varepsilon hh_t=0, \tag{6.139}$$

where the subscripts denote partial derivatives in which independent variables t and θ denote the time and the polar angle, respectively, and the dependent variables are the average velocity u and the level $h>0$ of the atmosphere perturbed from h_0, whereas $\varepsilon \ll 1$ is a small parameter.

We eliminate u_t and h_t from the terms of equations (6.138)–(6.139) with ε by substituting there

$$u_t=-uu_\theta-h_\theta-\varepsilon F_1,\quad h_t=-uh_\theta-hu_\theta-\varepsilon F_2,$$

where

$$F_1=\frac{1}{2}(3hu_t-uh_t-u^2h_\theta+2hh_\theta),\quad F_2=hh_t.$$

Due to the fact that ε is a small parameter, the terms of order ε^2 can be considered as small perturbations to the first- and zero-order terms (unperturbed model). Ignoring the terms with ε^2, equations (6.138)–(6.139) become

$$u_t+uu_\theta+h_\theta-\varepsilon h\left(uu_\theta+\frac{h_\theta}{2}\right)=0, \tag{6.140}$$

$$h_t+uu_\theta+hu_\theta-\varepsilon h(uh_\theta+hu_\theta)=0. \tag{6.141}$$

Our main concern is a simplified version of the model in which the perturbations (nonlinear terms at ε) are ignored. We eliminate u_t and h_t from the terms of equations (6.138)–(6.139) and ignore the terms with ε. This leads to the following unperturbed system:

$$u_t+uu_\theta+h_\theta=0, \tag{6.142}$$

$$h_t+uh_\theta+hu_\theta=0. \tag{6.143}$$

The perturbed model (6.140)–(6.141) will be considered in Section 6.8.

6.8 Invariant solution and Fibonacci spiral

By using the invariance method presented in the first part of the book, we can check that the system (6.142)–(6.143) admits the infinite-dimensional Lie algebra composed of the operators

$$X_1 = t\frac{\partial}{\partial t} + \frac{\partial}{\partial u}, \quad X_2 = \theta\frac{\partial}{\partial \theta} + 2h\frac{\partial}{\partial h}, \quad X_3 = t\frac{\partial}{\partial t} + \theta\frac{\partial}{\partial \theta},$$
$$X_4 = (2\theta - 6tu)\frac{\partial}{\partial t} + (6th - 3tu^2)\frac{\partial}{\partial \theta} + 4hu\frac{\partial}{\partial h} + (u^2 + 4h)\frac{\partial}{\partial u} \tag{6.144}$$

and

$$X^{\infty} = \xi^1(h,u)\frac{\partial}{\partial t} + \xi^2(h,u)\frac{\partial}{\partial \theta}, \tag{6.145}$$

where the functions $\xi^1(h,u)$ and $\xi^2(h,u)$ solve the system of first-order equations

$$\xi_h^2 + \xi_u^1 - \xi_h^1 = 0, \quad \xi_u^2 - u\xi_u^1 + \xi_h^1 = 0. \tag{6.146}$$

In fact, there is actually quite a bit of work involved in working out the groups of (6.144) since first extensions of groups involving two independent and two dependent variables are required. Moreover, the system (6.142)–(6.143) admits eight symmetries, one of which involves arbitrary functions, which complicates sorting through solutions of the determining equations.

In particular, the operator X_1 is admitted due to the invariance of the system (6.142)–(6.143) under the Galilean transformation group

$$\bar{\theta} = \theta + ta_1, \quad \bar{u} = u + a_1 \tag{6.147}$$

with an arbitrary parameter a_1. The operator X_2 indicates that the system (6.142)–(6.143) is invariant under the nonuniform scaling transformation group

$$\bar{\theta} = \theta e^{a_2}, \quad \bar{h} = he^{2a_2}, \quad \bar{u} = ue^{a_2}, \tag{6.148}$$

where a_2 is another arbitrary parameter and the operator X_4 is responsible for a specific nonscaling symmetry of the one-dimensional shallow water model (6.142)–(6.143).

The term *similarity solution* refers to invariant solutions based on scaling transformations (6.147). However, it is not necessary to restrict the term "similarity solution" to scaling groups. Bluman and Cole's original book [11], which is essentially Bluman's PhD thesis in book form, is titled *Similarity Methods for Differential Equations* and covers more than just scaling groups.

It can be checked by direct substitution that the shallow water model (6.142)–(6.143) is *invariant* under the transformation (6.147). One calculates the *invariant* $J(t,\theta,u,h)$ of the group X_2 by solving the first-order linear PDE

$$X_2 J = 0. \tag{6.149}$$

The latter equation (6.149) gives three functionally independent invariants corresponding to the transformation (6.147):

$$J_1 = t, \quad J_2 = \frac{u}{\theta}, \quad J_3 = \frac{h}{\theta^2}.$$

Accordingly, we look for the invariant solution of (6.142)–(6.143) in the form

$$u = \theta U(t), \quad h = \theta^2 H(t). \tag{6.150}$$

Direct substitution of the presentation (6.150) in the system (6.142)–(6.143) yields the following nonlinear ordinary differential equations:

$$\frac{dU}{dt} + U^2 + 2H = 0, \tag{6.151}$$

$$\frac{dH}{dt} + 3UH = 0. \tag{6.152}$$

In case when $H \neq 0$, equation (6.152) can be written as the following coupled equations:

$$H = \mathrm{e}^{-3W}, \quad U = \frac{dW}{dt}, \tag{6.153}$$

where W satisfies the nonlinear differential equation of second order

$$\frac{d^2W}{dt^2} + \left(\frac{dW}{dt}\right)^2 + 2e^{-3W} = 0. \tag{6.154}$$

Integration of equation (6.154) yields

$$\frac{dW}{dt} = 2e^{-\frac{3W}{2}} \sqrt{1 + k\,\mathrm{e}^W}, \tag{6.155}$$

where k is a constant, so W is given implicitly by the equation

$$\int \frac{\mathrm{e}^{3W/2}}{\sqrt{1 + k\,\mathrm{e}^W}}\, dW = \pm 2(t - t_0), \tag{6.156}$$

where t_0 is an arbitrary constant.

In particular, if the function H is known, the function u can be expressed in terms of H as

$$u = \pm 2\theta \sqrt{H} \sqrt{1 + kH^{-1/3}}. \tag{6.157}$$

6.8.1 Case studies

Equation (6.156) provides an implicit representation of the function $W(t)$ and hence the functions $U(t)$ and $H(t)$ are (6.150) due to equations (6.153). We calculate the integral and distinguish the following three cases:

Case $k = 0$

Example 6.6. When $k = 0$, equation (6.156) has the form

$$F_0(W) := \mathrm{e}^{\frac{3W}{2}} = \pm 3(t - t_0). \tag{6.158}$$

In this case, equations (6.150) provide us with the solution

$$H = \frac{1}{9(t - t_0)^2}, \quad U = \frac{2}{3(t - t_0)}, \tag{6.159}$$

and thus the exact solution of the system (6.142)–(6.143) is given by

$$u = \frac{2\theta}{3(t - t_0)}, \quad h = \frac{\theta^2}{9(t - t_0)^2}. \tag{6.160}$$

□

Case $k > 0$

Example 6.7. When $k > 0$, the functions U and H are given by

$$U = \pm 2\,\mathrm{e}^{-\frac{3W}{2}}\sqrt{1 + k\mathrm{e}^{W}}, \quad H = \mathrm{e}^{-3W}, \tag{6.161}$$

where $k > 0$ and W is given implicitly by the equation

$$F_+(W) := \mathrm{e}^{\frac{W}{2}}\sqrt{\frac{1}{k} + \mathrm{e}^{W}} - \frac{1}{k}\ln\left(\mathrm{e}^{\frac{W}{2}} + \sqrt{\frac{1}{k} + \mathrm{e}^{W}}\right) = \pm 2(t - t_0). \tag{6.162}$$

□

Case $k < 0$

Example 6.8. When $k < 0$, the functions U and H are given by

$$U = \pm 2\,\mathrm{e}^{-\frac{3W}{2}}\sqrt{1 + k\mathrm{e}^{W}}, \quad H = \mathrm{e}^{-3W}, \tag{6.163}$$

where $k < 0$ and W is given implicitly by the equation

$$F_-(W) := -\mathrm{e}^{\frac{W}{2}}\sqrt{-\frac{1}{k} - \mathrm{e}^{W}} - \frac{1}{k}\arcsin\left(\sqrt{-k}\,\mathrm{e}^{\frac{W}{2}}\right) = \pm 2(t - t_0). \tag{6.164}$$

The function $F_{-}(W)$ is defined subject to constraint

$$|k| < e^{-W}. \tag{6.165}$$

For example, the function $F_{-}(W)$ is not defined for $k = 0.1$. □

We will approximate the functions $F_{\pm}(W)$ for large and small values of W and also for large and small values of k.

6.8.2 Asymptotic analysis of invariant solutions

We will analyze the asymptotic behavior of the similarity solution (6.150) for large and small W as well as for large and small values of k.

Approximation of the similarity solution for large and small W

We next approximate the function $F_{+}(W)$ given by equation (6.162) in the limit $W \to \pm\infty$. We start with $F_{+}(W) = f(W) + g(W)$, where

$$f(W) = e^{\frac{W}{2}}\sqrt{\frac{1}{k} + e^{W}}, \quad g(W) = -\frac{1}{k}\ln\left(e^{\frac{W}{2}} + \sqrt{\frac{1}{k} + e^{W}}\right). \tag{6.166}$$

First, we consider the limiting behavior of $f(W)$,

$$\lim_{W\to-\infty} f(W) = 0, \tag{6.167}$$

and for $W \gg 0$, since $f(W) \sim e^{W}$, we look for $\lim_{W\to+\infty} f(W) - e^{W}$. Using $\frac{1}{u} = e^{\frac{W}{2}}$, we find

$$\lim_{W\to+\infty} f(W) - e^{W} = \lim_{u\to+\infty} \frac{1}{u}\sqrt{\frac{1}{k} + \frac{1}{u^2}} - \frac{1}{u^2} = \lim_{u\to+\infty} \frac{u\sqrt{\frac{u^2}{k} + 1} - u}{u^3}. \tag{6.168}$$

After applying the L'Hopital rule three times, we get

$$\lim_{W\to+\infty} f(W) = \frac{1}{2k}. \tag{6.169}$$

Next, with the limiting behavior of $g(W)$,

$$\lim_{W\to-\infty} g(W) = \frac{\ln k}{2k}. \tag{6.170}$$

For $W \gg 0$, since $g(W) \sim -\frac{1}{k}\ln(2e^{\frac{W}{2}})$, we get

$$\lim_{W\to+\infty} g(W) + \frac{\ln 2}{k} + \frac{W}{2k} = 0. \tag{6.171}$$

Using this, we can find two approximations for $F_+(W)$:

$$W \ll 0: \quad F_+(W) \approx \frac{\ln k}{2k}, \tag{6.172}$$

since

$$\lim_{W \to -\infty} \left(f(W) + g(W) - \left[\frac{\ln k}{2k} \right] \right) = 0, \tag{6.173}$$

and

$$W \gg 0: \quad F_+(W) \approx e^W + \frac{1}{k}\left(\frac{1}{2} - \ln 2 - \frac{W}{2} \right), \tag{6.174}$$

since

$$\lim_{W \to +\infty} \left(f(W) + g(W) - \left[e^W + \frac{1}{k}\left(\frac{1}{2} + \ln 2 - \frac{W}{2} \right) \right] \right) = 0. \tag{6.175}$$

This analysis shows that there are two approximations for $F_+(W)$, for very large W and for very small W. In particular, $\lim_{W \to -\infty} F_-(W) = 0$, but $\lim_{W \to +\infty} F_-(W)$ does not exist since $F_-(W)$ is subject to the constraint $W < -\ln(-k)$.

6.8.3 Approximation of the similarity solution for large and small *k*

We note that the function $F_+(W)$ represents a better approximation of $F_0(W)$ for larger values of $|k|$, whereas the function $F_-(W)$ represents a better approximation of $F_0(W)$ for smaller values of $|k|$, which is demonstrated in Figure 6.4. In particular, panel (a) shows the curve F_0 and F_+ for $k = 0.1$ and $k = 10$. It can be shown that $\lim_{k \to \infty} F_+ = F_0$. Panel (b) shows the curve F_0 and F_- for $k = 0.04$ and $k = 0.020$. It can be shown that $\lim_{k \to 0} F_- = F_0$, subject to the constraint (6.165). Finally, panel (c) shows F_0 with F_+ and F_- for $k = 0.02$. For this reason we call the exact solution (6.150) an *attractor solution* in the sense that F_0 is defined as the smallest unit which cannot be itself decomposed into two or more attractors with distinct basins of attraction. This restriction is necessary since, in general, a dynamical system may have multiple attractors, each with its own basin of attraction.

6.9 Nonlinear analysis

We start with analyzing the exact *attractor solution* $h(\theta, t)$ representing the deviation of the free boundary (see equation (6.150)) from the unperturbed boundary $r = 1 + \varepsilon$ for the case $k = 0$ at different values of time t. As shown in Figure 6.5, the attractor solution is a decreasing function of t for fixed values of θ but it is an increasing function of θ

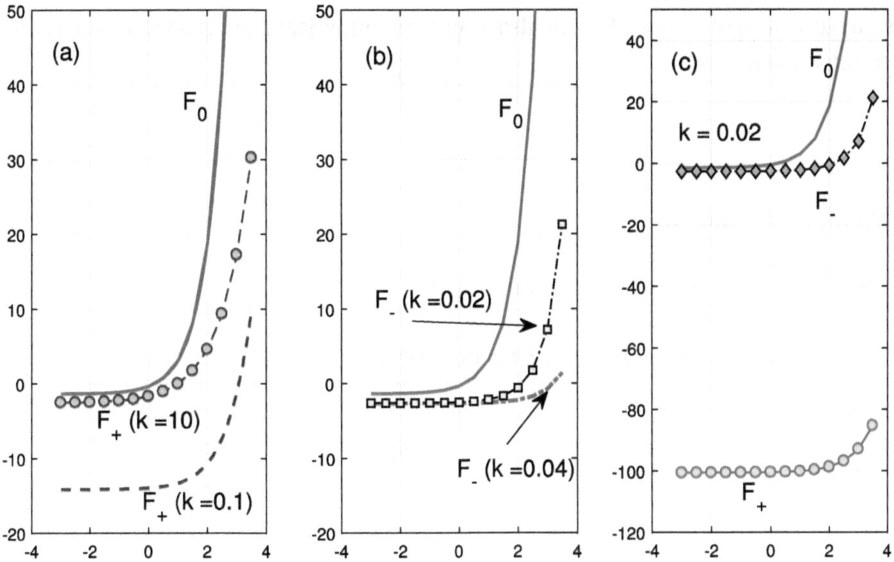

Figure 6.4: Approximation of $F_0(W)$ by $F_+(W)$ and $F_-(W)$ for large and small values of $|k|$. We observe that $\lim_{k\to\infty} F_+ = F_0$, as illustrated in panel (a), and $\lim_{k\to 0} F_- = F_0$, as illustrated in panel (b). Panel (c) shows the curves F_+ and F_- with F_0 for the value of $k = 0.02$.

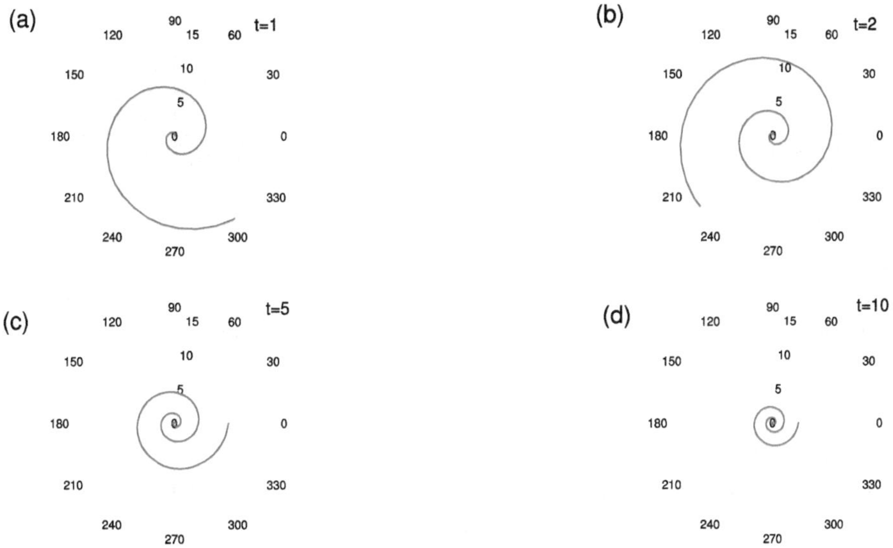

Figure 6.5: Attractor solution $h = \frac{\theta^2}{9(t-t_0)^2}$ for the case $k = 0$ and different values of time $t = 1$, $t = 2$, $t = 5$, and $t = 10$.

for fixed values of time t. We will also apply the numerical technique to compare the *attractor solution*

$$h = \frac{\theta^2}{9(t - t_0)^2} \tag{6.176}$$

with the exact solution

$$h(\theta, t) = \theta^2 e^{-3W(t)} \tag{6.177}$$

in both cases of $k > 0$ and $k < 0$. The results are shown in Figures 6.6 and 6.7.

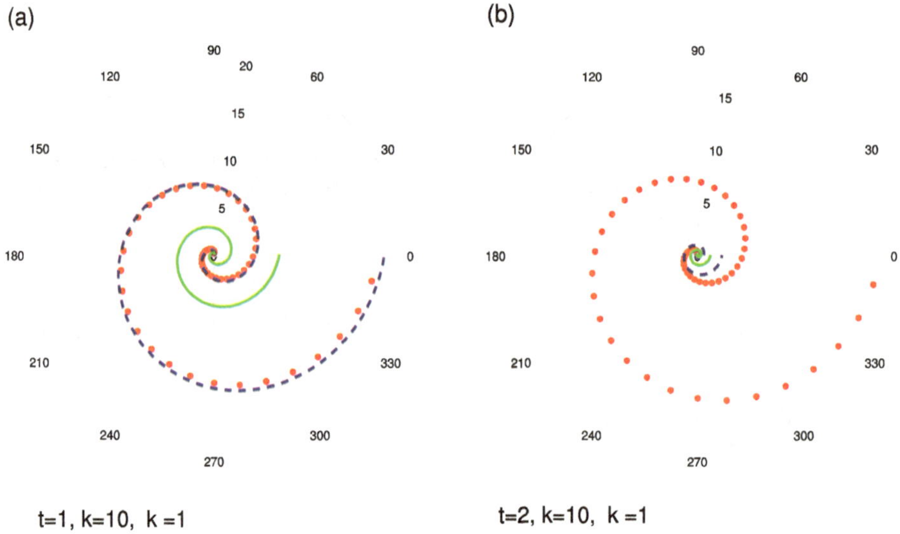

Figure 6.6: Comparison of the solution $h(\theta, t)$ for $k > 0$ with the attractor solution $h = \frac{\theta^2}{9(t-t_0)^2}$ at $t = 1$ and different values of k. Panel (a) shows the attractor solution (red line) and the solution $h(\theta, t)$ evaluated at $k = 10$ (blue line) and $h(\theta, t)$ evaluated at $k = 1$ (green line); panel (b) shows these solutions for $k = 10$ and 1 as well, but evaluated at $t = 2$.

In particular, Figure 6.6 shows the solution $h(\theta, t)$ given by (6.177) for $k > 0$ with the *attractor solution* given by (6.176) at $t = 1$ and different values of k. Panel (a) shows the *attractor solution* (red line) and the solution $h(\theta, t)$ given by (6.176) evaluated at $k = 10$ (blue line) and $k = 1$ (green line). We can observe that the *attractor solution* represents a better approximation of the exact solution given by (6.176) at $k = 10$ than at $k = 1$, as expected from the asymptotic analysis showing that $\lim_{k\to\infty} F_+ = F_0$. A similar conclusion holds for larger values of time t, as illustrated in panel (b) of this figure, showing the solutions evaluated at $t = 2$.

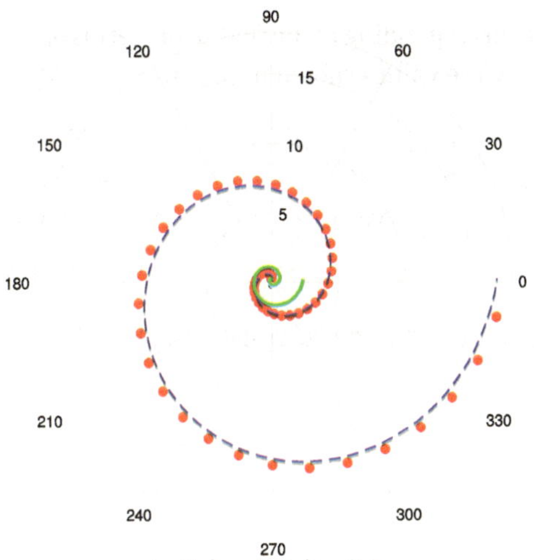

Figure 6.7: Comparison of the solution $h(\theta, t)$ for $k < 0$ with the attractor solution $h = \frac{\theta^2}{9(t-t_0)^2}$ at $t = 1$ and different values of k. The attractor solution is plotted as a red line and the solution $h(\theta, t)$ evaluated at $k = -0.4$ is shown as a blue line, whereas $h(\theta, t)$ evaluated at $k = -0.1$ is shown as a green line.

Similarly, Figure 6.7 shows the nonlinear solution $h(\theta, t) = \theta^2 e^{-3W(t,k)}$ given by (6.177) for $k < 0$ with the attractor solution $h = \frac{\theta^2}{9(t-t_0)^2}$ given by (6.176) at $t = 1$ and different values of k. In this plot, the attractor solution is plotted as a red line and the solution $h(\theta, t)$ evaluated at $k = -0.4$ is shown as a blue line, whereas $h(\theta, t)$ evaluated at $k = -0.1$ is shown as a green line. We note that the similarity solution $h(\theta, t)$ is defined for $k < 0$ and $0 < t < -\frac{\pi}{4k}$, which covers the whole range of $F_{-}(W)$.

6.10 Fibonacci spirals

We associate the parameter k with the Fibonacci sequence $\{k_n\}$ defined by $k_n = k_{n-1} - k_{n-2}$ with $k_0 = 1$ and $k_1 = 1$. Here each value k_n corresponds to the circular arc $\theta \in [\frac{m\pi}{2}, \frac{(m+1)\pi}{2}]$, where $m = 0, 1, 2, \ldots$, so that $k_1 = 1$ corresponds to the circular arc $\theta \in [0, \frac{\pi}{2}]$ with $m = 0$. Also, $k_2 = 1$ corresponds to the circular arc $\theta \in [\frac{\pi}{2}, \pi]$ with $m = 1$, $k_3 = 2$ corresponds to the circular arc $\theta \in [\pi, \frac{3\pi}{2}]$ with $m = 2$, and so on. Next, the parameter t_0 is chosen for each value of k such that the endpoints of the arc segments $\theta \in [\frac{m\pi}{2}, \frac{(m+1)\pi}{2}]$ connect smoothly with the next circular segment $\theta \in [\frac{(m+1)\pi}{2}, \frac{(m+2)\pi}{2}]$, for any given value of time t, so that $t - t_0 = \Delta_k t$, where we call $\Delta_k t$ *model hyperparameter* and it is tuned for the given predictive model represented by the Fibonacci spiral. We tune the hyperparameter in order to discover how the parameters k_n and t_0 of the model result in the given prediction.

For example, for $t = 1$, the values of $\Delta_k t$ corresponding to the first seven terms of the Fibonacci sequence $\{k_n\}$ are approximated by the values shown in the table in (6.178).

$k = 1$	$k = 1$	$k = 2$	$k = 3$	$k = 5$	$k = 8$	$k = 13$
$\theta \in \left[0, \frac{\pi}{2}\right]$	$\theta \in \left[\frac{\pi}{2}, \pi\right]$	$\theta \in \left[\frac{\pi}{2}, \frac{3\pi}{2}\right]$	$\theta \in \left[\frac{3\pi}{2}, 2\pi\right]$	$\theta \in \left[2\pi, \frac{5\pi}{2}\right]$	$\theta \in \left[\frac{5\pi}{2}, 3\pi\right]$	$\theta \in \left[3\pi, \frac{7\pi}{2}\right]$
$\Delta_1 t = 0$	$\Delta_1 t = 0$	$\Delta_2 t = -0.22$	$\Delta_3 t = -0.287$	$\Delta_5 t = -0.338$	$\Delta_8 t = -0.367$	$\Delta_{13} t = -0.385$

(6.178)

The values of $\Delta_k t$ are also shown in Figure 6.8 and the corresponding few circular arcs are shown in Figure 6.9. We observe that

$$\lim_{k \to \infty} \Delta_k \Big|_{t=1} = -0.41, \tag{6.179}$$

which is also seen from Figure 6.8.

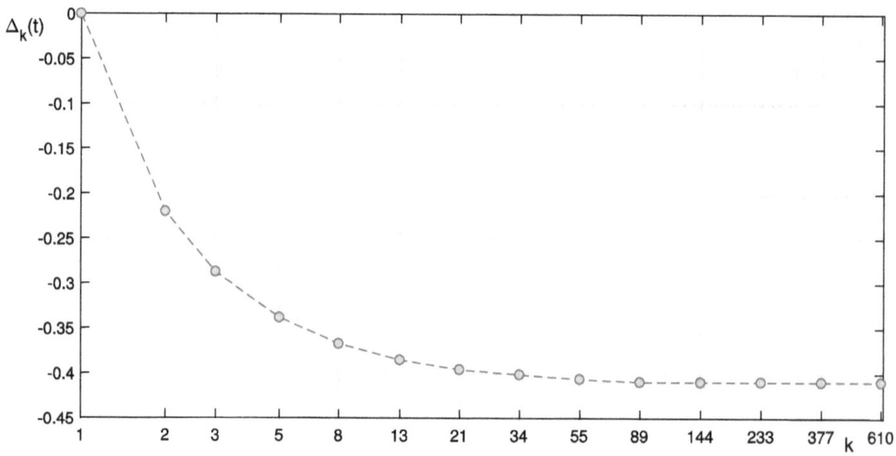

Figure 6.8: The values of $\Delta_k t$ calculated for the first 14 terms of the Fibonacci sequence.

In particular, based on the results for the hyperparameter $\Delta_k t$ shown in (6.178), we visualize our exact solution (6.150) for $h(\theta, t) = \theta^2 e^{-3W(t)}$ for $t = 1$ in Figure 6.9 as the sequence of circular arcs plotted at the values of time given by $t + \Delta_k t$. For example, when $k = 2$, we evaluate our exact solution as a function of θ at the value of time $1 + 0.22 = 1.22$. Similarly, when $k = 3$, we evaluate our exact solution as a function of θ at the value of time $1 + 0.287 = 1.287$, and so forth.

Figure 6.10 shows the attractor solution $h = \frac{\theta^2}{9(t-t_0)^2}$ with the numerical approximation of the perturbation $h(\theta, t) = \theta^2 e^{-3W(t)}$ for $k = 1$ and $\theta = \frac{\pi}{8}$ and $\theta = \frac{\pi}{4}$ as a function of time t. We note that $h(\theta, t)$ represents the deviation of the free boundary from the unperturbed state $r = 1 + \varepsilon$, so the feasible domain of h is given by

$$\Phi_h = \{h : \quad 0 \leqslant h \leqslant \varepsilon\}, \tag{6.180}$$

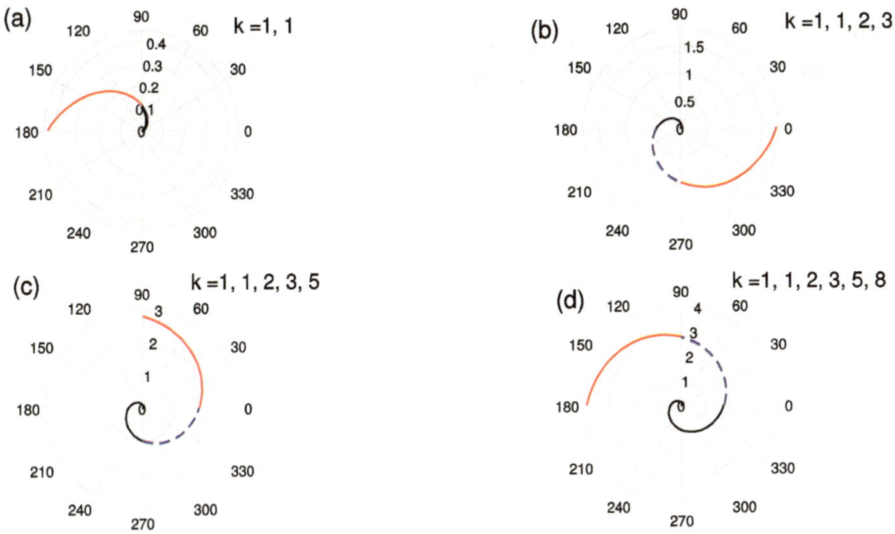

Figure 6.9: First few circular segments connected with each other for the first few terms of the Fibonacci sequence. The smoothness of connections is designed by choosing the corresponding values of the *hyper-parameter* $\Delta_k t$.

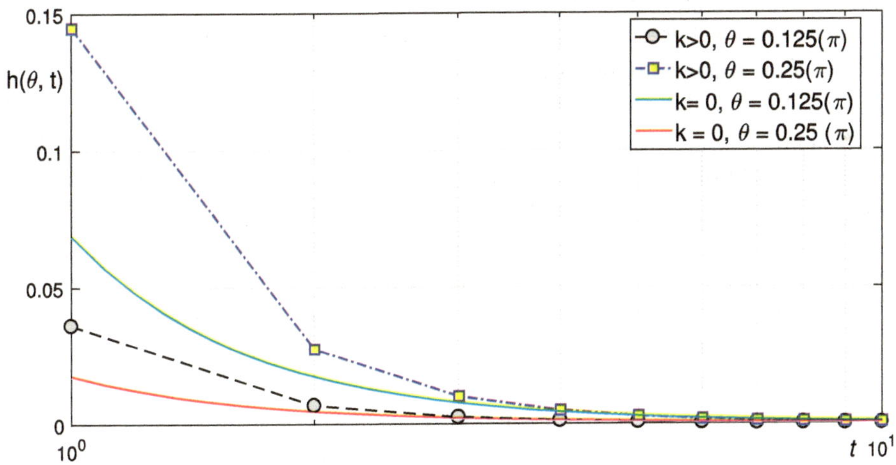

Figure 6.10: Comparison of the attractor solution with $k = 0$ with the numerical approximation of the perturbation $h(\theta, t)$ for $k = 1$ and $\theta = \frac{\pi}{8}$ and $\theta = \frac{\pi}{4}$.

where $\varepsilon < 1$ is a small parameter since it is assumed that the unperturbed level of atmospheric "depth" is small compared to the radius of the Earth. As shown in Figure 6.10, h is a decreasing function of time t, but it is an increasing function of polar angle θ, which means that our nontrivial solution is valid only for very small values of θ.

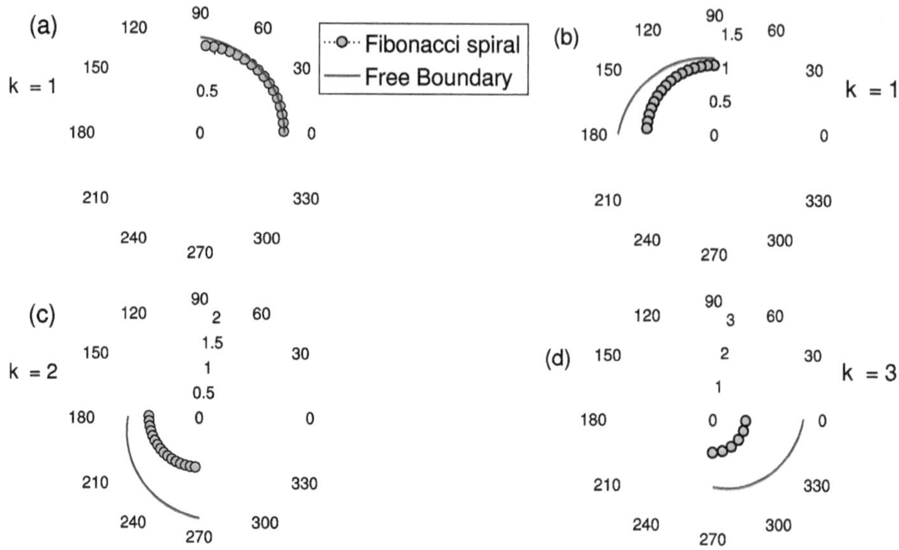

Figure 6.11: Comparison of the free boundary $\eta = 1 + \varepsilon h(\theta, t)$ and the Fibonacci spiral for the first three values of k.

Figure 6.11 shows the free boundary given by $\eta = 1 + h(\theta, t)$ and the Fibonacci spiral for the first four values of k, namely, for $k = 1, 1, 2, 3$. According to the values chosen as shown in (6.178), panel (a) shows the segment of the Fibonacci spiral (green circles) and the free boundary (blue solid line) evaluated at the values of $k = 1$, $t = 1$, and $t_0 = 0$. In panel (b), the segment of the Fibonacci spiral is compared with the free boundary evaluated at the values of $k = 1$, $t = 1$, and $t_0 = 0$. Panel (c) shows the segment of the Fibonacci spiral and the free boundary evaluated at the values of $k = 2$, $t = 1.22$, and $t_0 = -0.22$. Finally, panel (d) shows the segment of the Fibonacci spiral and the free boundary evaluated at the values of $k = 3$, $t = 1.287$, and $t_0 = -0.287$.

As discussed earlier, the solution for the free boundary perturbation $h(\theta, t)$ is only valid in the domain Φ_h defined by (6.180), i. e., in a very narrow band $|h| \leqslant \varepsilon < 1$. For example, if $\varepsilon = 0.15$, we can see from Figure 6.10 that our solution is valid for very small values of θ bounded by $\theta \in (0, \frac{\pi}{4})$. Thus, for example, if $\theta = \frac{\pi}{4}$, the solution is bounded by $|h| \leqslant 0.15$. So, in this domain, the Fibonacci spiral corresponding to $k = 1$ is a very good approximation of the nontrivial solution η, as shown in panel (a) of Figure 6.11. For larger values of k, we observe that the solution η diverges from the Fibonacci spiral.

6.11 Hodograph method

Let us write the operator (6.145) as an infinitesimal transformation

$$\bar{t} \approx t + \xi^1(h, u), \quad \bar{\theta} \approx \theta + \xi^2(h, u). \tag{6.181}$$

Equations (6.146) show that the infinitesimal transformation (6.181) changes the variables (t, θ) by adding to them an arbitrary solution of the system of linear differential equations

$$\theta_h + t_u - ut_h = 0, \quad \theta_u + ht_h - ut_u = 0 \tag{6.182}$$

and that the system (6.182) admits the infinitesimal transformation (6.181).

Hence, the operator (6.145) is admitted both by the nonlinear system (6.142)–(6.143) and by the linear system (6.182). This fact predicts a possibility to map the nonlinear model (6.142)–(6.143) to the linear system (6.182) by an appropriate change of variables. It is important to note that the linear system is homogeneous, i. e., invariant under the uniform dilation

$$\bar{t} = te^{a_5}, \quad \bar{\theta} = \theta e^{a_5}$$

produced by the operator X_3 from (6.144).

We conclude that the operator (6.145) and the operator X_3 from (6.144) are responsible for the possibility of mapping of the nonlinear system (6.142)–(6.143) to the linear homogeneous system (6.182).

Indeed, as will be shown here the systems (6.142)–(6.143) and (6.182) are connected by the hodograph transformation (the interchange of the independent and dependent variables):

$$(t, \theta) \Leftrightarrow (h, u). \tag{6.183}$$

6.11.1 Mapping the nonlinear shallow water model to a linear system

The hodograph transformation (6.183) for the system (6.142)–(6.143) means that the new independent variables $\bar{t}, \bar{\theta}$ and the new dependent variables $\bar{h}, \bar{u}$ are introduced by the equations

$$\bar{t} = h, \quad \bar{\theta} = u, \tag{6.184}$$

$$\bar{h} = t, \quad \bar{u} = \theta. \tag{6.185}$$

The differentiations in the new independent variables (6.184) are determined by the equations

$$\begin{aligned} D_t &= D_t(\bar{t})D_{\bar{t}} + D_t(\bar{\theta})D_{\bar{\theta}}, \\ D_\theta &= D_\theta(\bar{t})D_{\bar{t}} + D_\theta(\bar{\theta})D_{\bar{\theta}}. \end{aligned} \tag{6.186}$$

Whence, invoking the expressions for $\bar{t}, \bar{\theta}$ in equations (6.184), one has

$$D_t = h_t D_h + u_t D_u, \\ D_\theta = h_\theta D_h + u_\theta D_u. \tag{6.187}$$

Acting by the operators (6.186) on the first equation (6.185) and then on the second equation (6.185), we obtain

$$D_t(\bar{t})\bar{h}_{\bar{t}} + D_t(\bar{\theta})\bar{h}_{\bar{\theta}} = 1, \quad D_\theta(\bar{t})\bar{h}_{\bar{t}} + D_\theta(\bar{\theta})\bar{h}_{\bar{\theta}} = 0 \tag{6.188}$$

and

$$D_t(\bar{t})\bar{u}_{\bar{t}} + D_t(\bar{\theta})\bar{u}_{\bar{\theta}} = 0, \quad D_\theta(\bar{t})\bar{u}_{\bar{t}} + D_\theta(\bar{\theta})\bar{u}_{\bar{\theta}} = 1, \tag{6.189}$$

respectively. Due to the notation (6.184)–(6.185), equations (6.188)–(6.189) give

$$h_t\, t_h + u_t\, t_u = 1, \quad h_\theta\, t_h + u_\theta t_u = 0, \tag{6.190}$$

$$h_t\, \theta_h + u_t\, \theta_u = 0, \quad h_\theta\, \theta_h + u_\theta\, \theta_u = 1. \tag{6.191}$$

Solving equations (6.190) for t_h, t_u and equations (6.191) for θ_h, θ_u, we arrive at the following change of the first derivatives under the hodograph transformation (6.184)–(6.185):

$$t_h = -\frac{u_\theta}{\{u,h\}}, \quad t_u = \frac{h_\theta}{\{u,h\}}, \\ \theta_h = \frac{u_t}{\{u,h\}}, \quad \theta_u = -\frac{h_t}{\{u,h\}}, \tag{6.192}$$

where

$$\{u,h\} = u_t h_\theta - h_t u_\theta. \tag{6.193}$$

Solving equations (6.190)–(6.191) for h_t, h_θ, u_t, u_θ, one obtains the following expressions of the old derivatives through the new derivatives:

$$h_t = -\frac{\theta_u}{\{t,\theta\}}, \quad h_\theta = \frac{t_u}{\{t,\theta\}}, \\ u_t = \frac{\theta_h}{\{t,\theta\}}, \quad u_\theta = -\frac{t_h}{\{t,\theta\}}, \tag{6.194}$$

where

$$\{t,\theta\} = t_u\theta_h - t_h\theta_u. \tag{6.195}$$

The change of derivatives (6.194) maps the system (6.142)–(6.143) to the linear form:

$$\theta_h - u t_h + t_u = 0, \quad \theta_u + h t_h - u t_u = 0. \tag{6.196}$$

6.11.2 Reduction to a second-order linear equation

We further simplify the linear system (6.196) introducing the new dependent variables τ and χ defined by

$$\tau = ht, \quad \chi = \theta - ut. \tag{6.197}$$

Then the system (6.196) becomes

$$\tau_u + h\chi_h = 0, \quad \tau_h + \chi_u = 0 \tag{6.198}$$

and can be replaced by the single linear second-order equation

$$h\tau_{hh} - \tau_{uu} = 0. \tag{6.199}$$

In general, equation (6.199) is of the *mixed type*. It is hyperbolic when $h > 0$ and elliptic when $h < 0$. In our case equation (6.199) is hyperbolic because h is positive due to its physical meaning.

6.11.3 Characteristics

We dwell on the case $h > 0$ and consider the wave equation (6.199) with the variable coefficient. In this case the characteristic equation is written

$$h\omega_h^2 - \omega_u^2 = 0.$$

It splits in two linear PDEs,

$$\sqrt{h}\,\omega_h + \omega_u = 0$$

and

$$\sqrt{h}\,\omega_h - \omega_u = 0,$$

having the solutions

$$\omega_1(h,u) = 2\sqrt{h} - u$$

and

$$\omega_2(h,u) = 2\sqrt{h} + u,$$

respectively. Accordingly, equation (6.199) has two families of characteristic curves,

$$2\sqrt{h} - u = C_1 \tag{6.200}$$

and

$$2\sqrt{h} + u = C_2. \tag{6.201}$$

Signals governed by the wave equation (6.199) with the variable coefficient move along the characteristic curves (6.200) and (6.201). In particular, the characteristics carry essential information about singularities (e. g., discontinuity, shock waves, etc.) of the solutions. The next section is devoted to a discussion of the regular and characteristic Cauchy problems.

We introduce the characteristic variables

$$\alpha = 2\sqrt{h} - u, \quad \beta = 2\sqrt{h} + u \tag{6.202}$$

and rewrite the hyperbolic equation (6.199) in the form

$$\tau_{\alpha\beta} - \frac{\tau_\alpha + \tau_\beta}{2(\alpha + \beta)} = 0. \tag{6.203}$$

Remark 6.1. One can remove the first-order derivatives in equation (6.203) by the equivalence transformation

$$\tau = \sqrt{\alpha + \beta}\, T. \tag{6.204}$$

Then equation (6.203) takes the form

$$T_{\alpha\beta} - \frac{3\,T}{4(\alpha + \beta)^2} = 0. \tag{6.205}$$

6.12 Riemann's integration method

6.12.1 Preliminaries

In this section we apply Riemann's integration method [65] to the hyperbolic equation (6.203).

Riemann developed his method for investigating the propagation of plane waves of finite amplitude in aerodynamics. Later Riemann's method became a basic tool in solving the Cauchy problem

$$L[u] = f(x, y), \quad u|_\gamma = u_0(x), \quad u_y|_\gamma = u_1(x) \tag{6.206}$$

for the hyperbolic operator

$$L[u] = u_{xy} + a(x,y)u_x + b(x,y)u_y + c(x,y)u \tag{6.207}$$

with two independent variables x, y. The initial data in (6.206) are given on an arbitrary noncharacteristic curve γ in the (x,y)-plane.

Riemann's method reduces the solution of the problem (6.206) to construction of the function (known as *Riemann's function*)

$$R(M, M_0) \tag{6.208}$$

of two points $M(x,y)$, $M_0(x_0,y_0)$ satisfying the adjoint equation

$$L^*[R] = 0 \tag{6.209}$$

and the following conditions on the characteristics:

$$R|_{x=x_0} = e^{\int_{y_0}^{y} a(x_0,\,\eta)d\eta}, \quad R|_{y=y_0} = e^{\int_{x_0}^{x} b(\xi,\,y_0)d\xi}. \tag{6.210}$$

Here L^* is the adjoint operator for the operator L in (6.207) defined by

$$L^*[v] = v_{xy} - (av)_x - (bv)_y + cv. \tag{6.211}$$

If Riemann's function is known, the solution to the Cauchy problem (6.206) is given by an integral formula derived in [65, Section 8]. In the case of the homogeneous equation (when $f = 0$) the integral formula gives the following expression for the solution u at an arbitrary point $M_0(x_0,y_0) \notin \gamma$:

$$\begin{aligned} u(M_0) &= \frac{1}{2}[(uR)_A + (uR)_B] \\ &+ \int_A^B \left\{ \left[\frac{1}{2}Ru_x + \left(bR - \frac{1}{2}R_x\right)u\right]dx - \left[\frac{1}{2}Ru_y + \left(aR - \frac{1}{2}R_y\right)u\right]dy \right\}. \end{aligned} \tag{6.212}$$

Here A and B are the points where the curve γ meets the characteristics $y = y_0$ and $x = x_0$, respectively, the subscripts A and B indicate that the corresponding expressions are taken at points A and B, respectively, and the integral $\int_A^B$ is taken along γ between A and B.

The significance of Riemann's function (6.208) is that it carries essential information about singularities of the solutions of the Cauchy problem (6.206).

6.12.2 Toward Riemann's function via the invariance principle

The construction of Riemann's function (6.208) requires the integration of a PDE (6.211). In general, this problem is not simpler than integrating the original equation (6.207).

However, in certain particular cases Riemann's function can be found by reducing the PDE (6.209) to a second-order linear ODE with singularities provided that one can deduce dependence of Riemann's function (6.208) on a certain "good" combination of the variables x, y, x_0, y_0. This approach was used for the first time by Riemann [65, Section 9]. This ad hoc method has been used in the literature for several other cases.

We will use here the invariance principle for constructing Riemann's function.

Formulation of the invariance principle. Let L be a linear partial differential operator with independent variables $x = (x^1, \ldots, x^n)$. We consider the boundary value problem

$$L[u] = f(x), \quad u|_S = h(x) \tag{6.213}$$

with data on the manifold S. The problem (6.213) is said to be invariant under the group G of transformations in the space (x, u) if:

(1) the equation $L[u] = f(x)$ admits a group G,
(2) the manifold S carrying the boundary conditions is invariant under G,
(3) the boundary condition $u|_S = h(x)$ is invariant under the action $\widetilde{G}$ of the group G on the manifold S.

The invariance principle states that one should seek the solution to the invariant boundary value problem (6.213) *as an invariant solution under the group G.*

Now we turn to equation (6.203). In this case the adjoint equation (6.209) has the form

$$R_{\alpha\beta} + \frac{R_\alpha + R_\beta}{2(\alpha+\beta)} - \frac{R}{(\alpha+\beta)^2} = 0 \tag{6.214}$$

and the conditions (6.210) on the characteristics are written

$$R|_{\alpha=\alpha_0} = \sqrt{\frac{\alpha_0+\beta_0}{\alpha_0+\beta}}, \quad R|_{\beta=\beta_0} = \sqrt{\frac{\alpha_0+\beta_0}{\alpha+\beta_0}}. \tag{6.215}$$

The calculation shows that equation (6.214) admits the four-dimensional Lie algebra spanned by the operators

$$X_1 = \frac{\partial}{\partial\alpha} - \frac{\partial}{\partial\beta}, \quad X_2 = \alpha\frac{\partial}{\partial\alpha} + \beta\frac{\partial}{\partial\beta},$$
$$X_3 = R\frac{\partial}{\partial R}, \quad X_4 = 2\alpha^2\left(\frac{\partial}{\partial\alpha} - \frac{\partial}{\partial\beta}\right) + (\alpha-\beta)R\frac{\partial}{\partial R}. \tag{6.216}$$

We require the invariance of the characteristics $\alpha = \alpha_0$, $\beta = \beta_0$ and of equations (6.215) with respect to a linear combination of the operators (6.216) and obtain the following operator admitted by the problem (6.214)–(6.215):

$$X = 2(\alpha - \alpha_0)(\alpha + \beta_0)\frac{\partial}{\partial \alpha} - 2(\beta - \beta_0)(\beta + \alpha_0)\frac{\partial}{\partial \beta} + [(\beta - \beta_0) - (\alpha - \alpha_0)]R\frac{\partial}{\partial R}. \tag{6.217}$$

Hence, the boundary value problem (6.214)–(6.215) is invariant under the one-parameter group with the generator (6.217). The invariant solution is written in the form

$$R = \frac{\alpha_0 + \beta_0}{\sqrt{(\alpha + \beta_0)(\beta + \alpha_0)}} V(z) \tag{6.218}$$

with an unknown function V of the variable

$$z = \frac{(\alpha - \alpha_0)(\beta - \beta_0)}{(\alpha_0 + \beta_0)(\alpha + \beta)}. \tag{6.219}$$

Substituting expression (6.218) in equation (6.214) we obtain the hypergeometric equation

$$z(1+z)V'' + (1+z)V' - \left[1 + \frac{1}{4(1+z)}\right]V = 0. \tag{6.220}$$

Equation (6.220) has two singular points, $z = 0$ and $z = -1$. By the definition (6.219) of z, the point $z = 0$ corresponds to the characteristics

$$\alpha = \alpha_0, \quad \beta = \beta_0 \tag{6.221}$$

of equation (6.214). Accordingly, two conditions (6.215) on the characteristics lead to the single initial condition

$$V(0) = 1 \tag{6.222}$$

for equation (6.220).

Furthermore, noting that

$$1 + z = \frac{(\alpha + \beta_0)(\beta + \alpha_0)}{(\alpha_0 + \beta_0)(\alpha + \beta)},$$

we see that the singular point $z = -1$ corresponds to the characteristics

$$\alpha = -\beta_0, \quad \beta = -\alpha_0 \tag{6.223}$$

of equation (6.220).

6.13 Shock waves

Validity of equations (6.192) requires that the Jacobian (6.193) does not vanish. This requirement means that we consider the functionally independent solutions $u(t,\theta)$, $h(t,\theta)$ only. Let us relax this restriction and describe the solutions with vanishing Jacobian, i. e., satisfying the equation

$$u_t h_\theta - h_t u_\theta = 0. \tag{6.224}$$

We substitute in equation (6.224) the expressions of u_t, h_t given by the system (6.142)–(6.143) and obtain

$$(h_\theta)^2 = h(u_\theta)^2,$$

or

$$\frac{\partial u}{\partial \theta} = \pm \frac{1}{\sqrt{h}} \frac{\partial h}{\partial \theta}. \tag{6.225}$$

Equation (6.225) yields

$$u + f(t) = \pm 2\sqrt{h}.$$

Substitution of this expression for u in the first equation of the system (6.142)–(6.143) shows that $f(t) = \text{const}$. Thus, we have two possibilities: either

$$u = -\alpha + 2\sqrt{h}, \quad \alpha = \text{const.}, \tag{6.226}$$

or

$$u = -\beta - 2\sqrt{h}, \quad \beta = \text{const.} \tag{6.227}$$

Equations (6.226)–(6.227) show that the hodograph transformation (6.192) collapses on the characteristic curves (6.200) and (6.201).

Substituting (6.226) and (6.227) in the second equation of the system (6.142)–(6.143), we obtain

$$h_t + (3\sqrt{h} - \alpha)h_\theta = 0 \tag{6.228}$$

and

$$h_t - (3\sqrt{h} + \beta)h_\theta = 0, \tag{6.229}$$

respectively. Thus, the system (6.142)–(6.143) has been reduced to two different forms (6.228) and (6.229) of the single first-order nonlinear PDE known as the Hopf equation.

This allows to investigate the formation of shock waves similar to toppling of waves in the ocean.

The general solutions of the Hopf equations (6.228) and (6.229) are given implicitly by the equations

$$h = F[\theta - t(3\sqrt{h} - \alpha)] \tag{6.230}$$

and

$$h = \Phi[\theta + t(3\sqrt{h} + \beta)], \tag{6.231}$$

respectively. Here F and Φ are arbitrary functions of the respective arguments given in the square brackets.

Differentiating both sides of equation (6.230) with respect to t and θ, one obtains

$$h_t = -2\sqrt{h}\,\frac{(3\sqrt{h} - \alpha)F'}{2\sqrt{h} + 3tF'} \tag{6.232}$$

and

$$h_\theta = \frac{2\sqrt{h}\,F'}{2\sqrt{h} + 3tF'}, \tag{6.233}$$

respectively. Dealing likewise with equation (6.231), one obtains

$$h_t = 2\sqrt{h}\,\frac{(3\sqrt{h} + \beta)\Phi'}{2\sqrt{h} - 3t\Phi'}, \tag{6.234}$$

$$h_\theta = \frac{2\sqrt{h}\,\Phi'}{2\sqrt{h} - 3t\Phi'}. \tag{6.235}$$

The derivatives (6.232) and (6.233) collapse if

$$2\sqrt{h} + 3tF'[\theta - t(3\sqrt{h} - \alpha)] = 0. \tag{6.236}$$

The function

$$h = h(t, \theta) \tag{6.237}$$

given implicitly by equation (6.236) defines a surface where the solution (6.230) has a singularity. This fact is crucial for investigating the formation of shock waves in atmospheric dynamics. The similar singularity surface for the solution (6.231) is defined by the equation

$$2\sqrt{h} - 3t\Phi'[\theta + t(3\sqrt{h} + \beta)] = 0. \tag{6.238}$$

6.14 Perturbed system

Now we return to the perturbed system (6.140)–(6.141) and discuss the influence of the perturbation on the similarity solution based on the scaling symmetry X_2 from (6.144).

6.14.1 Hodograph transform and simplified perturbation

The hodograph transformation (6.184)–(6.185) maps the perturbed system (6.140)–(6.141) to the linear system

$$\begin{aligned} &\theta_h + \left(1 - \frac{\varepsilon}{2} h\right) t_u - (1 - \varepsilon h) u t_h = 0, \\ &\theta_u + (1 - \varepsilon h)(h t_h - u t_u) = 0. \end{aligned} \tag{6.239}$$

In the variables τ and χ defined by equations (6.197), the perturbed system (6.239) takes the form

$$\begin{aligned} &\left(1 - \frac{\varepsilon}{2} h\right) \tau_u + h(\chi + \varepsilon u \tau)_h - \varepsilon u \tau = 0, \\ &(1 - \varepsilon h)\tau_h + (\chi + \varepsilon u \tau)_u = 0. \end{aligned} \tag{6.240}$$

Note that here the perturbations appear due to the shifting

$$\chi \to \tilde{\chi} = \chi + \varepsilon u \tau$$

in the factors $(1-\varepsilon h)$ and $(1-\varepsilon h/2)$ as well as in the term $-\varepsilon u \tau$ in the first equation (6.240).

In what follows, we will dwell on motions with large mean velocities $u^2 \gg h$. In this case only the perturbation term $-\varepsilon u \tau$ is essential. Therefore, we will replace the perturbed system (6.240) with the simplified perturbed system

$$\tau_u + h\chi_h - \varepsilon u \tau = 0, \quad \tau_h + \chi_u = 0, \tag{6.241}$$

where $\tilde{\chi}$ has been denoted by χ.

6.14.2 Hodograph transformation of symmetries

After performing the hodograph transformation (6.184)–(6.185) and introducing the variables (6.197), the operators (6.144) and (6.145) provide the infinitesimal symmetries

$$X_1 = \frac{\partial}{\partial u}, \; X_2 = 2h\frac{\partial}{\partial h} + u\frac{\partial}{\partial u} + 2\tau\frac{\partial}{\partial \tau} + \chi\frac{\partial}{\partial \chi}, \; X_3 = \tau\frac{\partial}{\partial \tau} + \chi\frac{\partial}{\partial \chi},$$

$$X_4 = (u^2 + 4h)\frac{\partial}{\partial u} + 4hu\frac{\partial}{\partial h} + 2h\chi\frac{\partial}{\partial \tau} + 2(\tau - u\chi)\frac{\partial}{\partial \chi}, \tag{6.242}$$

and

$$X^{\infty} = f^0(u,h)\frac{\partial}{\partial \tau} + g^0(u,h)\frac{\partial}{\partial \chi} \tag{6.243}$$

for the linear unperturbed system (6.239). The functions f^0, g^0 in the operator (6.243) are related to the functions ξ^1, ξ^2 in the operator (6.145) by

$$f^0 = h\xi^1, \quad g^0 = \xi^2 - u\xi^1. \tag{6.244}$$

Accordingly, equations (6.146) become

$$f_u^0 + hg_h^0 = 0, \quad f_h^0 + g_u^0 = 0. \tag{6.245}$$

6.14.3 Approximate symmetry for the perturbed system

We will consider here the perturbed system (6.241) and discuss the perturbation of the exact symmetry X_2 from the list (6.226). We find the perturbation of the operator X_2 in the form

$$Y = Y^0 + \varepsilon Y^1, \tag{6.246}$$

where

$$\begin{aligned} Y^0 &\equiv X_2 = 2h\frac{\partial}{\partial h} + u\frac{\partial}{\partial u} + 2\tau\frac{\partial}{\partial \tau} + \chi\frac{\partial}{\partial \chi}, \\ Y^1 &= (\tau u^2 + 2h\tau - 2uh\chi)\frac{\partial}{\partial \tau} + (\chi u^2 + 2h\chi - 2u\tau)\frac{\partial}{\partial \chi}. \end{aligned} \tag{6.247}$$

The operator (6.246) with the components (6.247) is an approximate symmetry of the perturbed system (6.241).

The components Y^0, Y^1 of the approximate symmetry (6.246) are represented in the variables t, θ, u, h as follows (see X_2 from (6.144)):

$$\begin{aligned} Y^0 &= \theta\frac{\partial}{\partial \theta} + 2h\frac{\partial}{\partial h} + u\frac{\partial}{\partial u}, \\ Y^1 &= (3tu^2 + 2th - 2\theta u)\frac{\partial}{\partial t} + (2\theta h - \theta u^2 + 2tu^3 - 2tuh)\frac{\partial}{\partial \theta}. \end{aligned} \tag{6.248}$$

The approximate symmetry (6.246) with the components (6.247), or with their (t, θ, u, h) representations (6.248), can be used for constructing approximately invariant solutions.

6.14.4 Approximately invariant solution

Here we construct the approximately invariant solution of equations (6.241) using the approximate scaling symmetry (6.246)–(6.247) and compare the result with the similarity solution, which refers to invariant solutions based on scaling transformations (6.147).

For constructing approximately invariant solutions we have to find functionally independent approximate invariants

$$J(u,h,\tau,\chi,\varepsilon) = J^0(u,h,\tau,\chi) + \varepsilon J^1(u,h,\tau,\chi) \tag{6.249}$$

determined by the approximate equation

$$Y(J) \approx 0, \tag{6.250}$$

where the operator Y is given by equation (6.246). Equation (6.250) should be satisfied in the first order of precision with respect to ε.

For $\varepsilon = 0$ equation (6.250) reduces to the linear homogeneous first-order PDE

$$2h\frac{\partial J^0}{\partial h} + u\frac{\partial J^0}{\partial u} + 2\tau\frac{\partial J^0}{\partial \tau} + \chi\frac{\partial J^0}{\partial \chi} = 0 \tag{6.251}$$

given three functionally independent invariants

$$J_1^0 = \frac{u^2}{h}, \quad J_2^0 = \frac{\tau}{h}, \quad J_3^0 = \frac{\chi}{\sqrt{h}}. \tag{6.252}$$

The invariant J_1^0 depends only on the independent variables u, h. Therefore it will be convenient to take it as an independent variable in the approximately invariant solution. We will use for J_1^0 the special notation

$$\lambda = \frac{u^2}{h}. \tag{6.253}$$

Using the above invariants, we arrive at the following representation for the unperturbed invariant solution:

$$\begin{aligned} \tau|_{\varepsilon=0} &\equiv \tau^{(0)} = hT(\lambda), \\ \chi|_{\varepsilon=0} &\equiv \chi^{(0)} = \sqrt{h}K(\lambda), \end{aligned} \tag{6.254}$$

where $T(\lambda)$ and $K(\lambda)$ are arbitrary functions of λ. Then we solve the approximate equation (6.250) using the representation (6.254) of the unperturbed invariant solution and obtain the following three functionally independent approximate invariants of the form (6.249):

$$J_1 = \lambda, \quad J_2 = \frac{\tau}{h} - \varepsilon h\left[\left(1 + \frac{\lambda}{2}\right)T - \sqrt{\lambda}K\right],$$
$$J_3 = \frac{\chi}{\sqrt{h}} - \varepsilon h\left[\left(1 + \frac{\lambda}{2}\right)K - \sqrt{\lambda}T\right]. \tag{6.255}$$

Equations (6.255) provide the following representation of the approximately invariant solution:

$$\tau = h\tau^0(\lambda) + \varepsilon h^2\left[\left(1 + \frac{\lambda}{2}\right)T - \sqrt{\lambda}\,K\right],$$
$$\chi = \sqrt{h}\chi^0(\lambda) + \varepsilon h^{3/2}\left[\left(1 + \frac{\lambda}{2}\right)K - \sqrt{\lambda}\,T\right]. \tag{6.256}$$

Substituting (6.256) in equations (6.241), we obtain the linear system of two ODEs

$$\lambda\chi^0_\lambda - \frac{1}{2}\chi^0 - 2\sqrt{\lambda}\,\tau^0_\lambda = 0, \quad \lambda\tau^0_\lambda - \tau^0 - 2\sqrt{\lambda}\,\chi^0_\lambda = 0 \tag{6.257}$$

for τ^0 and χ^0 and the linear system of two ODEs

$$2(\lambda + 1)\sqrt{\lambda}\,T_\lambda - \lambda\left(1 + \frac{\lambda}{2}\right)K_\lambda - \frac{(2 - 7\lambda)K}{4} - \sqrt{\lambda}\,\tau^0 = 0,$$
$$\left(1 + \frac{\lambda}{2}\right)T - \left(3 + \frac{\lambda}{2}\right)\lambda T_\lambda + 2\sqrt{\lambda}\,(1 + \lambda)K_\lambda - \frac{\sqrt{\lambda}\,K}{2} = 0 \tag{6.258}$$

for T and K.

We eliminate χ^0_λ from equations (6.257) to obtain

$$\chi^0 = \sqrt{\lambda}[(\lambda - 4)\tau^0_\lambda - \tau^0] \tag{6.259}$$

and rewrite the system (6.257) as the second-order equation

$$\lambda(\lambda - 4)\tau^0_{\lambda\lambda} - 2\tau^0_\lambda = 0. \tag{6.260}$$

The solution of equations (6.260)–(6.259) is given by

$$\tau^0 = q_1 + q_0\sqrt{\lambda(\lambda - 4)} - 4q_0\ln(\sqrt{\lambda} + \sqrt{\lambda - 4}),$$
$$\chi^0 = -q_1\sqrt{\lambda} - 4q_0\sqrt{\lambda - 4} + 4q_0\sqrt{\lambda}\ln(\sqrt{\lambda} + \sqrt{\lambda - 4}). \tag{6.261}$$

This coincides with that given, for example, by (6.162) after the substitution

$$\lambda = 4[1 + k\exp(W)], \quad q_0 = \pm 1/(8k^{3/2}), \quad q_1 = t_0 \pm (1/(2k^{3/2}))\ln(2\sqrt{k}).$$

6.15 Concluding remarks

In this modeling scenario, we have analyzed and visualized the exact invariant solution of the nonlinear simplified version of the shallow water equations (6.142)–(6.143)

$$\begin{aligned} u_t + uu_\theta + h_\theta &= 0, \\ h_t + uh_\theta + hu_\theta &= 0, \end{aligned} \tag{6.262}$$

which are used to simulate equatorial atmospheric waves of planetary scales. Our model is represented by the Cauchy–Poisson free boundary problem on the nonstationary motion of a perfect incompressible fluid circulating around a vortex field approximated by a circle of a large radius and the gravity is directed to the center of the circle. The solution for the free boundary perturbation $h(\theta, t)$ is only valid in a very narrow band $|h| \leqslant \varepsilon < 1$, as also shown schematically in Figure 6.13. We have shown that within this band, the Fibonacci spiral corresponding to $k = 1$ is a very good approximation of the nontrivial solution η, as shown in Figure 6.12. However, we observe that the solution η diverges from the Fibonacci spiral for increasing values of k. In other words, the physically relevant part of the solution matches exactly the Fibonacci spiral.

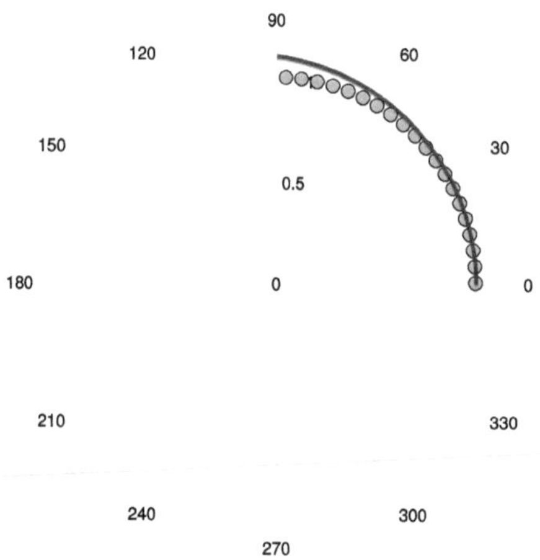

Figure 6.12: Comparison of the free boundary given by $h = 1 + \varepsilon\eta(\theta, t)$ and the Fibonacci spiral evaluated at the values of $k = 1$, $t = 1$, and $t_0 = 0$.

We note that the model (6.138)–(6.139) is invariant under the group of transformations

$$X^{\infty} = \xi^1(h, u)\frac{\partial}{\partial t} + \xi^2(h, u)\frac{\partial}{\partial \theta}, \tag{6.263}$$

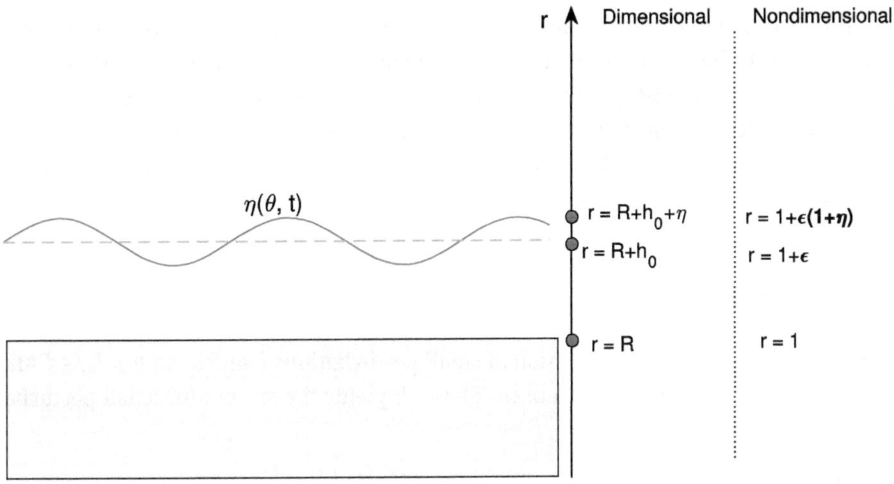

Figure 6.13: The model geometry.

where the functions $\xi^1(h,u)$, $\xi^2(h,u)$ solve the system of first-order equations

$$\xi^2_h + \xi^1_u - u\xi^1_h = 0, \quad \xi^2_u - u\xi^1_u + h\xi^1_h = 0. \tag{6.264}$$

If we write the operator (6.263) as an infinitesimal transformation

$$\bar{t} \approx t + \xi^1(h,u), \quad \bar{\theta} \approx \theta + \xi^2(h,u), \tag{6.265}$$

equations (6.264) show that the infinitesimal transformation (6.265) changes the variables (t,θ) by adding to them an arbitrary solution of the system of linear differential equations

$$\theta_h + t_u - ut_h = 0, \quad \theta_u + ht_h - ut_u = 0 \tag{6.266}$$

and that the system (6.266) admits the infinitesimal transformation (6.265).

Hence, the operator (6.263) is admitted both by the nonlinear system (6.138)–(6.139) and by the linear system (6.266). This fact predicts a possibility to map the nonlinear equations (6.138)–(6.139) to the linear system (6.266) by an appropriate change of variables. It is important to note that the linear system is homogeneous, i. e., invariant under the uniform dilation

$$\bar{t} = te^{a_5}, \quad \bar{\theta} = \theta e^{a_5}$$

produced by the operator X_3 from (6.144).

We conclude that the operator (6.263) and the operator X_3 from (6.144) are responsible for the possibility of mapping of the nonlinear system (6.138)–(6.139) to the linear homogeneous system (6.266).

It is also shown here that the hodograph transformation collapses on the solutions of two different forms of the single first-order nonlinear PDE known as the Hopf equation. This allows to investigate the formation of shock waves in atmospheric motion. The shock waves deserve particular attention because singularities in solutions of a mathematical model are observable in natural phenomena described by the considered mathematical model.

6.16 Exercises

Exercise 6.1. Show that introduction of small perturbations ζ and a via $h = h_0 + \zeta$ and $A = -gh_0t + a$ and linearization of (6.76)–(6.77) yields the system for small perturbations (6.79)–(6.80).

Exercise 6.2. Check that equation (6.86) is true if ω is given by (6.87).

Exercise 6.3. Show that the Boussinesq approximation (6.89)–(6.90) is obtained by substituting $h = h_0 + \varepsilon^2\zeta$ and $A = -gh_0t + \varepsilon^2 a$ into (6.76)–(6.77) and keeping the terms up to the order $0(\varepsilon^4)$.

Exercise 6.4. Check that equation (6.95) is satisfied if

$$2a_{\xi\tau} + 3a_\xi a_{\xi\xi} + \frac{1}{3}c_0 h_0^2 a_{\xi\xi\xi\xi} = 0,$$

where $c_0^2 = gh_0$.

Exercise 6.5. Check that

$$h_0 = 0, \psi_0 = -\frac{\Gamma}{2\pi}\log\left(\frac{r}{R}\right)$$

is a solution of the Cauchy–Poisson free boundary problem (6.124)–(6.128).

Exercise 6.6. Check that the unperturbed system model

$$u_t + uu_\theta + h_\theta = 0,$$
$$h_t + uh_\theta + hu_\theta = 0$$

admits the group generated by the following four infinitesimal operators:

$$X_1 = t\frac{\partial}{\partial t} + \frac{\partial}{\partial u}, \quad X_2 = \theta\frac{\partial}{\partial \theta} + 2h\frac{\partial}{\partial h}, \quad X_3 = t\frac{\partial}{\partial t} + \theta\frac{\partial}{\partial \theta},$$
$$X_4 = (2\theta - 6tu)\frac{\partial}{\partial t} + (6th - 3tu^2)\frac{\partial}{\partial \theta} + 4hu\frac{\partial}{\partial h} + (u^2 + 4h)\frac{\partial}{\partial u}.$$

Exercise 6.7. Check that the Galilean transformation group $\bar{\theta} = \theta + ta_1$, $\bar{u} = u + a_1$ is generated by the operator $X_1 = t\frac{\partial}{\partial t} + \frac{\partial}{\partial u}$.

Exercise 6.8. Check that the Galilean transformation group $\bar{\theta} = \theta + ta_1$, $\bar{u} = u + a_1$ has the following three functionally independent invariants:

$$J_1 = t, \quad J_2 = \frac{u}{\theta}, \quad J_3 = \frac{h}{\theta^2}.$$

7 Modeling scenario 4: Rossby nonlinear atmospheric waves along a spherical planet

This chapter is devoted to the modeling of nonlinear viscous fluid flows within a thin rotating spherical shell that serves as a simple mathematical description of an atmospheric circulation caused by the temperature difference between the equator and the poles of a planet. A particular stationary flow is also superimposed on the model by which, under the assumption of no friction and a distribution of temperature dependent only upon latitude, models the zonal West-to-East flows in the upper atmosphere between the Ferrel and polar cells, as shown in Figure 7.1.

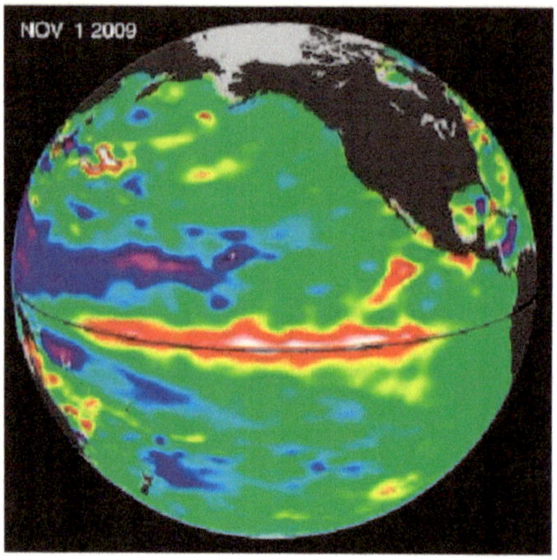

Figure 7.1: Sea-level height data from November 2009 showing the dynamics of warm water known as Kelvin waves that can be seen traveling eastward along the equator (black line) in the November 1, 2009 image. El Niños form when trade winds in the equatorial Western Pacific relax over a period of months, sending Kelvin waves eastward across the Pacific like a conveyor belt. Image credit: NASA/JPL.

Owing to the presence of the Rossby number in the model, the resulting achievable meteorological flows correspond to the asymptotically stable flows that are being translated along the equatorial plane. It will be shown here that application of Lie group analysis provides a new class of exact stationary and nonstationary solutions associated with the corresponding Euler (zero viscosity) and Navier–Stokes equations (nonzero viscosity) that can be presented in terms of elementary functions. It will be shown in this chapter that the energy balance is exactly the conservation law for one class of the solutions, whereas the second class of invariant solutions provides an asymptotic convergence of the energy balance to the conservation law.

Since nonsteady westerly winds in the Southern Hemisphere of our planet are an important force contributing to climate modulation [63, 1, 36], large-scale atmospheric

https://doi.org/10.1515/9783111387499-007

dynamics is currently a topic of extensive research. For example, recent studies [69] suggest that global warming has increased the temperature difference between the southern, mid, and high latitudes and that the loss of Antarctic ozone has contributed to this increase. Additionally, as has been shown in Refs. [34] and [6], owing to the Coriolis forces, the temperature difference between the equator and the poles of a sphere gives rise to waves advancing in the direction of the meridian and waves propagating along the circles of latitude. The importance of understanding the formation of cyclones and their time evolution in the Earth's atmosphere for the creation and distribution of weather systems throughout the world is summarized in Refs. [70, 50], and [51].

The problem forming our main focus of interest here is to consider long waves formed in the Earth's atmosphere, which are captured by the rotation of the Earth with a further goal to extend the previous analysis in [38] to the nonlinear case where the effects of Rossby waves are incorporated in the modeling equations with the goal to find more general (nonstationary in particular) exact solutions of the nonlinear Euler and Navier–Stokes equations modeling atmospheric waves in a thin rotating spherical shell. The new class of exact stationary and nonstationary solutions associated with the corresponding Euler (zero viscosity) and Navier–Stokes equations (nonzero viscosity) is presented in terms of elementary functions. The exact solutions are found by using Lie group methods.

The atmospheric motion modulated by the pressure distribution resulting from the combination of waves advancing in the direction of the meridian and waves propagating along the circles of latitude gives rise to cyclonic and anticyclonic phenomena, which are nowadays a paramount topic of research in atmospheric modeling. For example, the mechanisms of cyclone formation in the Earth's atmosphere have been recently studied in Ref. [6] by means of numerical modeling of the complete gas-dynamic equations. Recent laboratory experiments on cyclone and anticyclone formation in a rotating stratified fluid have been studied in Ref. [18]. Another point of interest in this chapter is the investigation of the asymptotic behavior of sources and sinks associated with the effects of westerly winds on the energy balance of atmospheric waves progressing in the Earth's atmosphere, which are captured by the rotation of the Earth.

7.1 Euler and Navier–Stokes equations

We address the nonlinear three-dimensional Euler (zero viscosity) and Navier–Stokes (nonzero viscosity) equations for an incompressible fluid in Lamb's form [47],

$$\frac{\partial \vec{u}}{\partial t} - \vec{u} \times \operatorname{rot} \vec{u} = -\nabla\left(\frac{\vec{u} \cdot \vec{u}}{2} + \frac{p}{\rho_0} + W\right) + 2\vec{u} \times \vec{\Omega} + \nu \Delta \vec{u}, \tag{7.1}$$

$$\nabla \cdot \vec{u} = 0 \tag{7.2}$$

in a thin spherical shell

$$\Pi = \{\vec{x} \in \mathbf{R}^3 : r_0 < |\vec{x}| < r_0 + \epsilon\}, \tag{7.3}$$

where $\epsilon \to 0$ is a small parameter for the thickness of the spherical shell, $\vec{u}$ is the velocity vector, p is the pressure, ρ_0 is the constant density, ν is the kinematic viscosity, Δ is the Laplacian operator, and W is the potential of the gravitational force which includes the Newtonian attraction. The term $2\vec{u} \times \vec{\Omega}$ is referred to as the Coriolis acceleration. Additionally, r_0 is a fixed altitude above a planet.

We employ the usual spherical coordinates (r, θ, ϕ) with the velocity vector $\mathbf{u} = u_r\mathbf{e}_r + u_\theta\mathbf{e}_\theta + u_\phi\mathbf{e}_\phi$, where $(\mathbf{e}_r, \mathbf{e}_\theta, \mathbf{e}_\phi)$ are basic orthonormal vectors along the spherical coordinates; θ denotes the polar (latitude) angle and ϕ denotes the azimuthal (longitude) angle, so that u_θ represents the velocity component along the meridian (which is positive when the velocity is directed to the South) and u_ϕ is associated with the velocity component along the circle of the latitude (which is positive when it is directed toward the East).

Since the velocity vector $\vec{u}$ and the pressure p are coupled together by the incompressibility constraint $\nabla \cdot \vec{u} = 0$, it is difficult to analyze the full set of three-dimensional equations (7.1)–(7.2). A common approach to simplify the nonlinear model in question is to use artificial methods such as pressure stabilization and projection methods [67]. However, as has been shown in Ref. [67], the error estimate of the pressure stabilization and projection methods is not mathematically precise. Instead, we employ the usual spherical coordinates (r, θ, ϕ) and use the reduction of equations (7.1)–(7.2) to the two-dimensional equations on the sphere

$$S = \{(\theta, \phi),\ 0 < \theta < \pi,\ 0 \leqslant \phi < 2\pi\} \tag{7.4}$$

which is obtained in the limit $\epsilon \to 0$. In our modeling, the depth is sent to zero in the limiting procedure for the nonfree surface model. The latter reduction relies on the theorem from the work *Temam & Ziane* [76]. Namely, as follows directly from our previous analysis in Ref. [38], the latter reduction is justified by Theorem B in [76], which states that the strong global solution of the three-dimensional Navier–Stokes equations (7.1)–(7.2) converges as $\epsilon \to 0$ to the strong global solution $\vec{u}(\theta, \phi, t)$ of the two-dimensional Navier–Stokes or Euler equations on a spherical surface (7.4), where

$$\vec{v}(\theta, \phi, t) = \lim_{\epsilon\to 0} \frac{1}{\epsilon r_0} \int_{r_0}^{r_0+\epsilon r_0} r\vec{u}(r, \theta, \phi, t)dr = (0, v_\theta, v_\phi). \tag{7.5}$$

The vector $\vec{v}(\theta, \phi, t)$ can thus be interpreted as the *average velocity* with respect to r.

In terms of physical interpretation, the above reduction can be interpreted as follows. Let the Earth be a sphere of radius r_0. Then, for $r = r_0$, we must set $u_r = 0$. In addition, for $r \to \infty$, we must assume that

$$\lim_{r\to\infty} \rho u_r = 0, \tag{7.6}$$

meaning that the atmosphere neither loses nor gains mass from outside. However, there exists a height for which ρu_r attains a maximum (the other possibility corresponds to the case when the vertical velocity $u_r = 0$), i. e., where

$$\left.\frac{\partial \rho u_r}{\partial r}\right|_{r=r_0} = 0.$$

It is thus postulated in the present work that at least one such level exists for the entire atmosphere of the planet. We call it the "mean level." In practice, in terms of atmospheric modeling of the Earth (see, e. g., [6, 8]), we can speak about it as a level of 3–5 km. Thus, in the limit $\epsilon \to 0$, the nonstationary three-dimensional viscous rotating fluid flow is confined to a sphere (7.4) of radius r_0 parametrized by the polar (latitude) angle θ and the azimuthal (longitude) angle ϕ.

It is useful to recast the model (7.1)–(7.2) to a nondimensional form by taking c_0 as the unit of velocity and r_0 as the unit of length and defining the Rossby number R_0 as follows:

$$R_0 = \frac{c_0}{2\Omega r_0}, \tag{7.7}$$

where Ω is the angular frequency of planetary rotation (for example, $\Omega = 2\pi$ rad/day $\approx 0.73 \times 10^{-4}\,\mathrm{s}^{-1}$ is the rate of the Earth's rotation).

After the above averaging procedure, the two-dimensional equations in spherical angles are written as follows (*Ibragimov & Pelinovsky* [38]):

$$\begin{aligned} &\frac{\partial v_\theta}{\partial t} + \frac{v_\phi \cos\theta}{R_o} + \frac{\partial p}{\partial \theta} + v_\theta \frac{\partial v_\theta}{\partial \theta} + \frac{v_\phi}{\sin\theta}\frac{\partial v_\theta}{\partial \phi} - v_\phi^2 \cot\theta \\ &\quad + \frac{1}{R_e}\left(\Delta_S v_\theta - \frac{v_\theta}{\sin^2\theta} - \frac{2\cos\theta}{\sin^2\theta}\frac{\partial v_\phi}{\partial \phi}\right) = 0, \end{aligned} \tag{7.8}$$

$$\begin{aligned} &\frac{\partial v_\phi}{\partial t} + \frac{v_\theta \cos(\theta)}{R_0} + \frac{1}{\sin\theta}\frac{\partial p}{\partial \phi} + v_\theta \frac{\partial v_\phi}{\partial \theta} + \frac{v_\phi}{\sin\theta}\frac{\partial v_\phi}{\partial \phi} + v_\theta v_\phi \cot\theta \\ &\quad + \frac{1}{R_e}\left(\Delta_S v_\phi - \frac{v_\phi}{\sin^2\theta} + \frac{2\cos\theta}{\sin^2\theta}\frac{\partial v_\theta}{\partial \phi}\right), \end{aligned} \tag{7.9}$$

$$\frac{\partial}{\partial \theta}(v_\theta \sin\theta) + \frac{\partial v_\phi}{\partial \phi} = 0, \tag{7.10}$$

where

$$\Delta_S = \frac{1}{\sin\theta}\frac{\partial}{\partial \theta}\left(\sin\theta \frac{\partial}{\partial \theta}\right) + \frac{1}{\sin^2\theta}\frac{\partial^2}{\partial \varphi^2} \tag{7.11}$$

is the Laplace–Beltrami operator in spherical angles and

$$R_e = \frac{c_o r_0}{\nu}$$

is the Reynolds number.

Thus, the case $R_e = \infty$ corresponds to the case of Euler equations (zero viscosity) and the case $R_e \neq \infty$ corresponds to the case of Navier–Stokes equations (nonzero viscosity).

Example 7.1 (Particular stationary zonal flow). One can check by direct differentiation that there exists an exact stationary solution to the two-dimensional Euler equations (7.8)–(7.10) with $R_e = \infty$ in spherical coordinates:

$$v_\theta = 0, \quad v_\phi = F(\theta), \quad p = \overline{p} = \int F \cos\theta \left(\frac{F}{\sin\theta} + \frac{1}{R_0} \right) d\theta + p_0, \quad p_0 = \text{const.}, \tag{7.12}$$

where $F(\theta)$ is an arbitrary function. Thus, to a certain extent, the exact solution (7.12) can be associated with Kelvin waves traveling eastward along the equator of the Earth, as shown in Figure 7.1.

We remark that in the case of Navier–Stokes flow (i. e., when if the effects of viscosity are included, $R_e \neq \infty$), the exact solution (7.12) is still valid provided that

$$F(\theta) = \frac{1}{\sin\theta}. \tag{7.13}$$

The exact stationary solution (7.12) has singularities in the pressure, which cannot be removed by a coordinate transformation. In spherical geometry, the curvature of the shape induces singularities in the velocity vector and in the pressure terms the stationary flows represent the basic flows that account the curvature of the sphere. If the spherical layer is transformed to a planar layer by a homotopy transformation, the stationary flows with singularities transform to a regular solution with a constant velocity field.

7.2 Nonlinear nonviscous flows in rotating reference frame

The inclusion of the Coriolis force creates a cyclonic rotation around the poles, i. e., West-to-East winds (see [3]). Namely, as has been indicated in Ref. [34], the temperature difference between the equator and the poles of a sphere gives rise to waves of two kinds. Waves of the first kind advance in the direction of the meridian; waves of the second kind advance in the direction of the circles of latitude. The atmospheric pressures and motions resulting from the combination of these two groups of intersecting waves give rise to the cyclonic and anticyclonic phenomena which are nowadays a paramount topic of research in atmospheric modeling. For example, the mechanisms of cyclone formation in the Earth's atmosphere have recently been studied in [6] with the help of numerical modeling using the complete system of gas-dynamic equations. The authors of [6] have shown that cyclones can appear in horizontal stratified shear flows of warm and wet air masses with a horizontal direction of gradients of the wind velocity components as a result of small disturbances of pressure which can be produced by Rossby waves. Cyclone and anticyclone formation in a rotating stratified fluid have been studied in a laboratory setting in [18]. The formation of cyclones and their time evolution in the Earth's atmosphere for the creation and distribution of weather systems throughout the world are summarized in [70].

In terms of applications to atmospheric sciences, it is useful to note atmospheric patterns at the North and South Poles spin around themselves in the counterclockwise sense at a rate of $\Omega = 2\pi$ rad/day, whereas atmospheric patterns in the domain $\theta \in [\theta_0, \pi - \theta_0]$ do not spin around themselves but simply translate provided $\theta_0 \in (0, \frac{\pi}{2})$. Owing to the Coriolis effects, the achievable meteorological flows rotating around the poles correspond to the flows that are being translated along the equatorial plane. This translational motion is well captured by the zonal flows considered in this chapter.

In view of the importance of zonal flows in terms of meteorological applications [6, 18, 34], we first focus on the system (7.8)–(7.10) on the sphere S in the limiting case $R_e \to \infty$ (nonviscous flow) and $R_o \neq \infty$ (nonzero rotation), with the superimposed stationary flows of the form (7.12) with F being an arbitrary function of θ. We thus look for a solution in the form

$$v_\theta = \mu \widehat{v}_\theta, \quad v_\phi = F(\theta) + \mu \widehat{v}_\phi, \quad p = \overline{p} + \mu \widehat{p}, \tag{7.14}$$

where $\mu > 0$ is a parameter and $\widehat{v}_\theta$, $\widehat{v}_\phi$, and $\widehat{p}$ are space- and time-dependent disturbances of the basic flow (7.12). Since we are interested in the exact solution of the nonlinear model, the case $\mu \geqslant 1$ is of special interest. The particular case $\mu \ll 1$ was studied in much detail in our previous work [38]. Within approximation (7.6), the Euler equations (7.8)–(7.10) with $R_e = \infty$ are equivalent to a single vorticity equation

$$\frac{\partial \zeta}{\partial t} + \frac{\mu}{\sin\theta} J(\psi, \zeta) + \frac{F}{\sin\theta}\frac{\partial \zeta}{\partial \phi} - \frac{1}{\sin\theta}\frac{\partial \psi}{\partial \phi} LF + \frac{1}{R_0}\frac{\partial \psi}{\partial \phi} = 0, \tag{7.15}$$

where the vorticity ζ is related to the stream function ψ through the Laplacian in spherical geometry, i. e.,

$$\zeta = \left[\frac{1}{\sin\theta}\frac{\partial}{\partial\theta}\left(\sin\theta\frac{\partial}{\partial\theta}\right) + \frac{1}{\sin^2\theta}\frac{\partial^2}{\partial\varphi^2}\right]\psi, \tag{7.16}$$

in which the stream function ψ is given by

$$v_\theta = -\frac{1}{\sin\theta}\frac{\partial\psi}{\partial\phi}, \quad v_\phi = \frac{\partial\psi}{\partial\theta}, \tag{7.17}$$

and L is the Sturm–Liouville operator for the associated Legendre functions:

$$L = \frac{1}{\sin\theta}\frac{d}{d\theta}\left(\sin\theta\frac{d}{d\theta}\right) - \frac{1}{\sin^2\theta}. \tag{7.18}$$

Additionally,

$$J(\psi,\zeta) = \frac{\partial\psi}{\partial\theta}\frac{\partial\zeta}{\partial\phi} - \frac{\partial\psi}{\partial\phi}\frac{\partial\zeta}{\partial\theta} \tag{7.19}$$

stands for the nonlinear Jacobian operator.

As shown in formula (7.12), $F(\theta)$ represents finite disturbances that are superimposed on a zonally averaged mean flow having only a latitudinal dependence. In particular, as has been discussed in [1], the presence of Rossby waves originates the general West-to-East flows caused by the planet rotation which are related to jet streams, i. e., zones of fast-moving West-to-East winds in the upper atmosphere between the Ferrel and polar cells. In terms of atmospheric modeling of the Earth, such flows exist due to the presence of the cold North and South Poles and the warm equator so that the pressure is low at high latitudes above the poles, high in the temperate zones, and even higher in the equatorial zones. Thus, under the assumption of no friction and a distribution of temperature dependent only upon latitude and altitude, the presence of $F(\theta)$ can be associated with a zonal flow directed from West to East (it can also serve as a mathematical descriptor of Kelvin waves demonstrated in Figure 7.1). The experimental results in [18] justify the conclusions on the existence of such zonal flows discussed in [34].

From a mathematical standpoint, the presence of $F(\theta)$ can be explained as follows. The set

$$\overline{u}_\theta = 0, \quad \overline{u}_\phi = F(\theta), \quad \overline{p} = \int F\left(\frac{F}{\sin\theta} + \frac{1}{R_0}\right)\cos\theta d\theta + p_a \tag{7.20}$$

represents the particular class of stationary exact solutions of the Euler equations (7.8)–(7.10) with $R_e = \infty$ and $p_a = \text{const}$. This ansatz follows directly from our previous work [38], in which the Euler flows on a sphere were investigated without the effects of rotation.

Equation (7.15) is written in a noninertial reference frame. Rossby waves are generated by pseudoforces, which support the West-to-East flows so that the fluid particles at the North and South Poles spin around themselves at a rate of $\Omega = 2\pi$ rad/day, whereas fluid particles in the spherical domain $\theta \in [\theta_0, \pi - \theta_0]$ do not spin around themselves but simply translate provided $0 \leqslant \theta_0 < \pi/2$. The complete sphere corresponds to the limit $\theta_0 \to 0$. In the latter case, the interval $[0, \pi]$ connects two singular points of the Sturm–Liouville operator (7.18). Owing to the Coriolis effects, the achievable meteorological flows correspond to the fluid rotation rotate around the poles, i. e., translation along the equatorial plane.

The exact solutions in terms of elementary functions for the nonlinear model (7.15) are found by using Lie group methods.

7.3 Invariant solutions of Euler equations

Solution of the determining equations for symmetries shows that the Lie algebra admitted by equation (7.15) with an arbitrary form of a zonal flow $F(\theta)$ and Rossby number R_0 is spanned by the following operators:

$$X_1 = \frac{\partial}{\partial t}, \quad X_2 = \frac{\partial}{\partial \phi}, \quad X_3 = \gamma(t)\frac{\partial}{\partial \psi}, \tag{7.21}$$

$$X_4 = 2R_0\mu t\frac{\partial}{\partial t} - \mu t\frac{\partial}{\partial \phi} + (\cos\theta - 2R_0H(\theta) - 2\mu R_0\psi)\frac{\partial}{\partial \psi}, \tag{7.22}$$

$$X_5 = \mu\sin\left(\phi + \frac{t}{2R_0}\right)\frac{\partial}{\partial \theta} + \mu\frac{\cos\theta}{\sin\theta}\frac{\partial}{\partial \phi} + \left(F(\theta) + \frac{\sin\theta}{2R_0}\right)\sin\left(\phi + \frac{t}{2R_0}\right)\frac{\partial}{\partial \psi}, \tag{7.23}$$

$$X_6 = \mu\cos\left(\phi + \frac{t}{2R_0}\right)\frac{\partial}{\partial \theta} + \mu\frac{\cos\theta}{\sin\theta}\sin\left(\phi + \frac{t}{2R_0}\right)\frac{\partial}{\partial \phi} - \left(F(\theta) + \frac{\sin\theta}{2R_0}\right)\cos\left(\phi + \frac{t}{2R_0}\right)\frac{\partial}{\partial \psi}, \tag{7.24}$$

where μ is a constant, $\gamma(t)$ in the operator X_3 given by (7.21) is an arbitrary function, and $H(\theta)$ in the operator X_4 given by (7.22) is the integral of $F(\theta)$, i. e., $H'(\theta) = F(\theta)$.

It is worth noting that symmetries (7.21)–(7.24) were obtained through analytical derivation, without the use of any numerical packages.

For example, the operator X_2 generates the obvious translation $\overline{\phi} = \phi + \beta_2$ of the angle ϕ, where β_2 is the group parameter (the subscript in the group parameter coincides with that of the group generator). Likewise, the operator X_4 generates the one-parameter group of a more complex form, namely,

$$\begin{aligned} &\bar{t} = te^{2\mu R_0\beta_4}, \quad \overline{\theta} = \theta, \quad \overline{\phi} = \phi + \frac{t}{2R_0}(1 - e^{2\mu R_0\beta_4}), \\ &\overline{\psi} = \psi e^{-2\mu R_0\beta_4} + \frac{\cos\theta - 2R_0H(\theta)}{2\mu R_0}(1 - e^{-2\mu R_0\beta_4}). \end{aligned} \tag{7.25}$$

Computing the commutators of the operators (7.21)–(7.24), one obtains the following commutator table:

Commutator table of the admitted symmetries (7.26)

	X_1	X_2	X_3	X_4	X_5	X_6
X_1	0	0	X_3'	$2\mu R_0X_1 - \mu X_2$	$\frac{1}{2R_0}X_6$	$-\frac{1}{2R_0}X_5$
X_2	0	0	0	0	X_6	$-X_5$
X_3	$-X_3'$	0	0	X_3''	0	0
X_4	$\mu X_2 - 2\mu R_0X_1$	0	$-X_3''$	0	0	0
X_5	$-\frac{1}{2R_0}X_6$	$-X_6$	0	0	0	μ^2X_2
X_6	$\frac{1}{2R_0}X_5$	X_5	0	0	$-\mu^2X_2$	0

Here

$$X_3' = \gamma'(t)\frac{\partial}{\partial\psi}, \quad X_3'' = -2\mu R_0[\gamma(t) + t\gamma'(t)]\frac{\partial}{\partial\psi}.$$

Example 7.2. Verify that the operators X_4, X_5, and X_6 are admitted by equation (7.15).

Solution. We denote the left-hand side of equation (7.15) by G, prolong the operators X_4 and X_5, and obtain after calculations

$$X_4(G) = -4\mu R_0 G, \quad X_5(G) = \mu \frac{\cos\theta}{\sin\theta} \sin\left(\phi + \frac{t}{2R_0}\right)G.$$

This proves that the operators X_4 and X_5 are admitted by equation (7.15). This fact for the operator X_6 follows from the commutator table in (7.26) because the commutator of any two admitted operators is also admitted. □

7.3.1 Invariant solution based on X_2 and X_4

The commutator table in (7.26) shows that X_2 and X_4 span an Abelian two-dimensional subalgebra. In order to find the invariant solution with respect to this subalgebra, we solve the equations

$$X_2 J(t, \phi, \theta, \psi) = 0, \quad X_4 J(t, \phi, \theta, \psi) = 0$$

to obtain two functionally independent invariants

$$\xi = \theta, \quad \mu = t\left(\psi + \frac{H(\theta)}{\mu} - \frac{\cos\theta}{2\mu R_0}\right).$$

Letting $\mu = \Phi(\xi)$, we obtain the following G/H factor system providing a general form for the invariant solutions:

$$\psi = \frac{\cos\theta}{2\mu R_0} - \frac{H(\theta)}{\mu} + \frac{1}{t}\Phi(\theta). \tag{7.27}$$

Substituting (7.27) in equation (2.101) we obtain the second-order ODE

$$(1 - \xi^2)\Phi'' - 2\xi\Phi' = 0, \tag{7.28}$$

whence

$$\Phi = C_1 + C_2 \ln|\tan(\theta/2)|.$$

Substituting this in (7.27) we obtain the following invariant solution:

$$\psi^{[1]} = \frac{\cos\theta}{2\mu R_0} - \frac{H(\theta)}{\mu} + \frac{C_1}{t} + \frac{C_2}{t}\ln|\tan(\theta/2)|, \tag{7.29}$$

where C_1 and C_2 are arbitrary constants.

Thus, the solution (7.29) is invariant under the translation of the angle ϕ (due to the independence on this angle) and the transformation (7.25).

Example 7.3. Determine the pressure p for the particular case $F = 0$ (no influence of zonal flows) and $\mu = 1$, which corresponds to the nonperturbed Euler equations (7.8)–(7.10).

Solution. The velocity components corresponding to the invariant presentation (7.29) for $\psi^{[1]}$ are written in terms of elementary functions as

$$v_\theta = 0, \quad v_\phi = \frac{C_2}{t\sin\theta} - \frac{\sin\theta}{2R_0}, \tag{7.30}$$

whereas the pressure is determined from equations (7.8)–(7.10) by means of the following compatible system:

$$\frac{\partial p}{\partial\theta} = \frac{C_2^2\cos\theta}{t^2\sin^3\theta} - \frac{\cos\theta\sin\theta}{4R_0^2}, \quad \frac{\partial p}{\partial\phi} = \frac{C_2}{t^2}. \tag{7.31}$$

This means that the pressure is also determined in terms of the elementary functions as

$$p = \frac{C_2}{t^2}\phi - \frac{C_2^2}{2t^2\sin^2\theta} - \frac{\sin^2\theta}{8R_0^2} + p_0(t), \tag{7.32}$$

where p_0 is an arbitrary function. □

In the more general case when F is a nonzero arbitrary function of θ and μ is an arbitrary parameter, the solution (v_θ, v_ϕ, p) for the perturbed Euler model (7.15) is expressed as

$$v_\theta = 0, \quad v_\phi = \frac{C_2}{t\sin\theta} - \frac{\sin\theta}{2\mu R_0} - \frac{F(\theta)}{\mu}, \tag{7.33}$$

$$p = \frac{C_2}{t^2}\phi - \frac{\mu C_2^2}{2t^2\sin^2\theta} + \frac{\cos^2\theta}{8\mu R_0^2} - \frac{1}{\mu}\int\left(\frac{\cos\theta}{R_0} + F\cot\theta\right)F d\theta + p_0(t). \tag{7.34}$$

7.3.2 Invariant solution based on X_4 and X_5

The operators X_4 and X_5 also span an Abelian subalgebra. In order to find a basis of invariants of this subalgebra, we find first the following three functionally independent invariants for X_4 as follows:

$$\theta, \quad \nu = t + 2R_0\phi, \quad \sigma = t\left(\psi + \frac{H(\theta)}{\mu} - \frac{\cos\theta}{2\mu R_0}\right).$$

Acting on the invariants by the operator X_5, we obtain

$$X_5(\theta) = \mu \sin\left(\phi + \frac{t}{2R_0}\right) = \mu \sin\frac{\nu}{2R_0},$$
$$X_5(\nu) = 2\mu R_0 \frac{\cos\theta}{\sin\theta}\cos\left(\phi + \frac{t}{2R_0}\right) = 2\mu R_0 \frac{\cos\theta}{\sin\theta}\cos\frac{\nu}{2R_0}\cdot$$

The reckoning also shows that $X_5(\sigma) = 0$. Hence, X_5 is written in terms of the invariants as follows:

$$X_5 = \mu \sin\frac{\nu}{2R_0}\frac{\partial}{\partial\theta} + 2\mu R_0 \frac{\cos\theta}{\sin\theta}\cos\frac{\nu}{2R_0}\frac{\partial}{\partial\nu}\cdot \tag{7.35}$$

Solving the equation $X_5 J(\theta, \nu, \sigma) = 0$ with the operator X_5 given by (7.35), one readily obtains the invariants σ and

$$\lambda = \sin\theta\cos\frac{\nu}{2R_0}.$$

Thus, the subalgebra spanned by X_4, X_5 has the following two functionally independent invariants:

$$\lambda = \sin\theta\cos\left(\phi + \frac{t}{2R_0}\right), \quad \sigma = t\left(\psi + \frac{H(\theta)}{\mu} - \frac{\cos\theta}{2\mu R_0}\right). \tag{7.36}$$

Letting $\sigma = \Phi(\lambda)$, we obtain the following general form for the invariant solutions:

$$\psi^{[2]} = \frac{\cos\theta}{2\mu R_0} - \frac{H(\theta)}{\mu} + \frac{1}{t}\Phi(\lambda). \tag{7.37}$$

Substituting the presentation (7.37) in the left side of (7.15), we obtain the following equation for $\Phi(\lambda)$:

$$(1 - \lambda^2)\Phi'' - 2\lambda\Phi' = 0. \tag{7.38}$$

Since $|\lambda| \leqslant 1$ by definition of λ, the solution of equation (7.38) is given by

$$\Phi = C_1 + C_2 \ln\frac{1-\lambda}{1+\lambda}, \tag{7.39}$$

where $|\lambda| \neq 1$ and C_1, C_2 are arbitrary constants.

Example 7.4. In the particular case $F = 0$ and $\mu = 1$, which corresponds to the non-perturbed Euler equations (7.8)–(7.10), the velocity components corresponding to the invariant presentation (7.37) for $\psi^{[2]}$ are written in terms of elementary functions as

$$v_\theta = \frac{\Phi'}{t}\sin\left(\phi + \frac{t}{2R_0}\right), \quad v_\phi = \frac{\Phi'}{t}\cos\theta\cos\left(\phi + \frac{t}{2R_0}\right) - \frac{\sin\theta}{2R_0}, \tag{7.40}$$

whereas the pressure is determined from equations (7.8)–(7.10) by means of the following compatible system:

$$\begin{aligned} &-\frac{1}{t^2}\Phi'\sin\left(\phi + \frac{t}{2R_0}\right) + \frac{\mu\lambda}{t^2}\Phi'^2\cos\theta\cos\left(\phi + \frac{t}{2R_0}\right) + \frac{F\cot\theta}{\mu}\left(F + \frac{1}{R_0\sin\theta}\right) \\ &\quad + \frac{\cos\theta\sin\theta}{4\mu R_0^2} = -\frac{\partial p}{\partial\theta} \end{aligned} \tag{7.41}$$

$$-\frac{1}{t^2}\Phi'\cos\theta\cos\left(\phi + \frac{t}{2R_0}\right) - \frac{\mu\lambda}{t^2}\Phi'^2\sin\left(\phi + \frac{t}{2R_0}\right) = -\frac{1}{\sin\theta}\frac{\partial p}{\partial\phi}. \tag{7.42}$$

□

7.4 Invariant solutions of Navier–Stokes equations

We now consider the Navier–Stokes equations obtained by disturbing the perturbed Euler equations (7.8)–(7.10) by an infinitesimal viscosity μ:

$$\begin{aligned} &\frac{\partial v_\theta}{\partial t} + \frac{\tilde{F}}{\sin\theta}\frac{\partial v_\theta}{\partial\phi} - 2\tilde{F}v_\phi\cot\theta - \frac{\cos\theta}{R_0}v_\phi + \frac{\partial p}{\partial\theta} \\ &\quad + \mu\left\{v_\theta\frac{\partial v_\theta}{\partial\theta} + \frac{v_\phi}{\sin\theta}\frac{\partial v_\theta}{\partial\phi} - v_\phi^2\cot\theta\right\} + \frac{1}{R_e}\left(\Delta_S v_\theta - \frac{v_\theta}{\sin^2\theta} - \frac{2\cos\theta}{\sin^2\theta}\frac{\partial v_\phi}{\partial\phi}\right) = 0, \end{aligned} \tag{7.43}$$

$$\begin{aligned} &\frac{\partial v_\phi}{\partial t} + \tilde{F}'v_\theta + \frac{\tilde{F}}{\sin\theta}\frac{\partial v_\phi}{\partial\phi} + \tilde{F}v_\theta\cot\theta + \frac{v_\theta\cos\theta}{R_0} + \frac{1}{\sin\theta}\frac{\partial p}{\partial\phi} \\ &\quad \mu\left\{v_\theta\frac{\partial v_\phi}{\partial\theta} + \frac{v_\phi}{\sin\theta}\frac{\partial v_\phi}{\partial\phi} + v_\theta v_\phi\cot\theta\right\} + \frac{1}{R_e}\left(\Delta_S v_\phi - \frac{v_\phi}{\sin^2\theta} + \frac{2\cos\theta}{\sin^2\theta}\frac{\partial v_\theta}{\partial\phi}\right) = 0, \end{aligned} \tag{7.44}$$

$$\frac{\partial}{\partial\theta}(v_\theta\sin\theta) + \frac{\partial v_\phi}{\partial\phi} = 0, \tag{7.45}$$

where $\tilde{F}(t,\theta,\phi)$ is the deviation from the basic state $F(\theta)$ caused by the viscous terms. We note that the form of the stationary flow solution $F(\theta)$ depends strongly on the values of R_e. Namely, the two limiting cases $R_e = 0$ and $R_e = \infty$ correspond to the infinitely viscous and the nonviscous fluid, respectively, whereas all other cases correspond to viscous fluids in which F is given by (7.13). However, F is allowed to be an arbitrary function of θ in the case $R_e = \infty$.

In fact, there are several available results on similar perturbation techniques based on the fact that the velocities associated with the Euler and Navier–Stokes equations are "close" to each other in the case of small viscosity and no boundaries (see, e. g., [32, 4]).

Example 7.5. We will consider now the particular case

$$\mu = 1, \quad F(\theta) = -\frac{\sin\theta}{2R_0}, \tag{7.46}$$

which arises in the case of the extension of our results to the Navier–Stokes equations (7.43)–(7.45) obtained by viscous perturbations of the perturbed Euler equations (7.8)–(7.10). In this case, a similar utilization of the invariant solution (7.37) yields the corresponding velocity components for the perturbed Euler equations by $F(\theta)$ given by (7.46) as

$$v_\theta = \frac{\Phi'}{t}\sin\left(\phi + \frac{t}{2R_0}\right), \quad v_\varphi = \frac{\Phi'}{t}\cos\theta\cos\left(\phi + \frac{t}{2R_0}\right). \tag{7.47}$$

Upon substituting (7.46) and (7.47) in the Euler equations (7.8)–(7.10), we obtain the following overdetermined system for the pressure p:

$$\frac{\partial p}{\partial\theta} = \frac{\Phi'}{t^2}\sin\left(\phi + \frac{t}{2R_0}\right) - \frac{\Phi'^2}{t^2}\lambda\cos\theta\cos\left(\phi + \frac{t}{2R_0}\right), \tag{7.48}$$

$$\frac{\partial p}{\partial\phi} = \frac{\Phi'}{t^2}\lambda\cos\theta + \frac{\Phi'^2}{t^2}\lambda\sin\theta\sin\left(\phi + \frac{t}{2R_0}\right). \tag{7.49}$$

The integrability condition

$$\frac{\partial^2 p}{\partial\phi\partial\theta} - \frac{\partial^2 p}{\partial\theta\partial\phi} = 0$$

of the system (7.48)–(7.49) is written

$$\frac{\sin\theta}{t^2}\left[\left(\cos^2\theta\cos^2\left(\phi + \frac{t}{2R_0}\right) + \sin^2\left(\phi + \frac{t}{2R_0}\right)\right)\Phi'' - 2\lambda\Phi'\right] = 0.$$

As expected, the latter equation is equivalent to equation (7.38). Substituting in equation (7.48) the expression

$$\Phi' = \frac{k}{1-\lambda^2}, \quad k = \text{const.}, \tag{7.50}$$

obtained by integrating once equation (7.38), we have

$$\frac{\partial p}{\partial\theta} = \frac{k\sin(\phi + \frac{t}{2R_0})}{t^2(1-\lambda^2)} - \frac{\lambda k^2\cos\theta\cos(\phi + \frac{t}{2R_0})}{t^2(1-\lambda^2)^2}. \tag{7.51}$$

Integrating (7.51) we have

$$p = \frac{k\sin(\phi + \frac{t}{2R_0})}{2t^2}\left[\int\frac{d\theta}{1+\lambda} + \int\frac{d\theta}{1-\lambda}\right] + \frac{k^2}{2t^2}\int\frac{d(1-\lambda^2)}{(1-\lambda^2)^2} + h(\phi, t). \tag{7.52}$$

Here $h(\phi, t)$ is an arbitrary function. It is determined by substituting the expression (7.52) for p in equation (7.49). Then the reckoning shows that

$$\frac{\partial p}{\partial \phi} = \frac{k\lambda \cos\theta}{t^2(1-\lambda^2)} + \frac{\lambda k^2 \sin\theta \cos(\phi + \frac{t}{2R_0})}{t^2(1-\lambda^2)^2} + \frac{\partial h}{\partial \phi}. \tag{7.53}$$

Substituting (7.53) and (7.50) in equation (7.49), we obtain $\frac{\partial h}{\partial \phi} = 0$. Working out the integrals in (7.52) one can readily obtain the expression for the pressure in terms of elementary functions,

$$p = \frac{k}{t^2}\Sigma(t, \theta, \phi) - \frac{k^2}{2t^2(1-\lambda^2)} + p_0(t),$$

where we denote

$$\Sigma(t,\theta,\phi) = \arctan\left[\cot\left(\phi + \frac{t}{2R_0}\right) + \frac{\tan\frac{\sin\theta}{2}}{\sin(\phi + \frac{t}{2R_0})}\right] - \arctan\left[\cot\left(\phi + \frac{t}{2R_0}\right) - \frac{\tan\frac{\sin\theta}{2}}{\sin(\phi + \frac{t}{2R_0})}\right].$$

□

We now look for solutions of the Navier–Stokes model (7.43)–(7.45) in the form

$$\psi_v^{[1]} = \frac{\cos\theta}{2\mu R_0} - \frac{1}{\mu}\int \widetilde{F} d\theta + \frac{C_1}{t} + \frac{C_2}{t}\ \ln\left|\tan(\theta/2)\right|, \tag{7.54}$$

$$\psi_v^{[2]} = \frac{\cos\theta}{2\mu R_0} - \frac{1}{\mu}\int \widetilde{F} d\theta + \frac{1}{t}\Phi(\lambda), \tag{7.55}$$

which are obtained by perturbing the solutions (7.29) and (7.37) by $\widetilde{F}$. Here we limit ourselves to the simplest case where $\widetilde{F} = \widetilde{F}(\theta)$. Then the comparison with the case of nonzero viscosity ($R_e \neq \infty$) shows that both assumed solutions (7.54) and (7.55) will solve the resulting Navier–Stokes equations (7.43)–(7.45) if the following two equations are satisfied simultaneously:

$$E_1 = \Delta_S v_\theta - \frac{v_\theta}{\sin^2\theta} - \frac{2\cos\theta}{\sin^2\theta}\frac{\partial v_\phi}{\partial \phi} = 0 \tag{7.56}$$

and

$$E_2 = \Delta_S v_\phi - \frac{v_\phi}{\sin^2\theta} + \frac{2\cos\theta}{\sin^2\theta}\frac{\partial v_\theta}{\partial \phi} = 0. \tag{7.57}$$

One can check by direct differentiation that $\psi_v^{[1]}$ and $\psi_v^{[2]}$ given by (7.54) and (7.55) solve both equations (7.56) and (7.57) if $\widetilde{F}(\theta)$ is a solution of the nonhomogeneous linear ODE $L\widetilde{F} = \sin\theta/R_0$, where the Sturm–Liouville operator L_1 for the associated Legendre functions was introduced by formula (7.18). Thus, $\widetilde{F}$ is given by $\widetilde{F} = -\sin\theta/R_0$, which coincides with the form (7.46) for $F(\theta)$ considered in Example 7.5.

In particular, the exact asymptotically stable steady solution (7.13) obtained in Ref. [38] is given by $v_\phi^{[IP]} = 1/\sin\theta$, whereas two invariant solutions

$$v_\phi^{[1]} = \frac{C_2}{t\sin\theta}, \quad v_\phi^{[2]} = -\frac{2C_2}{t(1-\lambda^2)}\cos\theta\cos\left(\phi + \frac{t}{2R_0}\right) \tag{7.58}$$

are obtained from the general invariant presentations (7.29) and (7.37) in the special case when $F(\theta)$ is given by (7.46) in Example 7.5 (see also formulae (7.30) and (7.47)).

7.5 Discussion of invariant solutions

Our main results on the effects of the Rossby and Reynolds numbers R_0 and R_e on the zonal flows $F(\theta)$ and invariant solutions (7.54) and (7.55) are summarized in Table 7.1. The remainder of this chapter is devoted to discussion and visualization of the tabulated results.

We next visualize the obtained invariant solutions in terms of the velocity and pressure fields associated with the stream functions $\psi_\nu^{[1]}$ and $\psi_\nu^{[2]}$ given by (7.54) and (7.55) (see also entry [7] in Table 7.1). For example, the exact solutions corresponding to the invariant solution $\psi_\nu^{[1]}$ for the velocity field and the pressure are written in terms of elementary functions as follows:

$$\begin{aligned} v_\theta^{[2]} &= -\frac{1}{\sin\theta}\frac{\partial\psi^{[2]}}{\partial\phi} = \frac{-2c_2}{t(1-\lambda^2)}\sin\left(\phi + \frac{t}{2R_o}\right), \\ v_\phi^{[2]} &= \frac{\partial\psi^{[2]}}{\partial\theta} = \frac{-2c_2}{t(1-\lambda^2)}\cos\theta\cos\left(\phi + \frac{t}{2R_o}\right), \\ p^{[2]} &= -\frac{c_2}{t^2}\Sigma(t,\theta,\phi) - \frac{4c_2^2}{2t^2(1-\lambda^2)}, \end{aligned}$$

where

$$\Sigma(t,\theta,\phi) = \arctan\left[\cot\left(\phi + \frac{t}{2R_0}\right) + \frac{\tan\frac{\sin\theta}{2}}{\sin(\phi + \frac{t}{2R_0})}\right] - \arctan\left[\cot\left(\phi + \frac{t}{2R_0}\right) - \frac{\tan\frac{\sin\theta}{2}}{\sin(\phi + \frac{t}{2R_0})}\right].$$

The exact solutions for the velocity field and the pressure corresponding to the second invariant solution $\psi_\nu^{[2]}$ can be written in terms of elementary functions likewise.

Figure 7.2 shows the convergence of $v_\phi^{[1]}$ and $v_\phi^{[2]}$ as time increases to a steady-state regime. To understand better the nature of a such convergence, we compare the latter exact solutions at a certain fixed angle ϕ versus latitude as shown in Figure 7.2 at $\phi = 5°$ as the particular example. Namely, as we observe from Figure 7.2, the convergence

Table 7.1: Summary of invariant solutions and limiting cases. Since we are interested in the exact solutions of strongly nonlinear equations, we set $\mu = 1$.

	Rotation	Viscosity	Zonal flows	Invariant solutions
[1]	$R_o = \infty$	$R_e = \infty$	$F = 0$	$v_\theta^{[1]} = 0, v_\phi^{[1]} = \frac{c_2}{t\sin\theta}$ $v_\theta^{[2]} = \frac{\Phi'(\lambda_0)}{t}\sin\phi$ $v_\phi^{[2]} = \frac{\Phi'(\lambda_0)}{t}\cos\theta\cos\phi$
[2]	$R_o = \infty$	$R_e \neq \infty$	$F = 0$	$v_\theta^{[1]} = 0, v_\phi^{[1]} = \frac{c_2}{t\sin\theta}$ $v_\theta^{[2]} = \frac{\Phi'(\lambda_0)}{t}\sin\phi$ $v_\phi^{[2]} = \frac{\Phi'(\lambda_0)}{t}\cos\theta\cos\phi$
[3]	$R_o \neq \infty$	$R_e = \infty$	$F = 0$	$v_\theta^{[1]} = 0, v_\phi^{[1]} = \frac{c_2}{t\sin\theta} - \frac{\sin\theta}{2R_o}$ $v_\theta^{[2]} = \frac{\Phi'(\lambda)}{t}\sin(\phi + \frac{t}{2R_o})$ $v_\phi^{[2]} = \frac{\Phi'(\lambda)}{t}\cos\theta\cos(\phi + \frac{t}{2R_o}) - \frac{\sin\theta}{2R_o}$
[4]	$R_o \neq \infty$	$R_e = \infty$	$F = F(\theta)$	$v_\theta^{[1]} = 0, v_\phi^{[1]} = \frac{c_2}{t\sin\theta} - \frac{\sin\theta}{2R_o} - F(\theta)$ $v_\theta^{[2]} = \frac{\Phi'(\lambda)}{t}\sin(\phi + \frac{t}{2R_o})$ $v_\phi^{[2]} = \frac{\Phi'(\lambda)}{t}\cos\theta\cos(\phi + \frac{t}{2R_o}) - \frac{\sin\theta}{2R_o} - F(\theta)$
[5]	$R_o = \infty$	$R_e = \infty$	$F = F(\theta)$	$v_\theta^{[1]} = 0, v_\phi^{[1]} = \frac{c_2}{t\sin\theta} - F(\theta)$ $v_\theta^{[2]} = \frac{\Phi'(\lambda_0)}{t}\sin\phi$ $v_\phi^{[2]} = \frac{\Phi'(\lambda_0)}{t}\cos\theta\cos\phi - F(\theta)$
[6]	$R_o = \infty$	$R_e \neq \infty$	$F = F(\theta)$	$v_\theta^{[1]} = 0, v_\phi^{[1]} = \frac{c_2}{t\sin\theta} - F(\theta)$ $v_\theta^{[2]} = \frac{\Phi'(\lambda_0)}{t}\sin\phi$ $v_\phi^{[2]} = \frac{\Phi'(\lambda_0)}{t}\cos\theta\cos\phi - F(\theta)$
[7]	$R_o \neq \infty$	$R_e \neq \infty$	$F = -\frac{\sin\theta}{2R_o}$	$v_\theta^{[1]} = 0, v_\phi^{[1]} = \frac{c_2}{t\sin\theta}$ $v_\theta^{[2]} = \frac{\Phi'(\lambda)}{t}\sin(\phi + \frac{t}{2R_o})$ $v_\phi^{[2]} = \frac{\Phi'(\lambda)}{t}\cos\theta\cos(\phi + \frac{t}{2R_o})$

occurs at the vicinity of the value $v_\phi = 100/\sin\theta$. This convergence can be explained as follows. We remark that the general solution of the homogeneous equation $LF = 0$ is given by

$$F(\theta) = \frac{\alpha}{\sin\theta} + \beta, \tag{7.59}$$

where α and β are arbitrary parameters. The stationary solution (7.59) is the exact solution of the Navier–Stokes model (7.43)–(7.43), a nonrotating reference frame ($R_o = \infty$) whose asymptotic stability has been studied earlier in our work [38]. Thus, the steady-state regime shown in Figure 7.2 corresponds to the asymptotically stable exact solution (7.13) in which we arbitrarily set $\alpha = 100$ and $\beta = 0$.

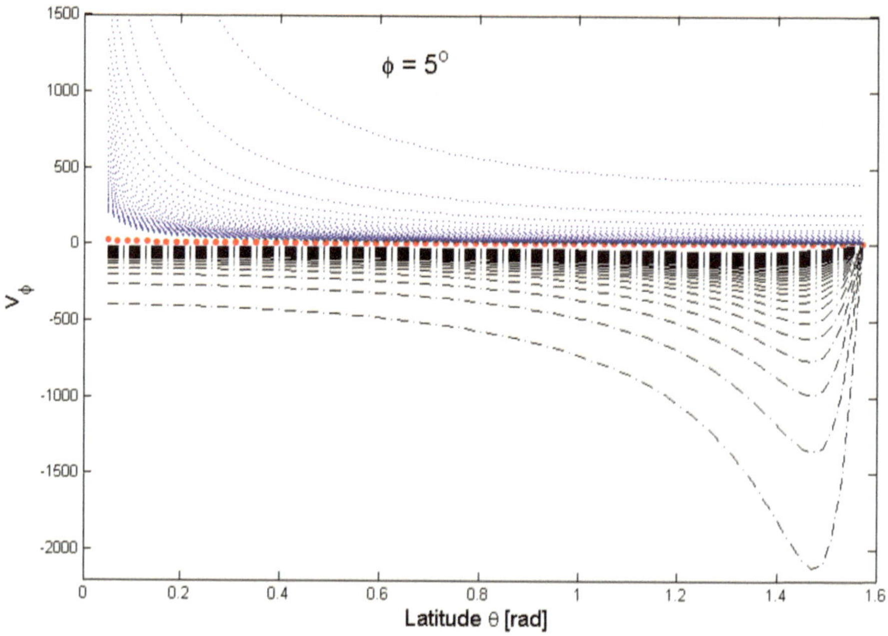

Figure 7.2: Comparison of exact solutions from entry [7] in Table 7.1 written in terms of the azimuthal velocity v_ϕ component plotted on a spherical surface at different values of time t. The red circles correspond to the stationary trivial solution $v_\phi^{[IP]} = 1/\sin\theta$ whose asymptotic stability was investigated in our previous studies in Ref. [38]. The dotted and dashed lines correspond to the invariant solutions $v_\phi^{[1]}$ and $v_\phi^{[2]}$, respectively, at $\phi = 5°$. We choose $R_0 = 68.493$ and $t \in [0.5, 10]$.

7.6 Effects of Rossby waves on the energy balance of zonal flows

Here we consider another application of invariant solutions $\psi^{[1]}$ and $\psi^{[2]}$ given by (7.29) and (7.37) for the nonlinear Euler model (7.15) describing nonviscous atmosphere dynamics in a thin rotating spherical shell affected by a presence of latitude-dependent zonal flows. Our main focus of interest here is to investigate the asymptotic behavior of sources and sinks associated with the effects of zonal flows (which can also serve as a model of westerly winds, which originate from the high-pressure areas in the horse latitudes and tend towards the poles and steer extratropical cyclones in this general manner) on the energy balance of atmospheric waves progressing in the Earth's atmosphere, which are captured by the rotation of the earth. The inquiry is motivated by dynamically significant zonal flows due to Coriolis forces in modeling of large-scale oceanic and atmospheric motion with applications to meteorology, climate variability models, the general atmospheric circulation model, weather prediction, and a variety of applications to large-scale atmosphere effects.

An energy balance refers to an equation which relates the increase or decrease of a quantity to terms representing supply or destruction. In a continuous medium, the typical form of a balance equation is written as (see, e. g., [40])

$$\frac{\partial P}{\partial t} + \operatorname{div} Q = S, \tag{7.60}$$

in which P is the quantity being conserved, Q is the flux (i. e., a transport of P within the medium), and S represents a source if $S > 0$ and a sink if $S < 0$. In the particular case when $S = 0$, the balance equation (7.60) yields the *conservation law*.

Let us write the Euler model (7.15) in the form (7.60). Since

$$Q = \left(\frac{\partial p}{\partial \theta} u_\theta + \frac{1}{\sin\theta} \frac{\partial p}{\partial \phi} u_\phi \right), \tag{7.61}$$

we can write the remaining terms P and S in equation (7.60) as

$$P = \frac{1}{2}(u_\theta^2 + u_\phi^2),$$

$$S = F\left(u_\theta \frac{\partial u_\theta}{\partial \phi} + u_\phi \frac{\partial u_\phi}{\partial \phi} - u_\theta u_\phi \cos\theta \right) + \frac{dF}{d\theta} u_\theta u_\phi \sin\theta \tag{7.62}$$

$$+ \left(\frac{\partial u_\theta}{\partial \phi} + \frac{\partial u_\phi}{\partial \phi} + \frac{\partial u_\phi}{\partial \theta} \sin\theta \right) u_\theta u_\phi + u_\theta^2 \frac{\partial u_\theta}{\partial \theta} \sin\theta, \tag{7.63}$$

which means that the source/sink terms consist exclusively of zonal flows $F(\theta)$ as well as of nonlinear terms.

The primary focus is to investigate the asymptotic behavior of the source/sink terms $S^{[1]}$ and $S^{[2]}$ as $t \to \infty$, where $S^{[1]}$ and $S^{[2]}$ are the source/sink terms associated with the first and second invariant solutions $\psi^{[1]}$ and $\psi^{[2]}$, respectively, of the nonlinear atmospheric model (7.15), describing incompressible fluid flows within an infinitely thin rotating spherical shell.

The main result can be formulated as follows:

$$\begin{array}{ll} S^{[1]} \equiv 0 \quad \Rightarrow & \textbf{Conservation law}, \\ S^{[2]} \neq 0; \text{ however,} & \begin{cases} S^{[2]} \underset{t\to\infty}{\to} 0, & \text{if } F(\theta) = -\frac{\sin\theta}{2R_0}, \\ S^{[2]} \underset{t\to\infty}{\to} S^*, & \text{if } F(\theta) \text{ is arbitrary,} \end{cases} \end{array} \tag{7.64}$$

where $S^* = \text{const}$.

This results means that, when evaluated on the first invariant solution $\psi^{[1]}$, the balance equation (7.60) represents the conservation law since $S^{[1]} \equiv 0$. But when evaluated on the second invariant solution $\psi^{[2]}$, the balance equation (7.60) represents the conservation law asymptotically in the sense that $\lim_{t\to\infty} S^{[1]} = 0$ for a very particular form of zonal flow,

$$F(\theta) = -\frac{\sin\theta}{2R_0}. \tag{7.65}$$

Finally, when $F(\theta)$ is arbitrary, the balance equation (7.60) evaluated on the second invariant solution $\psi^{[2]}$ does not provide the conservation law precisely, although it represents some "averaged" form of the conservation law in an asymptotic sense.

It would be interesting to give a physical interpretation to the question why, when evaluated on the first invariant solution $\psi^{[2]}$, the balance equation (7.60) can be associated with the conservation law only for that special class of zonal flows as given by (7.65). Thus, it would be an interesting task to better understand the role of the equatorial flows of the form $F(\theta)$ on the global atmospheric dynamics. One particular question of interest is to visualize the energy density on a surface of a sphere with different forms of the flows $F(\theta)$ and investigate if the particular form (7.65) can be used to get an analytic approximation of the equatorial Kelvin wave as can be seen traveling eastward along the equator, as shown in Figure 7.1.

7.7 Concluding remarks

In this modeling scenario, we have found the exact solutions to the Navier–Stokes equations in a thin rotating spherical shell in terms of elementary functions. Particularly, it has been shown here that the exact solutions were found as the extension of the invariant solutions for the corresponding Euler equations. Namely, as one can check by direct substitution, the governing equations (7.43)–(7.43) are invariant with respect to the obvious translation $\overline{\phi} = \phi + a$ of the angle ϕ, where the constant a is the group parameter. Apparently, it so happens that application of the *approximate equivalence transformations* shows that the Navier–Stokes equations are also invariant with respect to the one-parameter transformation group of a more complex form, namely,

$$\overline{t} = te^{2\mu R_0 a}, \quad \overline{\theta} = \theta, \quad \overline{\phi} = \phi + \frac{t}{2R_0}(1 - e^{2\mu R_0 a}),$$

$$\overline{\psi} = \psi e^{-2\mu R_0 a} + \frac{\cos\theta - 2R_0 H(\theta)}{2\mu R_0}(1 - e^{-2\mu R_0 a}).$$

The observation served to construct the exact solution presented in entry [7] of Table 7.1. We also remark that the above transformations are valid for the Euler model (7.15), which agrees with the general conclusions on the vanishing viscosity in Cauchy's problem for hydrodynamics equations (*Golovkin* [32]). The summary of the exact solutions corresponding to the limiting cases of zero/nonzero viscosity and zero/nonzero rotation is also shown schematically in Figure 7.3 for different choices of $F(\theta)$.

The exact solutions discussed here are used to describe physically relevant zonal flows. For example, the presence of the Coriolis force creates a cyclonic rotation around the poles, i. e., West-to-East winds. From a mathematical standpoint, the velocity and the pressure terms are unbounded in the neighborhood of the pole. Unbounded terms

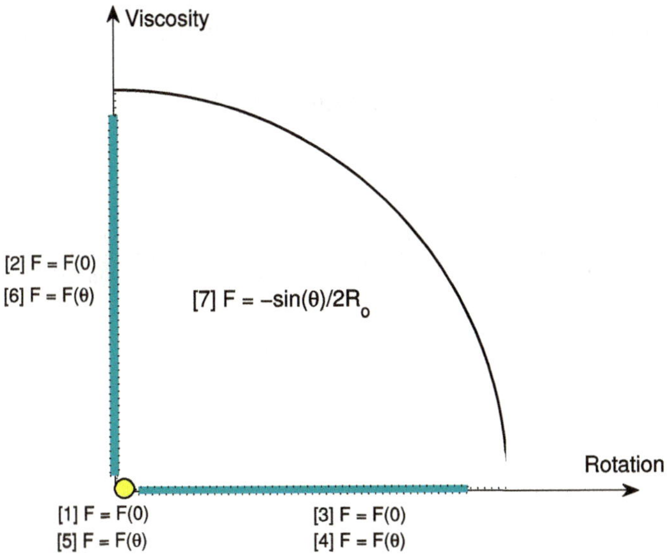

Figure 7.3: Discussion of the invariant solutions $\psi^{[1]}$ and $\psi^{[2]}$ corresponding to the entries in Table 7.1 depending on the limiting case.

are common in differential equations posed in spherical coordinates and it introduces a host of computational problems that are collectively known as the "pole problem" (see, e. g., [74]).

7.8 Exercises

Exercise 7.1. Show that two functionally independent invariants of

$$X_2 = \frac{\partial}{\partial \phi}, \quad X_4 = 2R_0\mu t\frac{\partial}{\partial t} - \mu t\frac{\partial}{\partial \phi} + (\cos\theta - 2R_0H(\theta) - 2\mu R_0\psi)\frac{\partial}{\partial \psi}$$

are given by

$$\xi = \theta, \quad \mu = t\left(\psi + \frac{H(\theta)}{\mu} - \frac{\cos\theta}{2\mu R_0}\right).$$

Exercise 7.2. Show that three functionally independent invariants of the operator

$$X_4 = 2R_0\mu t\frac{\partial}{\partial t} - \mu t\frac{\partial}{\partial \phi} + (\cos\theta - 2R_0H(\theta) - 2\mu R_0\psi)\frac{\partial}{\partial \psi}$$

are given by

$$\theta, \quad \nu = t + 2R_0\phi, \quad \sigma = t\left(\psi + \frac{H(\theta)}{\mu} - \frac{\cos\theta}{2\mu R_0}\right).$$

Exercise 7.3. Find a one-parameter group of transformations generated by the operator

$$X = 2R_0\mu t\frac{\partial}{\partial t} - \mu t\frac{\partial}{\partial \phi} + (\cos\theta - 2R_0H(\theta) - 2\mu R_0\psi)\frac{\partial}{\partial \psi}.$$

Exercise 7.4. Find the commutator $[X, Y]$, where

$$X = \frac{\partial}{\partial t}, \quad Y = \mu\sin\left(\phi + \frac{t}{2R_0}\right)\frac{\partial}{\partial \theta} + \mu\frac{\cos\theta}{\sin\theta}\frac{\partial}{\partial \phi} + \left(F(\theta) + \frac{\sin\theta}{2R_0}\right)\sin\left(\phi + \frac{t}{2R_0}\right)\frac{\partial}{\partial \psi}.$$

8 Modeling scenario 5: invariant solutions as ocean whirlpools

In this modeling scenario we will derive the solution of the linearized equations for internal gravity waves confined in a volume with a cylindrical sidewall. The analysis is applied to geophysical flows, where the effect of rotation due to Coriolis force is a prominent feature. The effects of the different aspect ratios (i. e., the radius and the height) of the cylindrical container on the internal wave field are presented in this chapter with the goal of extension to a fully nonlinear case. We will find the exact solutions of three-dimensional equations of motion for internal gravity waves in cylindrical coordinates in unbounded media by means of *approximate transformation groups* of equations with a small parameter. Introduction of the small parameter has been motivated by justifying the analogy of the Kelvin hypothesis on vanishing of the velocity component u_r normal to the wall for the rectilinear motion. In our case the radial component u_r of the velocity vector is nonzero and achieves its maximum in the interior, which also agrees with analytical predictions [28]. However, as we shall see from the linear analysis, u_r can be considered to be small in the limiting case when the aspect ratio $\sigma = H/r_0$ is small, in which H and r_0 are the basin's depth and radius, respectively. As a particular application to ocean and atmospheric modeling, in terms of linear modeling, the time series of the energy density will be visualized as spinning patterns that appear to be rotating in a counterclockwise sense when looking at the North Pole from above. Such spinning patterns could be compared with the flow around a low-pressure area that is usually being linked with modeling of hurricanes. In terms of zero-order approximate transformations, the invariant solutions were visualized as funnels having something in common with the geometric structure of oceanic whirlpools.

8.1 Preliminaries

The discovery of two giant whirlpools in the Atlantic Ocean, off the coast of Guyana and Suriname (shown in Figure 8.1), was sensational because this part of the ocean has been studied thoroughly, and no one expected anything like that to appear in the area.

To date, it is still not known where the whirlpools came from and what surprises they may bring to people. However, it is recognized that the oceanic whirlpools play a key role in global climate as well, transporting ocean heat from the equator northward and eventually feeding into the Gulf Stream system. Additionally, such whirlpools influence the atmosphere and form cyclonic air mass (source: www.TelevisionFanatic.com (2011)).

In this chapter, we will model oceanic and atmospheric whirlpools by means of internal waves confined by a cylindrical boundary affected by the rotation of the Earth. For example, such waves can be found in the lake *Guatavita* (Spanish: Laguna de Guatavita or Lago Guatavita), which is located in the Cordillera Oriental of the Colombian Andes

https://doi.org/10.1515/9783111387499-008

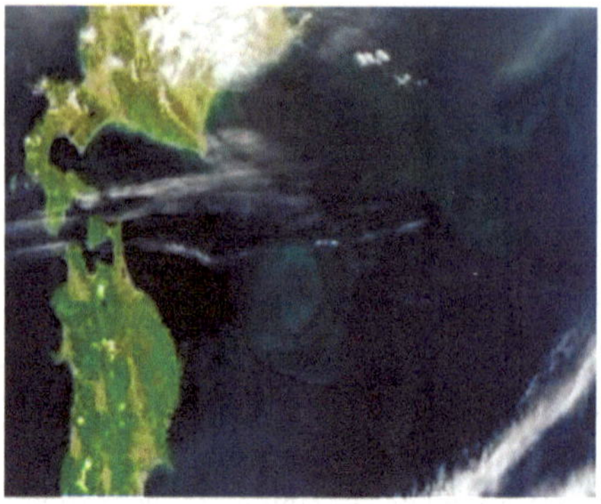

Figure 8.1: Two giant whirlpools in the Atlantic Ocean, off the coast of Guyana and Suriname. Source: watchers.news (2011).

Figure 8.2: Guatavita lagoon.

in the municipality of Sesquilé, in Almeidas Province, Cundinamarca, Colombia, 57 kilometers (35 miles) North-East of Bogotá, the capital of Colombia (see Figure 8.2). The lake is circular and has a surface area of 19.8 hectares. The crater's origin still remains unexplained. The lake is now a focus of ecotourism, and its association with the legend of El Dorado is also a major attraction.

Today the consensus is that internal waves with effects of the Earth's rotation play an important role in the dynamics of the ocean, especially in the large-scale general circulation, and they are responsible for a large fraction of the mixing and energy exchange occurring in the deep ocean. Understanding of internal waves is also essential since they are potentially hazardous to all subsea operations, including oil and gas drilling operations, and have already caused several costly and dangerous incidents. Particularly, the effects of internal waves on the Deepwater Horizon incident were studied recently in Ref. [36].

Fluids that are both rotating and stratified are known to possess a fascinating variety of wave motions. The applicability of our analysis is demonstrated by comparing the results with available analytical and numerical solutions reported in previous studies for similar situations of stratified flows. One of the purposes of the linear analysis is to characterize the effects of the basin's aspect ratios on rotating stratified fluid oscillations in order to justify the introduction of a small parameter. Also, as will be shown in the sequel, the analysis presented in this work allows to associate the linear model with the latitude-dependent whirlpools that do exist in the ocean and atmosphere. Particularly, the time series of the resulting energy density for baroclinic Kelvin waves were visualized as spinning patterns that appear to be rotating in a counterclockwise sense when looking at the North Pole from above. Such spinning patterns have something in common with the flow around a low-pressure area that is usually linked with modeling of hurricanes or whirlpools in the atmosphere, as shown in Figure 8.3. Surprisingly, there is also some similarity to the flow visualization as the azimuthal waves developed in the cylindrical container that was investigated in the experimental studies in [48], where the authors experimentally investigated the dynamics of surface waves excited by oscillations from a cylindrical sidewall. It is also demonstrated in this work that the zero-order invariant solutions of the nonlinear Boussinesq model can be visualized as

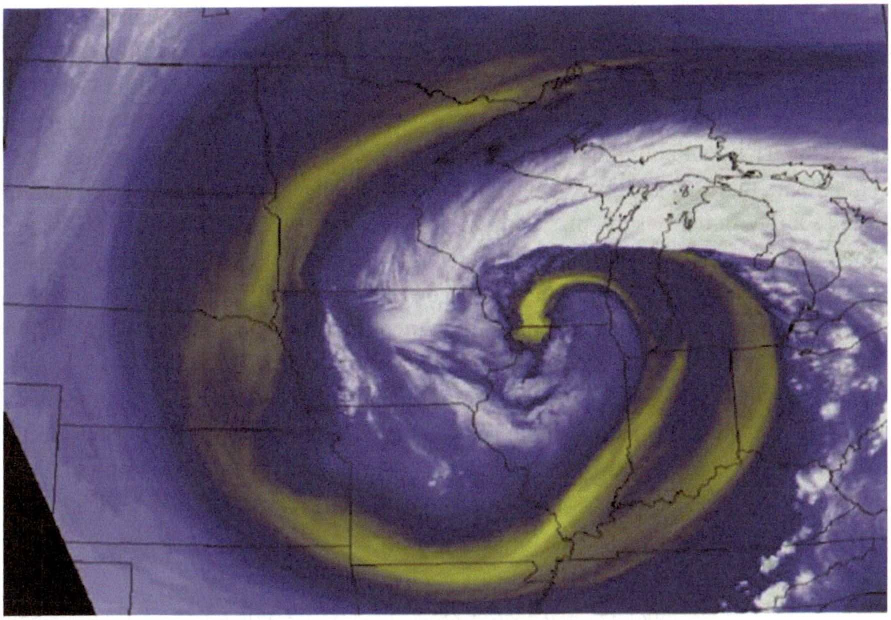

Figure 8.3: A monster whirlpool in the atmosphere. This is an image from the NOAA GOES satellite showing not clouds but water vapor (yellow, dry; white, moist) in a closed giant vortex over the Midwest. It is a huge whirlpool of air in the atmosphere. Some scientists believe these "cut off" circulations will occur more often in a changing climate.

funnels that can be associated with oceanic whirlpools. At present, it remains an open issue where the whirlpools came from and what surprises they may bring.

In this modeling scenario, the variety of exact solutions for nonlinear governing equations for internal waves in cylindrical coordinates will be found by means of *approximate transformation groups and approximate symmetries* of equations with small parameter. The algorithm used in this work is based on the ASP symmetry packages (*Jefferson & Carminati* [42]) and DESOLVII (*Vu, Jefferson, & Carminati* [77]) to find both the approximate symmetries and approximately invariant solutions of the governing equations in question. Particularly, the radial component of the velocity vector is assumed to be small, which is justified by linear analysis of baroclinic Kelvin waves contained in a cylindrical basin. Methods of classical group analysis allow to single out symmetries with remarkable properties among all equations of mathematical physics. Unfortunately, any small perturbation of an equation breaks the admissible group of transformations and reduces the applied value of these "refined" equations and group theoretical methods in general. Therefore, the development of group analysis methods that are stable with respect to small perturbations of differential equations has become vital.

8.2 Model equations of internal waves

We address the nonlinear equations of motion for internal waves affected by the Earth's rotation. The axes are x (assumed eastward for definiteness) and y (northward), and $\widehat{k}$ is the unit vector with z in the vertical direction (opposite gravity). The fluid velocity is $\vec{u} = (u, v, w)$ relative to the Cartesian coordinate system (x, y, z). Motion is considered to be outside frictional boundary layers, allowing viscosity and diffusion to be neglected. In the theory of ocean circulation and within meteorological applications, it is commonplace to apply Boussinesq approximation, in which the full variation of density is retained only in the buoyancy force in the vertical momentum balance. Wherever else it occurs, in the horizontal momentum balance and in the continuity equation, density is replaced by a constant (in the simplest implementation). In particular, this means that the three-dimensional velocity field is assumed to be solenoidal. The analysis is applied to geophysical flows, where the effect of rotation due to Coriolis force is a prominent feature. Thus, within the Boussinesq approximation (the density variations are neglected everywhere except in the gravitational term), the governing equations of motion for internal waves observed in a system of coordinates rotating with angular velocity $\vec{\Omega}$ are written in the form (see, e. g., [31])

$$\rho_0\left[\frac{\partial \vec{u}}{\partial t} + \vec{u}\cdot\nabla\vec{u} + 2\vec{\Omega}\times\vec{u}\right] = -\nabla p - g\rho\widehat{k}, \tag{8.1}$$

$$\frac{\partial \rho}{\partial t} + \vec{u}\cdot\nabla\rho + w\frac{d\bar{\rho}}{dz} = 0, \tag{8.2}$$

$$\nabla\cdot\vec{u} = 0, \tag{8.3}$$

where g is the acceleration due to gravity, the term $2\vec{\Omega} \times \vec{u}$ is referred to as the *Coriolis acceleration,* and the exact ocean density $\rho(x, y, z, t)$ is divided into a constant part ρ_0 (the *reference density*), a part $\overline{\rho}(z)$ (the *ambient equilibrium stratification*) that varies only with depth, and a perturbation $\rho'(x, y, z, t)$ which represents the density variation caused by the waves. We thus write

$$\rho = \rho_0 + \overline{\rho}(z) + \rho'(x, y, z, t), \tag{8.4}$$

in which ρ_0 is the constant reference density, $\overline{\rho}(z)$ is a background stable density profile with the associated buoyancy frequency N defined by

$$N^2 = -\frac{g}{\rho_0}\frac{d\overline{\rho}}{dz}, \tag{8.5}$$

and we require $\rho_0 + \overline{\rho}$ and $\overline{p}$ to be consistent with the state of rest, i. e.,

$$\frac{d\overline{p}}{dz} = -(\rho_0 + \overline{\rho})g. \tag{8.6}$$

The quantity N measures the degree of density stratification of a fluid with average potential density $\rho(z)$ and thus represents the frequency with which a vertically displaced fluid element would be expected to oscillate because of restoring buoyancy forces. If the displacement is not strictly vertical, as in the case of internal waves, the restoring force is less, so the frequency of oscillations is reduced.

Since typical oceanic values are

$$\rho_0 \sim 1.04\,\mathrm{g\,cm^{-3}}, \quad |\overline{\rho}| \sim 0.03\,\mathrm{g\,cm^{-3}}, \quad |\rho'| \sim 0.003\,\mathrm{g\,cm^{-3}},$$

we can assume $|\rho'|, |\overline{\rho}| \ll \rho_0$ and the Boussinesq approximation is quite accurate. Additionally, the total pressure is given by

$$p(x, z, t) = \overline{p}(z) + p'(x, z, t), \tag{8.7}$$

where p' is the pressure deviation from the basic state

$$\overline{p}(z) = P_0 - \rho_0 g z - g\int_0^z \overline{\rho}(\xi)d\xi, \tag{8.8}$$

in which $P_0 = \text{const}$. The traditional f-plane approximation is applied, whereby we take $2\vec{\Omega} = (0, 0, f)$, where f is the inertial frequency which depends on the rotation rate of the Earth (angular velocity $\Omega = 2\pi$ rad/day $\approx 0.73 \times 10^{-4}\,\mathrm{s^{-1}}$).

In summary, the Boussinesq approximation, rooted on the assumption that the density does not depart much from a mean value, has allowed us to replace the exact density ρ by its reference value ρ_0 everywhere, except in front of the gravitational acceleration and in the energy equation, which has become an equation governing density variations.

Let there be a cylindrical model basin with radius r_0 and depth H with $z \in [0, H]$. We employ the usual cylindrical coordinates (r, θ, z) via $x = r\cos\theta, y = r\sin\theta, z = z$ so that the rate-of-strain tensor is

$$e_{rr} = \frac{\partial u_r}{\partial r}, \quad e_{\theta\theta} = \frac{1}{r}\frac{\partial u_\theta}{\partial \theta}, \quad e_{r\theta} = \frac{r}{2}\frac{\partial}{\partial r}\left(\frac{u_\theta}{r}\right) + \frac{1}{2r}\frac{\partial u_r}{\partial \theta}.$$

For completeness, we write explicitly the three-dimensional nonlinear Euler equations of motion (8.1)–(8.3) in the cylindrical domain within the Boussinesq approximation:

$$\frac{\partial u_r}{\partial t} + u_r\frac{\partial u_r}{\partial r} + \frac{u_\theta}{r}\frac{\partial u_r}{\partial \theta} + w\frac{\partial u_r}{\partial z} - \frac{u_\theta^2}{r} - fu_\theta = -\frac{\partial p}{\partial r}, \tag{8.9}$$

$$\frac{\partial u_\theta}{\partial t} + u_r\frac{\partial u_\theta}{\partial r} + \frac{u_\theta}{r}\frac{\partial u_\theta}{\partial \theta} + w\frac{\partial u_\theta}{\partial z} + \frac{u_r u_\theta}{r} + fu_r = -\frac{1}{r}\frac{\partial p}{\partial \theta}, \tag{8.10}$$

$$\frac{\partial w}{\partial t} + u_r\frac{\partial w}{\partial r} + \frac{u_\theta}{r}\frac{\partial w}{\partial \theta} + w\frac{\partial w}{\partial z} = -\frac{\partial p}{\partial z} - \rho g, \tag{8.11}$$

$$\frac{\partial \rho}{\partial t} + u_r\frac{\partial \rho}{\partial r} + \frac{u_\theta}{r}\frac{\partial \rho}{\partial \theta} = \frac{N^2}{g}w, \tag{8.12}$$

$$\frac{1}{r}\frac{\partial (ru_r)}{\partial r} + \frac{1}{r}\frac{\partial u_\theta}{\partial \theta} + \frac{\partial w}{\partial z} = 0. \tag{8.13}$$

Our model is idealized by assuming N to be uniform over the extent of the fluid, which corresponds to a vertically linear density variation. While this simplification is commonly used in laboratory and theoretical studies and it is quite reasonable for the thermocline region, it is not common in the deep regions of the ocean except when considering short wavelengths in comparison with the scale of density changes. At low frequencies, close to f, rotational effects are important. Such internal waves are sometimes called inertial-internal waves.

We derive the exact solutions for the nonlinear model (8.9)–(8.13) in a cylindrical wave field under the assumption that $u_r \ll 1$.

8.2.1 Linear model and utilization of the Kelvin hypothesis for $u_r \ll 1$

In terms of linear modeling, we assume that the motions are small deviations from the motionless state, which can be regarded as only z-dependent for sufficiently small Froude number $F_R = f^2L/g$, where L is a characteristic horizontal length scale (see, e. g., [28]). Then the smallness of the deviation can be attributed as a Rossby number characterizing the magnitude of the initial perturbations. Further, in order to facilitate analytical solitons in a closed form, our model is idealized by assuming N to be uniform over the extent of the fluid. This corresponds to a vertically linear density variation. While this simplification is commonly used in laboratory and theoretical studies and it is quite reasonable for the deep ocean, it is not common in the thermocline region of the ocean except when considering short wavelengths relative to the scale of density changes. At low frequencies, close to f, rotational effects are important. Such internal waves are sometimes called inertial-internal waves. At high frequencies, close to N and far from f, rotational effects are negligible. So, under the assumption of a small Froude number F_R, the linearized Boussinesq model (8.15)–(8.25) is written as

$$\frac{\partial u_r}{\partial t} - fu_\theta = -\frac{\partial p}{\partial r}, \tag{8.14}$$

$$\frac{\partial u_\theta}{\partial t} + fu_r = -\frac{1}{r}\frac{\partial p}{\partial \theta}, \tag{8.15}$$

$$\frac{\partial w}{\partial t} = -\frac{\partial p}{\partial z} - \rho g, \tag{8.16}$$

$$\frac{\partial \rho}{\partial t} = \frac{N^2}{g} w, \tag{8.17}$$

$$\frac{1}{r}\frac{\partial (ru_r)}{\partial r} + \frac{1}{r}\frac{\partial u_\theta}{\partial \theta} + \frac{\partial w}{\partial z} = 0. \tag{8.18}$$

We write the bottom boundary conditions at $z = 0$ and the rigid lid approximation for the free surface boundary condition at $z = H$ in terms of p as

$$p_z(r, \theta, 0, t) = p_z(r, \theta, H, t) = 0. \tag{8.19}$$

Also, in order to model the Kelvin waves, we employ the boundary condition of no normal flow at the coastline located at $r = r_0$. We rewrite equations (8.14) and (8.15) as the following single PDE:

$$\left(\frac{\partial^2}{\partial t^2} + f^2\right) u_r = -\frac{f}{r}\frac{\partial p}{\partial \theta} - \frac{\partial^2 p}{\partial t \partial r} = 0 \quad \text{when } r = r_0. \tag{8.20}$$

Thus, we need to solve the system of PDEs (8.14)–(8.18) subject to the boundary conditions (8.19) and (8.20).

8.2.2 Effects of rotation on oscillatory modes

Elimination of the pressure terms from (8.14) and (8.15) and expressing $\partial w/\partial z$ from (8.17) yields

$$\frac{\partial \varphi}{\partial t} - \frac{frg}{N^2}\frac{\partial^2 \rho}{\partial t \partial z} = 0, \tag{8.21}$$

where the vorticity φ is defined as

$$\varphi = \frac{\partial(ru_\theta)}{\partial r} - \frac{\partial u_r}{\partial \theta}. \tag{8.22}$$

We next define

$$F = \varphi - \frac{frg}{N^2}\frac{\partial \rho}{\partial z}. \tag{8.23}$$

As follows from (8.21), F can be regarded as time-independent, i. e., $F = F(r, \theta, z)$. Although $F \neq 0$ for steady geostrophic flows, since we are looking for solutions periodic in time with zero mean, hereafter we set $F = 0$. In the frame of the present analysis, we recast the model (8.14)–(8.18) as a single equation for the pressure perturbation, which is commonly done in dynamical meteorology. The resulting equation is

$$\frac{\partial^2}{\partial t^2}\Delta p + N^2 \Delta_h p + f^2 \frac{\partial^2 p}{\partial z^2} = 0, \tag{8.24}$$

where

$$\Delta_h = \frac{1}{r}\frac{\partial}{\partial r}\left(r\frac{\partial}{\partial r}\right) + \frac{1}{r^2}\frac{\partial^2}{\partial \theta^2} \tag{8.25}$$

is the two-dimensional Laplace–Beltrami operator in polar angles and

$$\Delta = \Delta_h + \frac{\partial^2}{\partial z^2}. \tag{8.26}$$

We employ the usual normal mode analysis and allow solutions to be found as expansions of vertical modes $\psi_n(z)$ of the form

$$p = \sum_n p_n(r, \theta, t)\psi_n(z), \tag{8.27}$$

where the uniform mean vertical stratification $\overline{\rho}(z)$ is absorbed into the definition of the pressure and density so that p has units of (potential) energy per unit mass and ρ represents (dimensionless) density anomaly [57].

In view of the boundary condition (8.19), it is more convenient to look for a specific example of the idealized vertical profile $\psi_n(z)$, e. g.,

$$p = \sum_n \widehat{p}_n(r)\cos(n\theta - \omega_n t)\cos(m_n z), \tag{8.28}$$

where m_n is the vertical wave number vector. The first boundary condition (8.19) is then automatically satisfied with the choice of the sinusoidal vertical structure (8.28), whereas the second boundary condition (8.19) implies that there exist an infinite number of discrete internal wave modes satisfying the dispersion relation for internal waves

$$\omega_n^2 = \frac{\alpha_n^2 N^2 + m_n^2 f^2}{m_n^2 + \alpha_n^2}, \quad m_n H = n\pi, \quad n = \pm 1, \pm 2, .., \tag{8.29}$$

which also agrees with the asymptotic expression obtained by Krauss [45], who studied internal waves in rectangular geometry with zero rotation. Here α_n is a parameter having the dimension of the azimuthal wave number vector.

Substitution of the presentation (8.28) into equation (8.24) yields an ODE for $\widehat{p}_n(r)$:

$$\frac{d^2\widehat{p}_n}{dr^2} + \frac{1}{r}\frac{d\widehat{p}_n}{dr} + \left[\frac{m_n^2(\omega_n^2 - f^2)}{N^2 - \omega_n^2} - \frac{n^2}{r^2}\right]\widehat{p}_n = 0. \tag{8.30}$$

In the mid/high-latitude upper ocean, N is typically one or two orders of magnitude larger than f ($N \sim 10^{-3}$ to $10^{-2}\,\mathrm{s}^{-1}$; $f \sim 10^{-4}\,\mathrm{s}^{-1}$), so we limit our analysis to the case where

$$|f| < |\omega_n| < |N|, \tag{8.31}$$

which implies that

$$\frac{m_n^2(\omega_n^2 - f^2)}{N^2 - \omega_n^2} = \alpha_n^2 > 0. \tag{8.32}$$

We next define a new independent variable ξ via $\xi = \alpha_n r$ and regard $\widehat{p}$ as a function of ξ. Thus, equation (8.30) is written as the Bessel equation of order n with the single singular point at $\xi = 0$ [78]:

$$\xi^2\frac{d^2\widehat{p}_n}{d\xi^2} + \xi\frac{d\widehat{p}_n}{d\xi} + (\xi^2 - n^2)\widehat{p}_n = 0. \tag{8.33}$$

Correspondingly, the solution of equation (8.33) is given by the Bessel function of the first kind of order n:

$$\widehat{p}_n(\xi) = J_n(\xi) = \left(\frac{\xi}{2}\right)^n \sum_l \frac{(-1)^l}{l!(l+n)!}\left(\frac{\xi}{2}\right)^{2l}. \tag{8.34}$$

Thus, away of the coastal boundary $r = r_0$, the solution of equation (8.30) is written as

$$p(r, \theta, z, t) = \sum_n J_n(\alpha_n r)\cos(n\theta - \omega_n t)\cos(m_n z). \tag{8.35}$$

By substituting the solution (8.35) into equation (8.20) and evaluating at $r = r_0$, we obtain the following transcendental equation:

$$G_n := fnJ_n(\xi_n^0) - \xi_0\omega_n J_n'(\xi_n^0) = 0, \tag{8.36}$$

in which

$$\xi_n^0 = \alpha_n r_0. \tag{8.37}$$

Equation (8.36) guarantees that an infinite set of oscillations exists provided that ω_n and ξ_n^0 are related by the equation

$$\omega_{n,k} = \pm\left(\frac{N^2[\xi_{n,k}^0]^2/r_0^2 + m_n^2 f^2}{m_n^2 + [\xi_{n,k}^0]^2/r_0^2}\right)^{\frac{1}{2}} \tag{8.38}$$

for each solution number k. Thus, equation (8.36) serves as an additional condition for the frequencies $\omega_{n,k}$ of the solution (8.35) obeying the condition (8.31). The frequencies $\omega_{n,k}$ satisfying the dispersion relation (8.38) are called eigenfrequencies. These are the frequencies for which the solution of the model (8.14)–(8.18) exists subject to the boundary conditions (8.19).

The eigenfrequencies $\omega_{n,k}$ are determined as follows. First, we graphically find the roots $\xi_{n,k}^0$ ($k = 1, 2, 3, \ldots$) of the equation $G_n = G_{n,k} = 0$ given by (8.36). We remark that the numerical values of the roots $\xi_{n,k}^0$ given by (8.37) do not depend on the particular choice of the basin's depth H or the radius r_0, but only depend on the aspect ratio $\sigma = H/r_0$. In view of this remark, most of our further simulations were done for fixed σ. Particularly, without loss of generality we let hereafter $\sigma < 1$ by setting $H = 10^3$ [m] and $r_0 = 10^4$ [m] so that our further simulations were done for the aspect ratio $\sigma = 0.1$. However, the method presented here can also be easily adjusted for the aspect ratios $\sigma > 1$ or $\sigma = 1$. Additionally, for the purpose of visualization of the analytic solutions, we consider the case when $\xi \in [0, \xi_{\max}]$, and we further set $\xi_{\max} = 100$.

Next, for each root $\xi_{n,k}^0$, the corresponding eignefrequencies $\omega_{n,k}$ were obtained from the dispersion relation (8.38). In the case when $N^2 = 0$, the eigenvalues of the similar eigenvalue problem (8.36)–(8.38) have been obtained numerically in Kudlick [46]. Note that ω can either be positive or negative. For the purpose of distinctness, without loss of generality, hereafter we choose the positive sign in (8.38).

We remark that there is a certain mode number cutoff $n_{\max}$ for the given $\xi_{\max}$. For example, when $\xi_{\max} = 100$, we can find 32 solutions ($k \leqslant k_{\max} = 32$) provided that $n \leqslant n_{\max} \leqslant 83$. Furthermore, there is an explicit latitude-dependent relation between the maximum solution number $k_{\max}$ and the mode cutoff number $n_{\max}$. Next, the effects of the Earth's rotation on the solutions of equation (8.36) are demonstrated in Figure 8.4, showing the distribution of the obtained eignefrequencies $\omega_{n,k}$ versus ξ within the admissible range of $k_{\max}$ and $n_{\max}$ at different latitudes. Namely, panel (a) shows the eigenfrequency distribution at latitude $\phi = 1°$ North, whereas panels (b), (c), and (d) correspond to the cases $\phi = 10°$, $\phi = 20°$, and $\phi = 70°$ North, respectively. The range of the obtained eigenfrequencies $\omega_{n,k}$ associated with the obtained values $\xi_{n,k}^0$ is embedded by the Coriolis parameter f (red line) and the buoyancy frequency N (blue line), i. e., $|f| < |\omega_{n,k}| < |N|$, in which $\omega_{n,k} = f$ and $\omega_{n,k} = N$ correspond to the limiting (horizontal and vertical) cases of internal wave propagation. The obtained set of $\omega_{n,k}$ represents the bounded and monotonic (and thus convergent) sequence of the eigenfrequencies such that $\lim_{\xi\to\infty} \omega_{n,k} = N$ and $\lim_{n\to\infty} \omega_{n,k} = f$, as we observe from Figure 8.4. Particularly, the latter limit agrees with the conclusion on higher harmonic internal waves merging

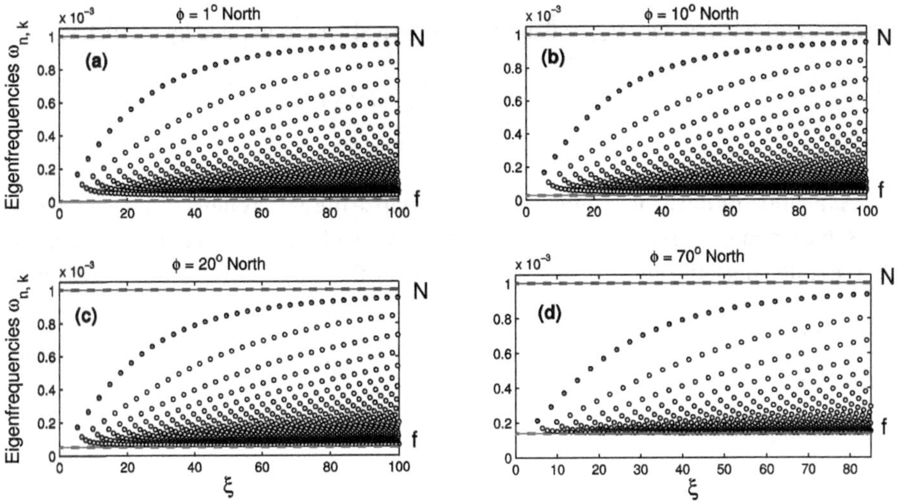

Figure 8.4: Distribution of the eigenfrequencies $\omega_{n,k}$ for $n \in [1, n_{\max}]$ and $k \in [1, k_{\max}]$ associated with $\xi_{\max} = 100$ at different latitudes. The domain of the existing eigenfrequencies is bounded by the Coriolis parameter f (red line) and the buoyancy frequency N (blue line) so that the eigenfrequencies band of internal Kelvin waves is limited, $|f| < |\omega_{n,k}| < |N|$.

when $\omega_{n,k} = f$ along the half-cones represented by possible internal wave group velocity vectors in terms of modeling of equatorial wave beams discussed in [52].

The corresponding eigenfunctions for the pressure are presented by

$$p(r, \theta, z, t) = \sum_{n=1}^{n_{\max}} \sum_{k=1}^{k_{\max}} J_n(\alpha_{n,k} r) \cos(n\theta - \omega_{n,k} t) \cos(m_n z), \tag{8.39}$$

where the eigenfrequencies are given by (8.38).

The truncated series is similar to (8.39), in which u_r, u_θ, w, and ρ are found as the corresponding eigenfunctions of the system (8.14)–(8.18), e. g.,

$$u_r(r, \theta, z, t) = \sum_{n=1}^{n_{\max}} \sum_{k=1}^{k_{\max}} \frac{\alpha_{n,k}}{\omega_{n,k}^2 - f^2} \left[\omega_{n,k} J_n' - \frac{fn}{\alpha_{n,k} r} J_n \right] \sin(n\theta - \omega_{n,k} t) \cos(m_n z), \tag{8.40}$$

$$u_\theta(r, \theta, z, t) = \sum_{n=1}^{n_{\max}} \sum_{k=1}^{k_{\max}} \frac{\alpha_{n,k}}{\omega_{n,k}^2 - f^2} \left[\frac{n\omega_{n,k}}{\alpha_{n,k} r} J_n - f J_n' \right] \cos(n\theta - \omega_{n,k} t) \sin(m_n z), \tag{8.41}$$

$$w(r, \theta, z, t) = \sum_{n=1}^{n_{\max}} \sum_{k=1}^{k_{\max}} \frac{m_n \omega_{n,k}}{N^2 - \omega_{n,k}^2} J_n \sin(n\theta - \omega_{n,k} t) \sin(m_n z), \tag{8.42}$$

$$\rho(r, \theta, z, t) = \sum_{n=1}^{n_{\max}} \sum_{k=1}^{k_{\max}} \frac{m_n N^2}{g(N^2 - \omega_{n,k}^2)} J_n \cos(n\theta - \omega_{n,k} t) \sin(m_n z), \tag{8.43}$$

for $k > 1$ and $n > 1$, where $a_{n,k} \neq 0$ are the numerical values defined by (8.32), i. e.,

$$a_{n,k} = +\left(\frac{m_n^2(\omega_{n,k}^2 - f^2)}{N^2 - \omega_{n,k}^2}\right)^{\frac{1}{2}}.$$

As an illustration to our simulations, we visualize the energy density associated with the Kelvin model (8.14)–(8.20). Within the framework of linear theory, in the absence of sources and sinks, the energy conservation law for the model (8.14)–(8.18) is written as

$$\frac{\partial}{\partial t}\left[\frac{1}{2}\left(u_r^2 + u_\theta^2 + w^2 + \frac{g^2\rho^2}{N^2}\right)\right] + \frac{1}{r}\frac{\partial}{\partial r}(ru_r p) + \frac{1}{r}\frac{\partial}{\partial \theta}(u_\theta p) + \frac{\partial}{\partial z}(wp) = 0. \tag{8.44}$$

The energy density

$$P = \frac{1}{2}\left(u_r^2 + u_\theta^2 + w^2 + \frac{g^2\rho^2}{N^2}\right) \tag{8.45}$$

is further associated with the truncated series similar to (8.40)–(8.43), in which u_r, u_θ, w, and ρ are found as the corresponding eigenfunctions of the system (8.14)–(8.18).

Figure 8.5 shows a snapshot of the energy density P versus depth z at different latitudes with the fixed parameter values $r = 5000$ [m] and $\theta = \pi/3$. The red line corresponds to the initial ($\tau = 0$) and the blue line corresponds to the final ($\tau = 100$) inertial period, where the ϕ-dependent inertial period is defined by $2\pi/f$ and the number of inertial periods is denoted by τ. First, we observe a convergence phenomenon on the time series at higher latitudes, that is, as τ increases from 0 to 100, the energy density tends to decrease monotonically from $P|_{\tau=0}$ to $P|_{\tau=100}$. It is also apparent that the magnitude of the energy density oscillation decreases at higher latitudes, that is, $P|_{\tau=0} \to P|_{\tau=100}$, as ϕ increases. We call this phenomenon the "*effect of rotational convergence.*" In addition to the phenomenon of rotational convergence, we observe the phenomenon of "*energy anomaly*" that manifests itself as the existence of the latitude-persistent region corresponding to approximately 600–700 [m] deep, where the energy density varies rapidly relative to the depth. In actuality, as follows from the further analysis, the latter "anomaly" is one of the fascinating effects of rotation and it can be explained by means of comparing the energy density at different latitudes ϕ versus z shown in Figure 8.5a and the maximum values of the energy density versus z at different values of polar angle θ shown in Figure 8.6b. For comparison purposes, the results in panels (a) and (b) were obtained from simulations using the same value of latitude, $\phi = 20°$ North. Particularly, we observe from panel (b) that the sharp energy variation (e. g., decrease) appears to have a circular (in a clockwise sense) nature.

Figure 8.7 shows the results of numerical simulations demonstrating the effect of the Earth's rotation on the time series of the energy density P versus radius. The energy densities at the initial ($\tau = 0$) and final ($\tau = 50$) inertial periods are plotted as red and blue lines, respectively. Particularly, panel (a) shows the energy density distribution at

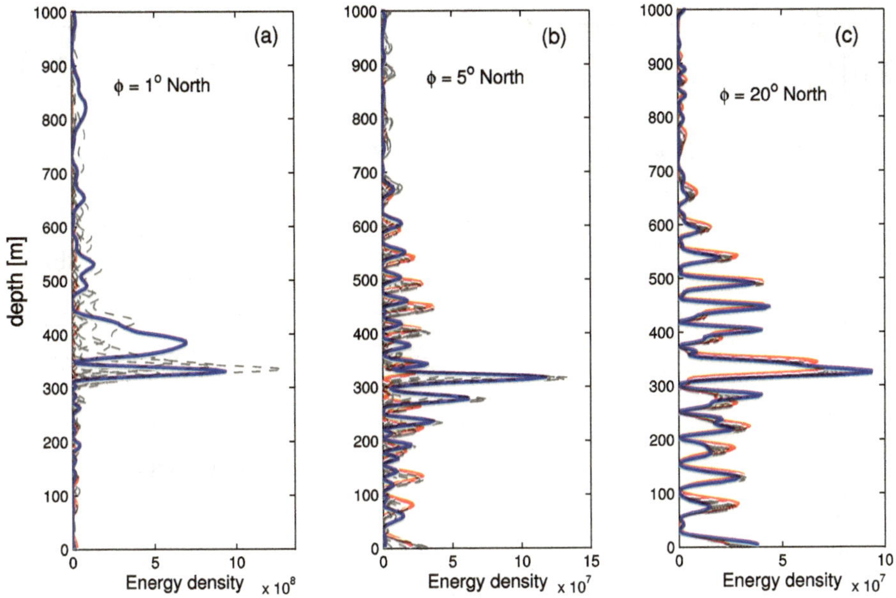

Figure 8.5: Time evolution of the energy density E versus depth z at different latitudes with the fixed parameter values $r = 5000$ m and $\theta = \pi/3$. The red line corresponds to the initial ($\tau = 0$) and the blue line corresponds to the final ($\tau = 100$) inertial period.

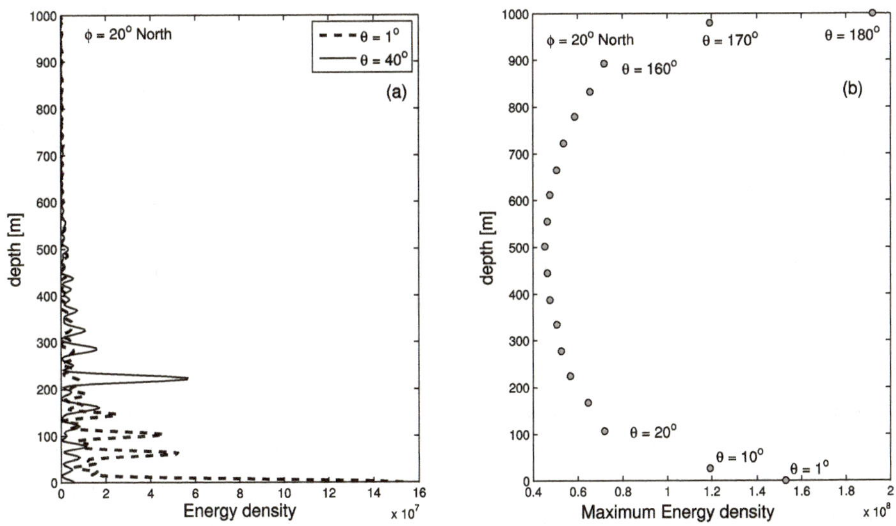

Figure 8.6: Prominent energy density peaks at different values of polar angle θ. Panel (a) shows the initial energy density (E) at values of $\theta = 1°$ (dashed line) and $\theta = 40°$ (solid line). The simulations were done at the value of latitude $\phi = 20°$ North for which we observed the convergence of the time series at initial and later times. Panel (b) shows the maximum values of the energy density at different values of θ varying monotonically from 1° to 180°.

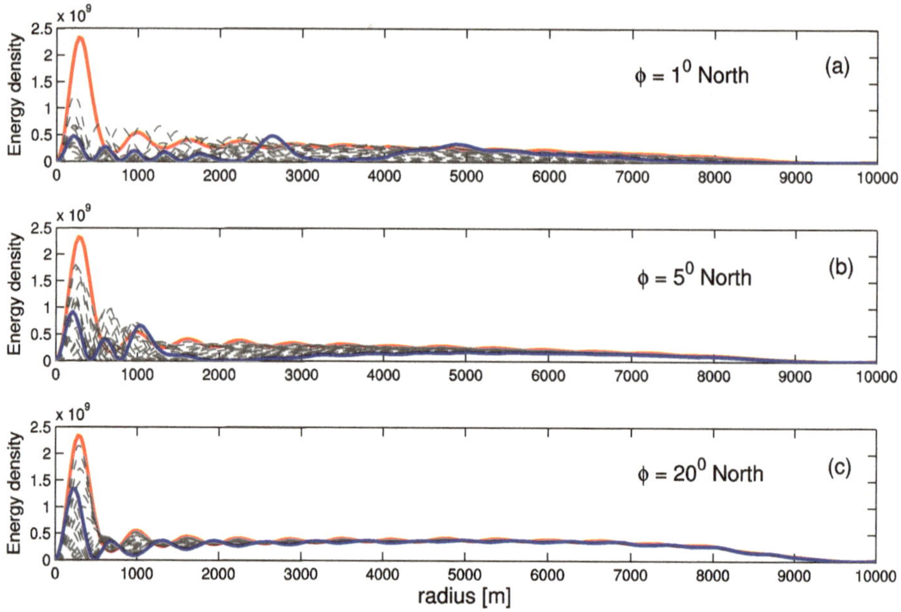

Figure 8.7: Time evolution of the energy density (P) versus radius r at different latitudes with the fixed parameter values $z = 333$ [m] and $\theta = \pi/3$. The red line corresponds to the initial ($\tau = 0$) and the blue line corresponds to the final ($\tau = 100$) inertial period.

latitude $\phi = 1°$ North, whereas panels (b) and (c) correspond to higher latitudes, i. e., $\phi = 5°$ and $\phi = 20°$ North, respectively. We observe that the most energetic region is closer to the origin. To some extent, such energetic peak can be associated with hole punch clouds (also known as punch hole clouds), as shown in Figure 8.8, showing a fallstreak hole, which represents a large circular gap that can appear in cirrocumulus or altocumulus clouds. It is believed that such holes are formed when the water temperature in the clouds is below freezing but the water has not frozen yet due to the lack of ice nucleation particles. When a portion of the water does start to freeze it will set off a domino effect, due to the Bergeron process, which is a process of ice crystal growth that occurs in mixed-phase clouds (containing a mixture of supercooled water and ice) in regions where the ambient vapor pressure falls between the saturation. This leaves a large, often circular, hole in the cloud. It is believed that a disruption in the stability of the cloud layer, such as that caused by a passing jet, may induce the domino process of evaporation which creates the hole. Because of the rarity and unusual appearance of a such cloud phenomena, fallstreak holes are often mistaken for unidentified flying objects or attributed to alien activity. As another illustration to our modeling, the results of numerical simulations presented in Figure 8.9 are used to demonstrate the contour plots showing snapshots of the equipotential curves of the energy density P. Similar spinning patterns and the development of a string of billows confined in a circular basin shown

Figure 8.8: Hole punch cloud: a fallstreak hole representing a large circular gap that can appear in cirrocumulus or altocumulus clouds (source: http://www.crystalinks.com/cloudholepunch.html).

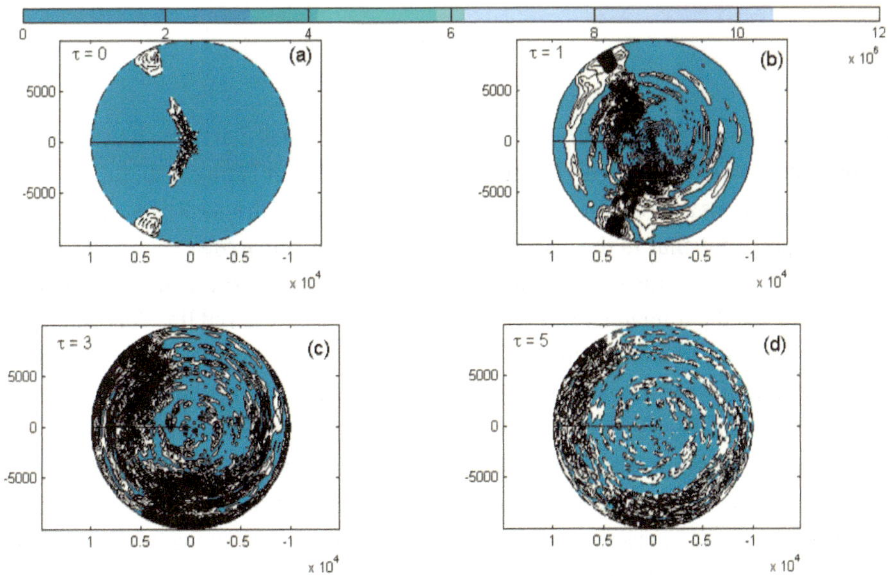

Figure 8.9: Snapshots of the contour plot of the energy density at fixed depth $z = 300$ [m] and fixed latitude $\phi = 20°$ North. Panel (a) shows the energy density at the initial inertial period $\tau = 0$ and panels (b), (c), and (d) correspond to $\tau = 1, 3$, and 5.

Figure 8.10: Schematic representation of flow around a low-pressure area (in this case, Hurricane Isabel) in the Northern Hemisphere. The picture can be found, e. g., at http://en.wikipedia.org/wiki/Hurricane_Isabel.

in Figure 8.9 can also be observed in the atmosphere. Particularly, such atmospheric motion can be associated with hurricanes, as illustrated in Figure 8.10, in which the pressure gradients are represented by blue arrows and the Coriolis acceleration, which is always perpendicular to the velocity, is represented by red arrows.

Figure 8.11 shows the corresponding time series for the eigenfunction u_r versus radius for the fixed parameter values $z = 333$ m and $\theta = \pi/3$. The red line corresponds to the initial ($\tau = 0$) and the blue line corresponds to the final ($\tau = 100$) inertial period. We remark that in the rectilinear trapped-wave solution, the radial component of velocity is identically zero (see, e. g., [54], which was also suggested by Lord Kelvin in his presentation in 1879, namely, the vanishing of the velocity component normal to the wall suggested the possibility that it be zero everywhere), whereas in our case u_r achieves its maximum in the interior, which agrees with analytical predictions in [28]. Similarly to the situation shown in Figure 8.5 for the energy density, we observe the phenomenon of the "*effect of rotational convergence*," namely, the time series tends to converge at higher latitudes, that is, as τ increases from 0 to 10, the radial component u_r increases monotonically from $u_r|_{\tau=0}$ to $u_r|_{\tau=50}$.

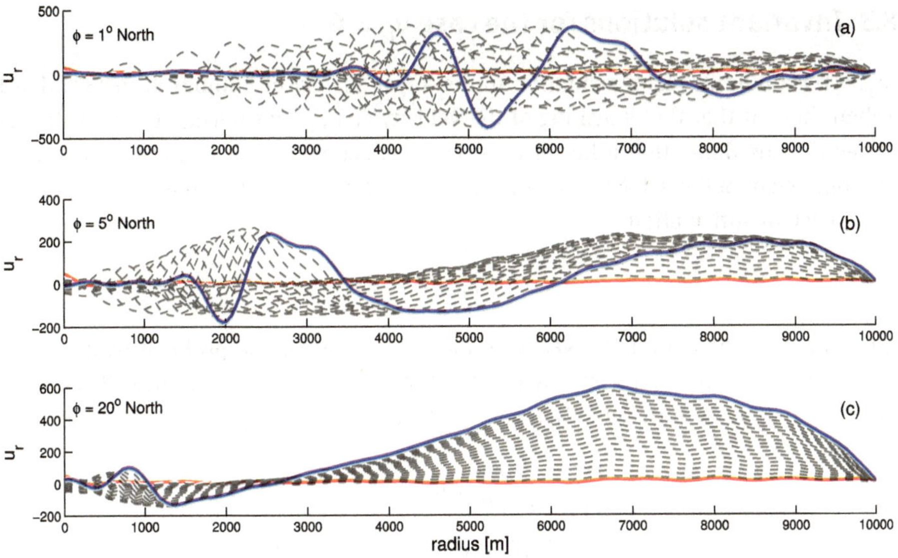

Figure 8.11: Snapshots of the truncated series (8.40) for the eigenfunction u_r versus the radius r at different latitudes and fixed parameter values $z = 333$ [m], $\theta = \pi/3$. The red line corresponds to the initial $\tau = 0$ and the blue line corresponds to the final $\tau = 100$ inertial period.

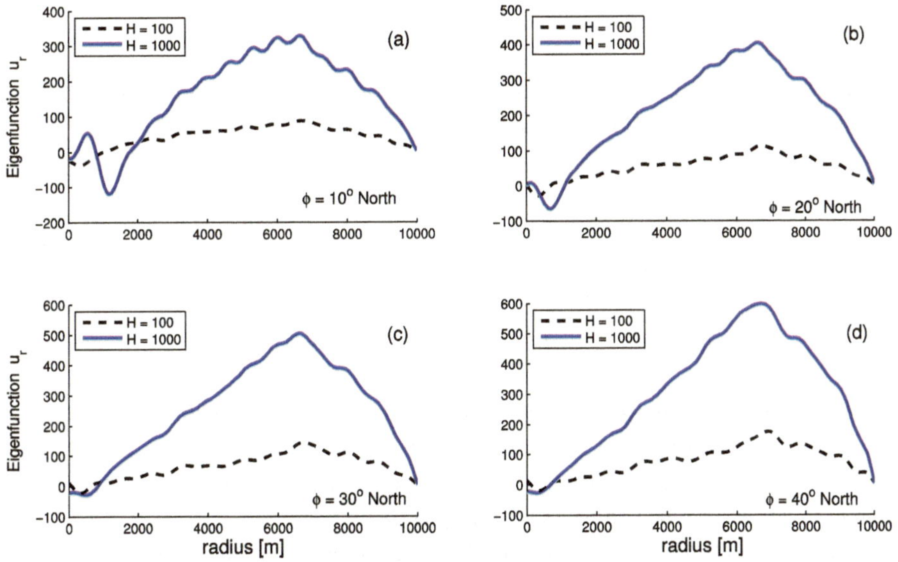

Figure 8.12: Justification of the Kelvin hypothesis for the cylindrical domain in the case $\sigma = H/r_0 \rightarrow 0$ by evaluating and comparing the eigenfunction u_r versus the radius r at $\tau = 400$ at different latitudes for the cases when $H = 1000$ and $H = 100$ and fixed parameter values $z = 333$ [m] and $\theta = \pi/3$.

8.3 Invariant solutions for the case $u_r = 0$

As he recounted in his presentation to the Royal Society of Edinburgh in 1879, Lord Kelvin thought that the vanishing of the velocity component normal to the wall suggested the possibility that it be zero everywhere. Following this hypothesis, we allow the component of the velocity that is normal to the lateral circular boundary to be zero, i. e., we let, in anticipation,

$$u_r = 0 \tag{8.46}$$

throughout the domain and investigate the consequences. The problem which forms the main focus of interest here is to obtain exact solutions of the nonlinear Boussinesq model (8.9)–(8.13) within Kelvin's hypothesis (8.46):

$$\frac{\partial u_\theta}{\partial t} + \frac{u_\theta}{r}\frac{\partial u_\theta}{\partial \theta} + w\frac{\partial u_\theta}{\partial z} + \frac{1}{r}\frac{\partial p}{\partial \theta} = 0, \tag{8.47}$$

$$\frac{\partial w}{\partial t} + \frac{u_\theta}{r}\frac{\partial w}{\partial \theta} + w\frac{\partial w}{\partial z} + \frac{\partial p}{\partial z} + \rho g = 0, \tag{8.48}$$

$$\frac{\partial p}{\partial r} - \frac{u_\theta^2}{r} - f\,u_\theta = 0, \tag{8.49}$$

$$\frac{\partial \rho}{\partial t} + \frac{u_\theta}{r}\frac{\partial \rho}{\partial \theta} - \frac{N^2}{g}\,w = 0, \tag{8.50}$$

$$\frac{1}{r}\frac{\partial u_\theta}{\partial \theta} + \frac{\partial w}{\partial z} = 0. \tag{8.51}$$

The symmetries of equations (8.47)–(8.51) contain two arbitrary functions, $\varphi(z)$ and $\psi(t)$, and are spanned by the following operators:

$$\begin{aligned}
&X_1 = \frac{\partial}{\partial t}, \quad X_2 = \frac{\partial}{\partial \theta}, \quad X_3 = \frac{\partial}{\partial z}, \quad X_\varphi = g\varphi(z)\frac{\partial}{\partial p} - \varphi'(z)\frac{\partial}{\partial \rho}, \quad X_\psi = \psi(t)\frac{\partial}{\partial p},\\
&X_4 = r\frac{\partial}{\partial r} + z\frac{\partial}{\partial z} + u_\theta\frac{\partial}{\partial u_\theta} + w\frac{\partial}{\partial w} + 2p\frac{\partial}{\partial p} + \rho\frac{\partial}{\partial \rho},\\
&X_5 = 2(ft + 2\theta)\frac{\partial}{\partial \theta} - 4r\frac{\partial}{\partial r} + 2fr\frac{\partial}{\partial u_\theta} + f^2r^2\frac{\partial}{\partial p}.
\end{aligned} \tag{8.52}$$

There are no group extensions in the case of homogeneous fluid ($N = 0$) and in the physically irrelevant case for the ocean $f = N$. For example, in the mid/high-latitude upper nonhomogeneous ocean, N is typically one or two orders of magnitude larger than f; N varies from 10^{-3} to $10^{-2}\,\mathrm{s}^{-1}$, while $f \sim 10^{-4}\,\mathrm{s}^{-1}$.

The symmetries (8.52) can be used to obtain exact solutions of the system (8.47)–(8.51) by computing the *invariant solutions*. In order to obtain all possible invariant solutions one has to construct *optimal systems of subalgebras* of the Lie algebra with the basis (8.52). Note that invariant solutions based on three-dimensional subalgebras are of particular interest because they are described by systems of ODEs.

8.3.1 Well-known invariant internal oscillation solution

We will first construct the invariant solution based on the subalgebra $L_2 = \langle X_2, X_3\rangle$. Invariance with respect to the translations in θ and z generated by X_2 and X_3, respectively, requires that the dependent variables are functions of t and r only. Accordingly, the system (8.47)–(8.51) reduces to the form

$$\frac{\partial u_\theta}{\partial t} = 0, \tag{8.53}$$

$$\frac{\partial p}{\partial r} = \frac{u_\theta^2}{r} + f u_\theta, \tag{8.54}$$

$$\frac{\partial w}{\partial t} = -\rho g, \tag{8.55}$$

$$\frac{\partial \rho}{\partial t} = \frac{N^2}{g} w. \tag{8.56}$$

Integration of equations (8.53)–(8.56) gives the invariant solution

$$u_\theta = U(r), \tag{8.57}$$

$$p = \int \left[\frac{1}{r} U^2(r) + f U(r)\right] dr + V(t - t_0), \tag{8.58}$$

$$w = W_1(r) \cos N(t - t_0) + W_2(r) \sin N(t - t_0), \tag{8.59}$$

$$\rho = \frac{N^2}{g}[W_1(r) \sin N(t - t_0) - W_2(r) \cos N(t - t_0)] \tag{8.60}$$

with arbitrary functions $U(r)$, $W_1(r)$, $W_2(r)$, and $V(t)$ and arbitrary constant t_0.

Particularly, by setting $U = V = 0$, if $W_1(r)$ and $W_2(r)$ are constants, we arrive at the well-known internal oscillation solution

$$u_\theta = 0, \quad p = 0, \quad w = A \cos N(t - t_0), \quad g\rho = AN \sin N(t - t_0), \tag{8.61}$$

where A and t_0 are arbitrary constants.

8.3.2 Invariant nonstationary solution

Let us construct the invariant solutions based on the three-dimensional subalgebra spanned by the operators X_3, X_4, X_5 from (8.52). According to the theorem on representation of nonsingular invariant manifolds ([59], Section 14.3), the invariant solutions can be represented via invariants $J(t, \theta, r, z, u_\theta, w, p, \rho)$ of the operators X_3, X_4, X_5. The invariance under X_3 requires that J does not depend on z. Thus, we have to solve the system

$$X_4(J) = r\frac{\partial J}{\partial r} + u_\theta\frac{\partial J}{\partial u_\theta} + w\frac{\partial J}{\partial w} + 2p\frac{\partial J}{\partial p} + \rho\frac{\partial J}{\partial \rho} = 0,$$
$$X_5(J) = 2(ft + 2\theta)\frac{\partial J}{\partial \theta} - 4r\frac{\partial J}{\partial r} + 2fr\frac{\partial J}{\partial u_\theta} + f^2r^22\frac{\partial J}{\partial p} = 0 \tag{8.62}$$

for the function $J = J(t, \theta, r, u_\theta, w, p, \rho)$. Integration of the system (8.62) gives the following basis of invariants:

$$J_1 = t, \quad J_2 = \frac{2u_\theta + fr}{(ft + 2\theta)r}, \quad J_3 = \frac{w}{(ft + 2\theta)r},$$
$$J_4 = \frac{8p + f^2r^2}{(ft + 2\theta)^2r^2}, \quad J_5 = \frac{\rho}{(ft + 2\theta)r}. \tag{8.63}$$

The representation of the invariant solutions via the basic invariants (8.63) is obtained by assuming that J_2, J_3, J_4, J_5 are unknown functions of J_1. This yields the following candidates for the invariant solutions:

$$u_\theta = -\frac{fr}{2} + (ft + 2\theta)rF(t), \quad w = (ft + 2\theta)rG(t),$$
$$p = -\frac{f^2r^2}{8} + (ft + 2\theta)^2r^2H(t), \quad \rho = (ft + 2\theta)rK(t). \tag{8.64}$$

Substituting (8.64) into equations (8.47)–(8.51) and solving the resulting equations for the unknown functions $F(t)$, $G(t)$, $H(t)$, and $K(t)$, we obtain

$$F(t) = 0, \quad G(t) = A\cos(Nt) + B\sin(Nt),$$
$$H(t) = 0, \quad K(t) = \frac{N}{g}[A\sin(Nt) - B\cos(Nt)], \tag{8.65}$$

where A and B are arbitrary constants. Substituting (8.65) in (8.64) we obtain the following nonstationary solution of the system (8.47)–(8.51):

$$u_\theta = -\frac{fr}{2}, \quad w = (ft + 2\theta)r[A\cos(Nt) + B\sin(Nt)],$$
$$p = -\frac{f^2r^2}{8}, \quad \rho = \frac{N}{g}(ft + 2\theta)r[A\sin(Nt) - B\cos(Nt)]. \tag{8.66}$$

8.3.3 Invariant stationary solution

Let us construct the invariant solutions based on the three-dimensional subalgebra spanned by the operators X_1, X_3, X_4 from (8.52). Proceeding as in the above, we obtain the following candidates for the invariant solutions:

$$u_\theta = r\,F(\theta), \quad w = r\,G(\theta), \quad p = r^2\,H(\theta), \quad \rho = r\,K(\theta). \tag{8.67}$$

Substituting (8.67) in equations (8.47)–(8.51), we obtain

$$
\begin{aligned}
&F(\theta) = F_0, \quad G(t) = C_1 \cos(k\theta) + C_2 \sin(k\theta), \\
&H(t) = \frac{F_0}{2}(F_0 + f), \quad K(t) = \frac{N}{g}[C_1 \sin(kt) - C_2 \cos(kt)],
\end{aligned} \tag{8.68}
$$

where C_1 and C_2 are arbitrary constants, F_0 is an arbitrary constant different from zero, and

$$k = \frac{N}{F_0}.$$

If $F_0 = 0$, the solution collapses to the trivial solution $u_\theta = w = p = \rho = 0$.

Substituting (8.68) into equations (8.47)–(8.51), the presentation (8.67) yields the following steady-state solution of the system (8.47)–(8.51):

$$
\begin{aligned}
&u_\theta = F_0\, r, \quad w = r\,[C_1 \cos(k\theta) + C_2 \sin(k\theta)], \\
&p = \frac{1}{2}\,(F_0^2 + fF_0)\, r^2, \quad \rho = \frac{N}{g}\, r[C_1 \sin(k\theta) - C_2 \cos(k\theta)].
\end{aligned} \tag{8.69}
$$

8.3.4 Invariant solutions as nonlinear whirlpools

Constructing as above the invariant solutions based on the three-dimensional subalgebra spanned by the operators X_2, X_4, X_5 from (8.52), we obtain the following candidates for the invariant solutions:

$$u_\theta = -\frac{f\,r}{2} + z\,F(t), \quad w = z\,G(t), \quad p = -\frac{f^2\,r^2}{8} + z^2\,H(t), \quad \rho = z\,K(t).$$

Substituting these expressions in equations (8.47)–(8.51), we obtain

$$F(t) = 0, \quad G(t) = 0, \quad K(t) = k = \text{const.}, \quad H(t) = -\frac{k}{2}\,g.$$

Thus, we have the following solution of the system (8.47)–(8.51):

$$u_\theta = -\frac{f\,r}{2}, \quad w = 0, \quad p = -\frac{f^2\,r^2}{8} - \frac{k}{2}\,gz^2, \quad \rho = kz, \tag{8.70}$$

where k is an arbitrary constant having the dimension $[\mathrm{m}^{-1}]$. If we take the coordinate system so that z is directed upward, then k should be negative. In this coordinate system the pressure deviation p vanishes on the surface of the funnel (whirlpool) with the apex on the bottom of the basin,

$$r = \frac{2}{f}\,\sqrt{-gk}\,z, \tag{8.71}$$

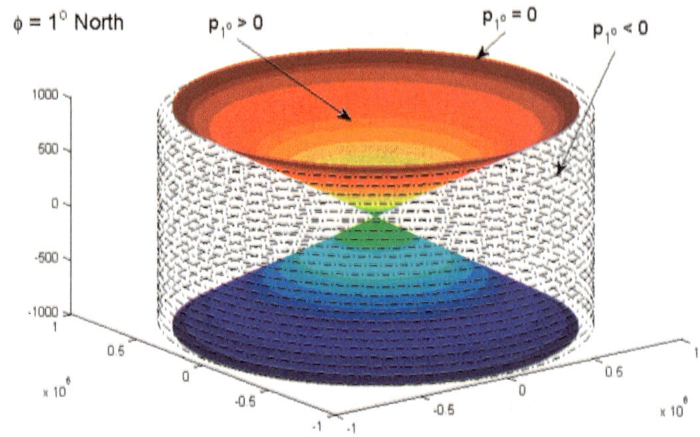

Figure 8.13: Nonlinear whirlpool in a water column of width 1000 km. The choice of the parameter $k = -9.5^{-7}$ is constrained by the requirement of the existence of the lowest-latitude whirlpool at latitude $\phi = 1°$ North.

and is positive inside of the funnel, i. e., under the condition

$$r < \frac{2}{f}\sqrt{-gk\,z}. \tag{8.72}$$

This is illustrated in Figure 8.13, showing the solution (8.70) in the form of the stationary funnel expanding/narrowing with depth. For this reason, we call the invariant solution (8.70) the *whirlpool* associated with nonlinear baroclinic Kelvin waves embedded in the rotating column of stratified fluid. The pressure deviation from the mean state is zero on the surface of the whirlpool, positive inside, and negative outside of the whirlpool. The intensity of rotation of fluid particles inside the whirlpool is monotonically increasing from zero at the center of the whirlpool to its maximum value at the whirlpool's shell. The existence of a whirlpool in a water column of fixed radius r_0 relies on the choice of the parameter k. For example, for $r_0 = 1000$ km and a fixed water column level $H = 1$ km, the choice of k satisfying the inequality $|k| < 1.653 \times 10^{-7}$ [m^{-1}] guarantees the existence of a whirlpool lying entirely in the cylindrical water column within the latitude range $\phi \in [1°, 90°]$ North.

In the world ocean, whirlpools result from the currents that move in a rotating direction, produced by rising and falling tides. Some of these exhibit a downward pull, called a vortex. Figure 8.14 shows the strait between Naruto and Awaji Island, which has a width of about 1.3 km (0.81 miles). The strait is one of the connections between the Pacific Ocean and the Inland Sea, a body of water separating Honshū and Shikoku, two of the main islands of Japan. In most cases, the craters, or rings, as scientists call them, are formed as a result of so-called vertical currents. The latter, in their turn, appear because of differences in water density which appear because of differences in the temperature of water layers. It is an open secret that cold water is thicker and heavier,

Figure 8.14: Photograph of the Naruto whirlpools, which are tidal whirlpools in the Naruto Strait, a channel between Naruto in Tokushima and Awaji Island in Hyōgo, Japan (the photograph is available at www.TelevisionFanatic.com).

so it goes down, to below the masses of warm water, which is lighter. This is the reason why warm currents in the world ocean always flow more closely to the surface, whereas colder currents flow more closely to the bottom.

8.4 Approximately invariant solutions for the case $u_r = \varepsilon v_r$

In this section we are interested in constructing the approximately invariant solutions of the complete nonlinear model (8.9)–(8.13) using the approximate symmetry method as proposed by Fushchych and Shtelen [29]. In terms of nonlinear modeling, we use the assumption that $u_r \ll 1$ throughout the domain. As has been discussed earlier, in the rectilinear trapped-wave solution, the radial component of velocity is identically zero. But in our case of the cylindrical domain, u_r, is nonzero and achieves its maximum in the interior, which also agrees with analytical predictions [28]. However, as we observe from Figure 8.12, we can assume that $u_r \to 0$ as $\sigma \to 0$. We remark that there are two possibilities for σ to become smaller: (i) when r_0 is fixed and $H \to 0$ and (ii) when H is fixed and $r_0 \to \infty$. Our conjecture, demonstrated in Figure 8.12, is based on the first possibility (in actuality, the second possibility will bring us to the same conclusion).

We now consider the case $u_r = \varepsilon v_r$ for some function v_r and small parameter $\varepsilon \ll 1$ such that as $\varepsilon \to 0$, $u_r \to 0$. Hence, we are interested in finding approximate symmetries and invariant solutions of the complete nonlinear model (8.9)–(8.13) with $u_r = \varepsilon v_r$ which is given to $o(\varepsilon)$ by

$$\varepsilon \frac{\partial v_r}{\partial t} + \varepsilon \frac{u_\theta}{r} \frac{\partial v_r}{\partial \theta} + \varepsilon w \frac{\partial v_r}{\partial z} - \frac{u_\theta^2}{r} - f u_\theta = -\frac{\partial p}{\partial r}, \tag{8.73}$$

$$\frac{\partial u_\theta}{\partial t} + \varepsilon v_r \frac{\partial u_\theta}{\partial r} + \frac{u_\theta}{r} \frac{\partial u_\theta}{\partial \theta} + w \frac{\partial u_\theta}{\partial z} + \varepsilon \frac{v_r u_\theta}{r} + \varepsilon f v_r + = -\frac{1}{r} \frac{\partial p}{\partial \theta}, \tag{8.74}$$

$$\frac{\partial w}{\partial t} + \varepsilon v_r \frac{\partial w}{\partial r} + \frac{u_\theta}{r} \frac{\partial w}{\partial \theta} + w \frac{\partial w}{\partial z} = -\frac{\partial p}{\partial z} - \rho g, \tag{8.75}$$

$$\frac{\partial \rho}{\partial t} + \varepsilon v_r \frac{\partial \rho}{\partial r} + \frac{u_\theta}{r} \frac{\partial \rho}{\partial \theta} = \frac{N^2}{g} w, \tag{8.76}$$

$$\varepsilon \frac{v_r}{r} + \varepsilon \frac{\partial v_r}{\partial r} + \frac{1}{r} \frac{\partial u_\theta}{\partial \theta} + \frac{\partial w}{\partial z} = 0. \tag{8.77}$$

We used the MAPLE computer algebra symmetry packages ASP (*Jefferson & Carminati* [42]) and DESOLVII (*Vu, Jefferson, & Carminati* [77]) to find both the approximate symmetries and approximately invariant solutions of (8.73)–(8.77). Only the key steps are described here; for further details on the method and algorithms, please refer to Ref. [42]. Moreover, in order to make clear the distinction between *classical* and *approximate* Lie point symmetries, we shall refer to the former as the *exact* symmetries of a given differential system.

Let us consider the system (8.73)–(8.77) with perturbation parameter ε. To find approximate symmetries we use the ASP routine ApproximateSymmetry. The routine first expands the dependent variables to first order in ε such that

$$v_r = v_{r0} + \varepsilon v_{r1} + o(\varepsilon^2), \tag{8.78}$$

$$u_\theta = u_{\theta 0} + \varepsilon u_{\theta 1} + o(\varepsilon^2), \tag{8.79}$$

$$p = p_0 + \varepsilon p_1 + o(\varepsilon^2), \tag{8.80}$$

$$w = w_0 + \varepsilon w_1 + o(\varepsilon^2), \tag{8.81}$$

$$\rho = \rho_0 + \varepsilon \rho_1 + o(\varepsilon^2). \tag{8.82}$$

The variables v_{ri}, $u_{\theta i}$, p_i, w_i, and ρ_i for $i = 0, 1, 2$ are the "new" dependent variables and are considered to be some smooth functions of r, t, z, and θ. However, note that as we are concerned only with expressions $o(\varepsilon)$, when we substitute (8.78) back into equations (8.73)–(8.77), all instances of v_{r1} will vanish from the system. Since the solution in the original variables will be given as $u_r = \varepsilon v_r \equiv \varepsilon v_{r0} + \varepsilon^2 v_{r1}$, we find that the function v_{r1} will disappear to $o(\varepsilon)$ anyway. Hence, for the final approximately invariant solutions, we need only substitute the solved new dependent variables back into

$$u_r = \varepsilon v_{r0}, \quad u_\theta = u_{\theta 0} + \varepsilon u_{\theta 1}, \quad p = p_0 + \varepsilon p_1, \quad w = w_0 + \varepsilon w_1, \quad \rho = \rho_0 + \varepsilon \rho_1. \tag{8.83}$$

The ApproximateSymmetry routine next substitutes (8.78)–(8.82) into the system (8.73)–(8.77) and splits the system at each order of ε to give the coupled system:

Order 1

$$\frac{\partial p_0}{\partial r} r - fru_{\theta 0} - u_{\theta 0}^2 = 0, \tag{8.84}$$

$$\frac{\partial p_0}{\partial \theta} + \frac{\partial u_{\theta 0}}{\partial t} r + u_{\theta 0}\frac{\partial u_{\theta 0}}{\partial \theta} + w_0 \frac{\partial u_{\theta 0}}{\partial z} r = 0, \tag{8.85}$$

$$\frac{\partial w_0}{\partial t} r + \rho_0 g r + w_0 \frac{\partial w_0}{\partial z} r + \frac{\partial p_0}{\partial z} r + u_{\theta 0}\frac{\partial w_0}{\partial \theta} = 0, \tag{8.86}$$

$$\frac{\partial \rho_0}{\partial t} gr - N^2 w_0 r + u_{\theta 0}\frac{\partial \rho_0}{\partial \theta} g = 0, \tag{8.87}$$

$$\frac{\partial u_{\theta 0}}{\partial \theta} + \frac{\partial w_0}{\partial z} r = 0, \tag{8.88}$$

Order ε

$$\frac{\partial p_1}{\partial r} r - fru_{\theta 1} - 2u_{\theta 0}u_{\theta 1} + w_0 \frac{\partial v_{r0}}{\partial z} r + \frac{\partial v_{r0}}{\partial t} r + u_{\theta 0}\frac{\partial v_{r0}}{\partial \theta} = 0, \tag{8.89}$$

$$frv_{r0} + v_{r0}\frac{\partial u_{\theta 0}}{\partial r} r + \frac{\partial p_1}{\partial \theta} + u_{\theta 0}\frac{\partial u_{\theta 1}}{\partial \theta} + u_{\theta 1}\frac{\partial u_{\theta 0}}{\partial \theta} + u_{\theta 0}v_{r0} \tag{8.90}$$

$$+ \frac{\partial u_{\theta 1}}{\partial t} r + rw_0 \frac{\partial u_{\theta 1}}{\partial z} + rw_1 \frac{\partial u_{\theta 0}}{\partial z} = 0, \tag{8.91}$$

$$\frac{\partial p_1}{\partial z} r + \frac{\partial w_1}{\partial t} r + \rho_1 g r + v_{r0}\frac{\partial w_0}{\partial r} r + rw_0 \frac{\partial w_1}{\partial z}$$

$$+ rw_1 \frac{\partial w_0}{\partial z} + u_{\theta 0}\frac{\partial w_1}{\partial \theta} + u_{\theta 1}\frac{\partial w_0}{\partial \theta} = 0, \tag{8.92}$$

$$\frac{\partial \rho_1}{\partial t} gr - N^2 w_1 r + v_{r0}\frac{\partial \rho_0}{\partial r} gr + gu_{\theta 0}\frac{\partial \rho_1}{\partial \theta} + guu_{\theta 1}\frac{\partial \rho_0}{\partial \theta} = 0, \tag{8.93}$$

$$\frac{\partial u_{\theta 1}}{\partial \theta} + r\frac{\partial v_{r0}}{\partial r} + v_{r0} + \frac{\partial w_1}{\partial z} r = 0. \tag{8.94}$$

These functions are stored by ASP and may be accessed using the command OrderFunctions();. As per the theory proposed in Ref. [29], the *approximate* symmetries of the original system (8.73)–(8.77) correspond to the *exact* symmetries of the split, coupled system (8.84)–(8.94). Hence, ASP calculates the determining equations for the split coupled system using DESOLVII routines and solves these using the DESOLVEII PDE solver pdecolv.

The exact symmetries of the coupled system (8.84)–(8.94), and thus the approximate symmetries of equations (8.73)–(8.77), are spanned by the following operators:

$$Y_1 = \frac{\partial}{\partial t}, \quad X_2 = \frac{\partial}{\partial \theta}, \quad Y_3 = \frac{\partial}{\partial z}, \quad Y_\psi = \psi(t)\frac{\partial}{\partial p_0}, \quad Y_\sigma = \sigma(t)\frac{\partial}{\partial p_1},$$

$$Y_\gamma = \gamma'(z)\frac{\partial}{\partial \rho_0} - g\gamma(z)\frac{\partial}{\partial p_0}, \quad Y_\varphi = \varphi'(z)\frac{\partial}{\partial \rho_1} - g\varphi(z)\frac{\partial}{\partial p_1}, \tag{8.95}$$

$$Y_4 = p_1\frac{\partial}{\partial p_1} + u_{\theta 1}\frac{\partial}{\partial u_{\theta 1}} + v_{r0}\frac{\partial}{\partial v_{r0}} + w_1\frac{\partial}{\partial w_1} + \rho_1\frac{\partial}{\partial \rho_1},$$
$$Y_5 = 2p_0\frac{\partial}{\partial p_0} + p_1\frac{\partial}{\partial p_1} + r\frac{\partial}{\partial r} + u_{\theta 0}\frac{\partial}{\partial u_{\theta 0}} + w_0\frac{\partial}{\partial w_0} + z\frac{\partial}{\partial z} + \rho_0\frac{\partial}{\partial \rho_0},$$

with arbitrary functions $\psi(t)$, $\sigma(t)$, $\gamma(z)$, and $\varphi(z)$.

As for classical Lie symmetries, we may use the operators (8.95), in any linear combination, to obtain exact solutions of equations (8.84)–(8.94) by computing group invariant solutions (see [42] for details). In particular, consider the operator

$$\bar{Y}_4 \equiv Y_4 + Y_5 = r\frac{\partial}{\partial r} + z\frac{\partial}{\partial z} + 2p_0\frac{\partial}{\partial p_0} + 2p_1\frac{\partial}{\partial p_1} + u_{\theta 0}\frac{\partial}{\partial u_{\theta 0}} + u_{\theta 1}\frac{\partial}{\partial u_{\theta 1}} + v_{r0}\frac{\partial}{\partial v_{r0}} + w_0\frac{\partial}{\partial w_0} + w_1\frac{\partial}{\partial w_1} + \rho_0\frac{\partial}{\partial \rho_0} + \rho_1\frac{\partial}{\partial \rho_1}, \tag{8.96}$$

which is equivalent to the symmetry generator X_4 from (8.52) for the case where $u_r = 0$. We shall later use (8.96) to retrieve some of the invariant solutions found using the exact symmetry generators found when $u_r = 0$.

The approximately invariant solutions of the original system (8.73)–(8.77) are determined by substituting the exact invariant solutions for the variables v_{r0}, $u_{\theta i}$, p_i, w_i, and ρ_i for $i = 0, 1$ into (8.83). These approximately invariant solutions will satisfy equations (8.73)–(8.77) to $O(\varepsilon^2)$.

We now seek to retrieve the approximate solutions analogous to the four solution types found in the previous section with $u_r = 0$ such that the solution for u_r is nonzero.

8.4.1 Well-known approximately invariant internal oscillation solution

Let us construct the invariant solution based on the subalgebra $L_2 = \langle Y_2, Y_3\rangle$, which is equivalent to that used for the reduction when $u_r = 0$ (refer to the previous section). The generators Y_2 and Y_3 are translations in θ and z, respectively, and hence imply that the dependent variables are functions of t and r only. The system (8.84)–(8.94) thus reduces to

$$\frac{\partial u_{\theta 0}}{\partial t} r = 0, \tag{8.97}$$
$$r\frac{\partial v_{r0}}{\partial r} + v_{r0} = 0, \tag{8.98}$$
$$\frac{\partial w_0}{\partial t} r + \rho_0 g r = 0, \tag{8.99}$$
$$\frac{\partial \rho_0}{\partial t} g r - N^2 r w_0 = 0, \tag{8.100}$$
$$\frac{\partial p_0}{\partial r} r - f r u_{\theta 0} - u_{\theta 0}^2 = 0, \tag{8.101}$$

$$\frac{\partial w_1}{\partial t} r + \rho_1 g r + v_{r0} \frac{\partial w_0}{\partial r} r = 0, \tag{8.102}$$

$$\frac{\partial \rho_1}{\partial t} g r - N^2 w_1 r + v_{r0} \frac{\partial \rho_0}{\partial r} g r = 0, \tag{8.103}$$

$$\frac{\partial p_1}{\partial r} r - f r u_{\theta 1} - 2 u_{\theta 0} u_{\theta 1} + \frac{\partial v_{r0}}{\partial t} r = 0, \tag{8.104}$$

$$f r v_{r0} + v_{r0} r \frac{\partial u_{\theta 0}}{\partial r} + u_{\theta 0} v_{r0} + \frac{\partial u_{\theta 1}}{\partial t} r = 0. \tag{8.105}$$

Since we seek to retrieve the internal oscillation solutions of the form (8.57)–(8.60) as found when $u_r = 0$ (and hence $\varepsilon = 0$ in equations (8.73)–(8.77)), we will have the following solutions for the unperturbed dependent variables:

$$u_{\theta 0} = U(r), \tag{8.106}$$

$$p_0 = \int \left[\frac{1}{r} U^2(r) + f U(r) \right] dr + V(t), \tag{8.107}$$

$$w_0 = W_2(r) \sin N(t - t_0) + W_1(r) \cos N(t - t_0), \tag{8.108}$$

$$\rho_0 = \frac{N}{g} [W_1(r) \sin N(t - t_0) - W_2(r) \cos N(t - t_0)], \tag{8.109}$$

with arbitrary functions $U(r)$, $V(t)$, $W_1(r)$, and $W_2(r)$ and arbitrary constant t_0. Substituting equations (8.106)–(8.109) into the reduced system (8.97)–(8.105), integration using MAPLE yields the invariant solution for the variables v_{r0}, $u_{\theta 1}$, p_1, w_1, and ρ_1. Subsequent substitution of these solutions and the internal oscillation type solutions (8.106)–(8.109) back into (8.83) gives the following approximately invariant solution:

$$u_r = \varepsilon \frac{G(t)}{r}, \tag{8.110}$$

$$u_\theta = U(r) - \varepsilon \frac{1}{r^2} \left[f r \int G(t)\, dt + U'(r) r \int G(t)\, dt + U(r) \int G(t)\, dt - K(r) r^2 \right], \tag{8.111}$$

$$\begin{aligned} p = {} & \int \left[\frac{1}{r} U^2(r) + f U(r) \right] dr + V(t) + \varepsilon \left\{ F(t) - \int \left[\frac{1}{r^3} \left(-f r^3 K(r) \right. \right. \right. \\ & + f^2 r^2 \int G(t)\, dt + f r^2 U'(r) \int G(t)\, dt + 3U(r) f r \int G(t)\, dt + r^2 G'(t) \\ & \left. \left. \left. + 2U(r) U'(r) r \int G(t)\, dt + 2U^2(r) \int G(t)\, dt - 2U(r) K(r) r^2 \right) dr \right] \right\}, \end{aligned} \tag{8.112}$$

$$\begin{aligned} w = {} & W_2(r) \sin N(t - t_0) + W_1(r) \cos N(t - t_0) + \varepsilon \frac{1}{rN} \Big\{ \sin(Nt) F_2(r) N r \\ & - \sin(Nt) \int \cos(Nt) [G'(t) W_1'(r) \cos(N(t - t_0)) + 2G(t) W_2'(r) N \cos(N(t - t_0)) \\ & - 2G(t) W_1'(r) N \sin(N(t - t_0)) + G'(t) W_2'(r) \sin(N(t - t_0))]\, dt \\ & + \cos(Nt) F_1(r) N r + \cos(Nt) \int \sin(Nt) [\cos(N(t - t_0)) G'(t) W_1'(r) \\ & + 2\cos(N(t - t_0)) G(t) W_2'(r) N - 2\sin(N(t - t_0)) G(t) W_1'(r) N \end{aligned} \tag{8.113}$$

$$+ \sin(N(t-t_0))G'(t)W_2'(r)]\,dt\Big\},$$

$$\rho = \frac{N}{g}[W_1(r)\sin N(t-t_0) - W_2(r)\cos N(t-t_0)] - \varepsilon\frac{1}{gr}\Big\{\cos(Nt)NF_2(r)r \tag{8.114}$$
$$- \cos(Nt)\int[\cos(Nt)[\cos(N(t-t_0))W_1'(r)G'(t)$$
$$+ 2\cos(N(t-t_0))NW_2'(r)G(t) + \sin(N(t-t_0))W_2'(r)G'(t)$$
$$- 2\sin(N(t-t_0))NW_1'(r)G(t)]\,dt - \sin(Nt)NF_1(r)r$$
$$- \sin(Nt)\int\sin(Nt)[\cos(N(t-t_0))W_1'(r)G'(t) + 2\cos(N(t-t_0))NW_2'(r)G(t)$$
$$+ \sin(N(t-t_0))W_2'(r)G'(t) - 2\sin(N(t-t_0))NW_1'(r)G(t)]\,dt$$
$$+ W_2'(r)G(t)\sin(N(t-t_0)) + W_1'(r)G(t)\cos(N(t-t_0))\Big\},$$

for arbitrary functions $G(t), K(r), F(t), F_1(r)$, and $F_2(r)$, which satisfies the original equations (8.73)–(8.77) to $O(\varepsilon^2)$. Moreover, when $\varepsilon = 0$ (and thus $u_r = 0$), we retain the internal oscillation solution (8.57)–(8.60) as expected.

8.4.2 Approximately invariant nonstationary solution

The invariant nonstationary type solution, as found for the case $u_r = 0$, was based on the three-dimensional subalgebra spanned by the operators X_3, X_4, X_5 in (8.52). Unfortunately, we do not have a candidate symmetry operator or linear combination of operators from (8.95) which matches X_5 in form. Hence, we use only the other two equivalent operators Y_3 and $\bar{Y}_4$, the first of which implies that the dependent variables in equations (8.84)–(8.94), and basis invariants, be independent of z. Using the DESOLVII routine reduceVar to reduce the system using the remaining symmetry operator $\bar{Y}_4$, we have the following basis of invariants:

$$K_1 = t, \quad K_2 = \theta, \quad W_1 = \frac{p_0}{r^2}, \quad W_2 = \frac{p_1}{r^2}, \quad W_3 = \frac{u_{\theta 0}}{r}, \quad W_4 = \frac{u_{\theta 1}}{r}, \tag{8.115}$$
$$W_5 = \frac{v_{r0}}{r}, \quad W_6 = \frac{w_0}{r}, \quad W_7 = \frac{w_1}{r}, \quad W_8 = \frac{\rho_0}{r}, \quad W_9 = \frac{\rho_1}{r},$$

where the reduced system is found by using DESOLVII to re-express the derivatives of the dependent variables $v_{r0}, u_{\theta 1}, p_1, w_1$, and ρ_1 in terms of the basic invariants and substitute these back into the split coupled system (8.84)–(8.94), now independent of z. The invariant solutions of this now re-expressed system are found by assuming that the basic invariants $W_i, i = 1, .., 9$, are unknown functions of K_1.

In order to retrieve the nonstationary solution found for the case $u_r = 0$, we assume the following solutions from (8.66) for the unperturbed dependent variables:

$$u_{\theta 0} = -\frac{1}{2}fr, \tag{8.116}$$

$$p_0 = -\frac{1}{8}f^2r^2, \tag{8.117}$$

$$w_0 = r(ft + 2\theta)(A\cos(Nt) + B\sin(Nt)), \tag{8.118}$$

$$\rho_0 = \frac{rN}{g}(ft + 2\theta)(A\sin(Nt) - B\cos(Nt)), \tag{8.119}$$

with arbitrary constants f, A, and B. Equations (8.116)–(8.119) supply the solutions for the basic invariants W_1, W_3, W_6, and W_8. Substituting these solutions into the re-expressed system and integrating for the unknown functions using MAPLE, we find the invariant solution for the variables v_{r0}, $u_{\theta 1}$, p_1, w_1, and ρ_1. Substituting these and the nonstationary solutions (8.116)–(8.119) back into (8.83) yields the following approximately invariant solution for equations (8.73)–(8.77):

$$u_r = -\varepsilon\frac{r}{2}F_3'\left(\frac{1}{2}ft + \theta\right),$$

$$u_\theta = -\frac{1}{2}fr + \varepsilon rF_3\left(\frac{1}{2}ft + \theta\right),$$

$$p = -\frac{f^2r^2}{8},$$

$$\begin{aligned}
w = {} & r(ft + 2\theta)(A\cos(Nt) + B\sin(Nt)) + \varepsilon\frac{r}{2N}\left[-2\theta A\sin(Nt)F_3'\left(\frac{1}{2}ft + \theta\right)\right.\\
& + \cos(Nt)AF_3'\left(\frac{1}{2}ft + \theta\right)ft^2N + \sin(Nt)BF_3'\left(\frac{1}{2}ft + \theta\right)ft^2N + 2\sin(Nt)BF_3'\left(\frac{1}{2}ft + \theta\right)\theta tN\\
& + 2\cos(Nt)AF_3'\left(\frac{1}{2}ft + \theta\right)\theta tN - 4\sin(Nt)BF_3\left(\frac{1}{2}ft + \theta\right)tN - 4\cos(Nt)AF_3\left(\frac{1}{2}ft + \theta\right)tN\\
& + 2\cos(Nt)F_5\left(\frac{1}{2}ft + \theta\right)g + 4A\sin(Nt)F_3\left(\frac{1}{2}ft + \theta\right) - 2\sin(Nt)F_4\left(\frac{1}{2}ft + \theta\right)g\\
& \left. - ftA\sin(Nt)F_3'\left(\frac{1}{2}ft + \theta\right)\right],
\end{aligned}$$

$$\begin{aligned}
\rho = {} & \frac{rN}{g}(ft + 2\theta)(A\sin(Nt) - B\cos(Nt)) + \varepsilon\frac{r}{2g}\left[-\cos(Nt)BNF_3'\left(\frac{1}{2}ft + \theta\right)ft^2\right.\\
& - 2\cos(Nt)BNF_3'\left(\frac{1}{2}ft + \theta\right)\theta t + 2\cos(Nt)F_4\left(\frac{1}{2}ft + \theta\right)g + ft^2A\sin(Nt)F_3'\left(\frac{1}{2}ft + \theta\right)N\\
& - 4A\sin(Nt)F_3\left(\frac{1}{2}ft + \theta\right)Nt - 4AF_3\left(\frac{1}{2}ft + \theta\right)\cos(Nt) + 2\sin(Nt)F_5\left(\frac{1}{2}ft + \theta\right)g\\
& + 4\cos(Nt)BNF_3\left(\frac{1}{2}ft + \theta\right)t + ftAF_3'\left(\frac{1}{2}ft + \theta\right)\cos(Nt) + 2\theta AF_3'\left(\frac{1}{2}ft + \theta\right)\cos(Nt)\\
& \left. + 2\theta A\sin(Nt)F_3'\left(\frac{1}{2}ft + \theta\right)Nt\right],
\end{aligned}$$

for arbitrary functions $F_3(\frac{1}{2}ft+\theta)$, $F_4(\frac{1}{2}ft+\theta)$, and $F_5(\frac{1}{2}ft+\theta)$.

We note that, for space considerations, we have chosen a *particular* solution for the variables v_{r0}, $u_{\theta 1}$, p_1, w_1, and ρ_1, simply to demonstrate that we have regained the desired nonstationary solution with a nonzero solution for u_r such that substitution into the original equations (8.73)–(8.77) is $O(\varepsilon^2)$ as required. There is, for example, the more general invariant solution which we show (for brevity) for the first three dependent variables only:

$$\begin{aligned}
u_r &= \varepsilon\frac{r}{2}\Bigg[4\int^t (F_1(\tau)(\sin(f\tau)\cos(ft+2\theta)-\cos(f\tau)\sin(ft+2\theta))\,d\tau) \\
&\quad -4\int^t (F_2(\tau)(\cos(f\tau)\cos(ft+2\theta)+\sin(f\tau)\sin(ft+2\theta))\,d\tau) - F_3'\left(\frac{1}{2}ft+\theta\right)\Bigg], \\
u_\theta &= -\frac{fr}{2}+\varepsilon\frac{r}{2}\Bigg[-4\int^t (F_1(\tau)(\cos(f\tau)\cos(ft+2\theta)+\sin(f\tau)\sin(ft+2\theta))\,d\tau) \\
&\quad -4\int^t (F_2(\tau)(\sin(f\tau)\cos(ft+2\theta)-\cos(f\tau)\sin(ft+2\theta)\,d\tau) + 2F_3\left(\frac{1}{2}ft+\theta\right)\Bigg], \\
p &= -\frac{1}{8}r^2f^2+\varepsilon r^2[2F_1(t)\sin(\theta)\cos(\theta)+2F_2(t)\cos^2(\theta)-F_2(t)].
\end{aligned}$$

8.4.3 Approximately invariant stationary solution

We construct the invariant stationary solution, equivalent to that found for $u_r=0$, based on the three-dimensional subalgebra spanned by the operators Y_1, Y_3, $\bar{Y}_4$. The generators Y_1 and Y_2 imply that the dependent variables in equations (8.84)–(8.94), and basis invariants, are independent of t and z. Proceeding as above, using reduceVar to reduce the system with the remaining symmetry operator $\bar{Y}_4$, we have the following basis of invariants:

$$\begin{aligned}
&K_1=\theta,\quad W_1=\frac{p_0}{r^2},\quad W_2=\frac{p_1}{r^2},\quad W_3=\frac{u_{\theta 0}}{r},\quad W_4=\frac{u_{\theta 1}}{r}, \\
&W_5=\frac{u_{r0}}{r},\quad W_6=\frac{w_0}{r},\quad W_7=\frac{w_1}{r},\quad W_8=\frac{\rho_0}{r},\quad W_9=\frac{\rho_1}{r}.
\end{aligned} \tag{8.120}$$

To retrieve the stationary solution found for the case $u_r=0$, we have the following solutions from (8.69) for the unperturbed dependent variables:

$$u_{\theta 0}=F_0 r, \tag{8.121}$$

$$p_0=\frac{1}{2}r^2(F_0^2+fF_0), \tag{8.122}$$

$$w_0=r[C_1\cos(k\theta)+C_2\sin(k\theta)], \tag{8.123}$$

$$\rho_0=\frac{Nr}{g}[C_1\sin(k\theta)-C_2\cos(k\theta)], \tag{8.124}$$

for arbitrary constants C_1, C_2, and F_0 and where $k = N/F_0$, which we substitute into the re-expressed system. Hence, integrating using MAPLE and solving for the unknown functions, we find the invariant solution for the variables v_{r0}, $u_{\theta 1}$, p_1, w_1, and ρ_1. Substituting these solutions and the stationary solutions (8.121)–(8.124) back into (8.83), we have the following approximately invariant solution:

$$\begin{aligned}
u_r &= \varepsilon r[C_7 \sin(2\theta) - C_6 \cos(2\theta)],\\
u_\theta &= rF_0 + \varepsilon r[C_5 + C_6 \sin(2\theta) + \cos(2\theta)C_7],\\
p &= \frac{1}{2}r^2(F_0^2 + fF_0) + \varepsilon\frac{r^2}{2}[C_6 f \sin(2\theta) + C_7 f \cos(2\theta) + fC_5 + 2F_0C_5],\\
w &= r[C_1 \cos(k\theta) + C_2 \sin(k\theta)] + \varepsilon\frac{r}{2F_0^2}[2N \sin(k\theta)C_1C_5\theta - 2N \cos(k\theta)C_2C_5\theta\\
&\quad + NC_1C_7 \sin(k\theta)\sin(2\theta) - NC_2C_7 \cos(k\theta)\sin(2\theta) + NC_2C_6 \cos(k\theta)\cos(2\theta)\\
&\quad + C_1C_6 \cos(k\theta)F_0 \sin(2\theta) + \sin(k\theta)C_2C_5F_0 + 2\cos(k\theta)C_3F_0^2\\
&\quad + F_0C_1C_7 \cos(k\theta)\cos(2\theta) - NC_1C_6 \sin(k\theta)\cos(2\theta) + C_2C_6 \sin(k\theta)F_0 \sin(2\theta)\\
&\quad + F_0C_2C_7 \sin(k\theta)\cos(2\theta) + \cos(k\theta)C_1C_5F_0 + 2\sin(k\theta)C_4F_0^2],\\
\rho &= \frac{Nr}{g}[C_1 \sin(k\theta) - C_2 \cos(k\theta)] + \varepsilon\frac{Nr}{2gF_0^2}[-2N \sin(k\theta)C_2C_5\theta - 2N \cos(k\theta)C_1C_5\theta\\
&\quad + NC_2C_6 \sin(k\theta)\cos(2\theta) + NC_1C_6 \cos(k\theta)\cos(2\theta) - C_2C_5 \cos(k\theta)F_0\\
&\quad - NC_1C_7 \cos(k\theta)\sin(2\theta) + F_0C_1C_7 \sin(k\theta)\cos(2\theta) - 2\cos(k\theta)C_4F_0^2\\
&\quad - NC_2C_7 \sin(k\theta)\sin(2\theta) - C_2C_6 \cos(k\theta)F_0 \sin(2\theta) - F_0C_2C_7 \cos(k\theta)\cos(2\theta)\\
&\quad + 2\sin(k\theta)C_3F_0^2 + C_1C_6 \sin(k\theta)F_0 \sin(2\theta) + C_1C_5 \sin(k\theta)F_0],
\end{aligned}$$

for arbitrary constants C_3, C_4, C_5, C_6, and C_7.

8.4.4 Approximately invariant nonlinear whirlpools

Finally, we construct the invariant solution based on the subalgebra spanned by the operators Y_1, Y_2, $\bar{Y}_4$, which are equivalent to the operators used to calculate the nonlinear whirlpool solution when $u_r = 0$. As above, the generators Y_1 and Y_2 imply that the dependent variables in equations (8.84)–(8.94), and basis invariants, are independent of t and θ. Using reduceVar and $\bar{Y}_4$, we have the following basis of invariants:

$$\begin{aligned}
&K_1 = \frac{r}{z}, \quad W_1 = \frac{p_0}{z^2}, \quad W_2 = \frac{p_1}{z^2}, \quad W_3 = \frac{u_{\theta 0}}{z}, \quad W_4 = \frac{u_{\theta 1}}{z},\\
&W_5 = \frac{v_{r0}}{z}, \quad W_6 = \frac{w_0}{z}, \quad W_7 = \frac{w_1}{z}, \quad W_8 = \frac{\rho_0}{z}, \quad W_9 = \frac{\rho_1}{z}.
\end{aligned} \tag{8.125}$$

Proceeding as before, to retrieve the nonlinear whirlpool solution found for the case $u_r = 0$, we have the following solutions for the unperturbed dependent variables as given in (8.70):

$$u_{\theta 0} = -\frac{1}{2}fr, \quad p_0 = -\frac{1}{8}f^2r^2 - \frac{1}{2}kgz^2, \quad w_0 = 0, \quad \rho_0 = kz, \tag{8.126}$$

for arbitrary constant k having the dimension m^{-1}. Substituting (8.126) into the re-expressed system, we first find that $u_{\theta 1}$ disappears from the system and hence may be considered an arbitrary function of the basis invariant r/z. Integrating and solving for the remaining unknown functions using MAPLE, we find the invariant solution for the variables v_{r0}, $u_{\theta 1}$, p_1, w_1, and ρ_1. Subsequent substitution of these solutions and equations (8.126) back into equations (8.83) gives the following approximately invariant solution:

$$u_r = \varepsilon\frac{C_2z^2}{r}, \tag{8.127}$$

$$u_\theta = -\frac{1}{2}fr + \varepsilon F\left(\frac{r}{z}\right), \tag{8.128}$$

$$p = -\frac{1}{8}f^2r^2 - \frac{1}{2}kgz^2 + \varepsilon C_1z^2, \tag{8.129}$$

$$w = 0, \tag{8.130}$$

$$\rho = kz - \varepsilon\frac{2C_1z}{g}, \tag{8.131}$$

for arbitrary function $F(\frac{r}{z})$ and arbitrary constants C_1 and C_2.

8.5 Discussions of invariant and approximately invariant solutions

The exact solutions of three-dimensional equations of motion for internal gravity waves in cylindrical coordinates in unbounded media were found by means of *approximate transformation groups* of nonlinear governing equations for internal waves with a small parameter. Introduction of the small parameter has been suggested by studying the eigenfrequencies of the linear model of uniformly stratified fluid motion confined in a rigid cylindrical basin.

As a particular application to ocean and atmospheric modeling, in terms of linear analysis, the time series of the energy density were visualized as spinning patterns that appear to be rotating in a counterclockwise sense when looking at the North Pole from above. Such spinning patterns were compared with the flow around a low-pressure area that is usually linked with models of hurricanes. In terms of nonlinear modeling, the conjecture on vanishing of the radial component of velocity throughout the domain as $\sigma \to 0$ brought us to construct a new exact solution depending on the r-

and z-coordinates, namely, the zero-order approximate transformations correspond to *exact symmetries* of the governing equations with $u_r = 0$, i. e.,

$$\frac{\partial u_\theta}{\partial t} + \frac{u_\theta}{r}\frac{\partial u_\theta}{\partial \theta} + w\frac{\partial u_\theta}{\partial z} + \frac{1}{r}\frac{\partial p}{\partial \theta} = 0, \tag{8.132}$$

$$\frac{\partial w}{\partial t} + \frac{u_\theta}{r}\frac{\partial w}{\partial \theta} + w\frac{\partial w}{\partial z} + \frac{\partial p}{\partial z} + \rho g = 0, \tag{8.133}$$

$$\frac{\partial p}{\partial r} - \frac{u_\theta^2}{r} - f\, u_\theta = 0, \tag{8.134}$$

$$\frac{\partial \rho}{\partial t} + \frac{u_\theta}{r}\frac{\partial \rho}{\partial \theta} - \frac{N^2}{g}w = 0, \tag{8.135}$$

$$\frac{1}{r}\frac{\partial u_\theta}{\partial \theta} + \frac{\partial w}{\partial z} = 0, \tag{8.136}$$

whereas the approximation $u_r = \varepsilon v_r$ provides us with approximately invariant solutions corresponding to the modeling equations at order $o(\varepsilon)$ given by

$$\varepsilon\frac{\partial v_r}{\partial t} + \varepsilon\frac{u_\theta}{r}\frac{\partial v_r}{\partial \theta} + \varepsilon w\frac{\partial v_r}{\partial z} - \frac{u_\theta^2}{r} - fu_\theta = -\frac{\partial p}{\partial r}, \tag{8.137}$$

$$\frac{\partial u_\theta}{\partial t} + \varepsilon v_r\frac{\partial u_\theta}{\partial r} + \frac{u_\theta}{r}\frac{\partial u_\theta}{\partial \theta} + w\frac{\partial u_\theta}{\partial z} + \varepsilon\frac{v_r u_\theta}{r} + \varepsilon f v_r + = -\frac{1}{r}\frac{\partial p}{\partial \theta}, \tag{8.138}$$

$$\frac{\partial w}{\partial t} + \varepsilon v_r\frac{\partial w}{\partial r} + \frac{u_\theta}{r}\frac{\partial w}{\partial \theta} + w\frac{\partial w}{\partial z} = -\frac{\partial p}{\partial z} - \rho g, \tag{8.139}$$

$$\frac{\partial \rho}{\partial t} + \varepsilon v_r\frac{\partial \rho}{\partial r} + \frac{u_\theta}{r}\frac{\partial \rho}{\partial \theta} = \frac{N^2}{g}w, \tag{8.140}$$

$$\varepsilon\frac{v_r}{r} + \varepsilon\frac{\partial v_r}{\partial r} + \frac{1}{r}\frac{\partial u_\theta}{\partial \theta} + \frac{\partial w}{\partial z} = 0. \tag{8.141}$$

For example, the exact symmetries of equations (8.47)–(8.51) contain two arbitrary functions, $\varphi(z)$ and $\psi(t)$, and are spanned by the following operators:

$$\begin{aligned}
&X_1 = \frac{\partial}{\partial t}, \quad X_2 = \frac{\partial}{\partial \theta}, \quad X_3 = \frac{\partial}{\partial z}, \quad X_\varphi = g\varphi(z)\frac{\partial}{\partial p} - \varphi'(z)\frac{\partial}{\partial \rho}, \quad X_\psi = \psi(t)\frac{\partial}{\partial p},\\
&X_4 = r\frac{\partial}{\partial r} + z\frac{\partial}{\partial z} + u_\theta\frac{\partial}{\partial u_\theta} + w\frac{\partial}{\partial w} + 2p\frac{\partial}{\partial p} + \rho\frac{\partial}{\partial \rho},\\
&X_5 = 2(ft + 2\theta)\frac{\partial}{\partial \theta} - 4r\frac{\partial}{\partial r} + 2fr\frac{\partial}{\partial u_\theta} + f^2 r^2\frac{\partial}{\partial p}\cdot
\end{aligned} \tag{8.142}$$

There are no group extensions in the case of homogeneous fluid ($N = 0$) and in the physically irrelevant case for the ocean $f = N$.

The symmetries (8.52) can be used to obtain exact solutions of the reduced (i. e., when $\varepsilon = 0$) system (8.47)–(8.51) and the complete (i. e., when $\varepsilon \ll 1$) system (8.9)–(8.13) by computing the *invariant and partially invariant solutions*. In order to obtain all possible invariant and partially invariant solutions, one has to construct *optimal systems of subalgebras* of the Lie algebra with the basis (8.52). Note that invariant solutions based on three-dimensional subalgebras are of particular interest because they are described by systems of ODEs.

For example, when $\varepsilon = 0$, we obtain the following exact solutions of the reduced system (8.47)–(8.51):

1. *Internal oscillation solution:*

$$u_\theta = U(r), \tag{8.143}$$

$$p = \int \left[\frac{1}{r} U^2(r) + fU(r)\right] dr + V(t - t_0), \tag{8.144}$$

$$w = W_1(r) \cos N(t - t_0) + W_2(r) \sin N(t - t_0), \tag{8.145}$$

$$\rho = \frac{N^2}{g} [W_1(r) \sin N(t - t_0) - W_2(r) \cos N(t - t_0)], \tag{8.146}$$

with arbitrary functions $U(r)$, $W_1(r)$, $W_2(r)$, and $V(t)$ and arbitrary constant t_0.

Particularly, by setting $U = V = 0$, if $W_1(r)$ and $W_2(r)$ are constants, we arrive at the well-known internal oscillation solution

$$u_\theta = 0, \quad p = 0, \quad w = A \cos N(t - t_0), \quad g\rho = AN \sin N(t - t_0), \tag{8.147}$$

where A and t_0 are arbitrary constants.

2. *Nonstationary solution:*

$$\begin{aligned} u_\theta &= -\frac{fr}{2}, \quad w = (ft + 2\theta)\, r\, [A \cos(Nt) + B \sin(Nt)], \\ p &= -\frac{f^2 r^2}{8}, \quad \rho = \frac{N}{g}\, (ft + 2\theta)\, r [A \sin(Nt) - B \cos(Nt)], \end{aligned} \tag{8.148}$$

where A and B are arbitrary constants.

3. *Stationary solution:*

$$\begin{aligned} u_\theta &= F_0\, r, \quad w = r\, [C_1 \cos(k\theta) + C_2 \sin(k\theta)], \\ p &= \frac{1}{2}\, (F_0^2 + fF_0)\, r^2, \quad \rho = \frac{N}{g}\, r[C_1 \sin(k\theta) - C_2 \cos(k\theta)], \end{aligned} \tag{8.149}$$

where C_1 and C_2 are arbitrary constants, F_0 is an arbitrary constant different from zero, and

$$k = \frac{N}{F_0}.$$

4. *Nonlinear whirlpools:*

$$u_\theta = -\frac{f\, r}{2}, \quad w = 0, \quad p = -\frac{f^2\, r^2}{8} - \frac{k}{2}\, gz^2, \quad \rho = kz, \tag{8.150}$$

where k is an arbitrary constant.

The existence of a whirlpool in a water column of fixed radius r_0 relies on the choice of the parameter k. For example, for $r_0 = 1000\,\text{km}$ and a fixed water column level $H = 1\,\text{km}$, the choice of k satisfying the inequality $|k| < 1.653 \times 10^{-7}\ [\text{m}^{-1}]$ guarantees the existence of a whirlpool lying entirely in the cylindrical water column within the latitude range $\phi \in [1°, 90°]$ North. Figure 8.15 shows the intensities of the whirlpools at different latitudes for the given radius $r_0 = 1000\,\text{km}$ with $k = -9.5^{-7}\,\text{m}^{-1}$. We observe that the domain of the positive pressure is narrowing to zero (center of the fluid column) at higher latitudes. The velocity for this solution only has an azimuthal component, which is a linearly growing function of r. The pressure is a function quadratic with respect to r and z, while the density deviation linearly depends on z. Because of the growing dependence of this solution at large r, one of the possible suggestions is the following: the obtained exact solution can be realized in the column of some radius R only; outside this column solution is trivial. The latter means that at the column boundary the velocity undergoes a jump. At first glance, one can suggest instability of the nonlinear whirlpool due to Kelvin–Helmholtz instability and the growth rate of this instability will linearly increase with the wave number for the plane interface. So in the given case for short perturbations with wavelength much less than R, one should expect instability. However, the preliminary analysis of stability/instability of the whirlpools reveals a latitude-dependent nontrivial stability criterion for the vertical structure of the model. Therefore, one of our goals for further studies is to perform asymptotic stability analysis of the resulting nonlinear whirlpools with the aim to relate them with whirlpools observed in nature. For example, in the world ocean, as a rule, whirlpools are observed in the presence of external flows which, for instance, can be connected with tides.

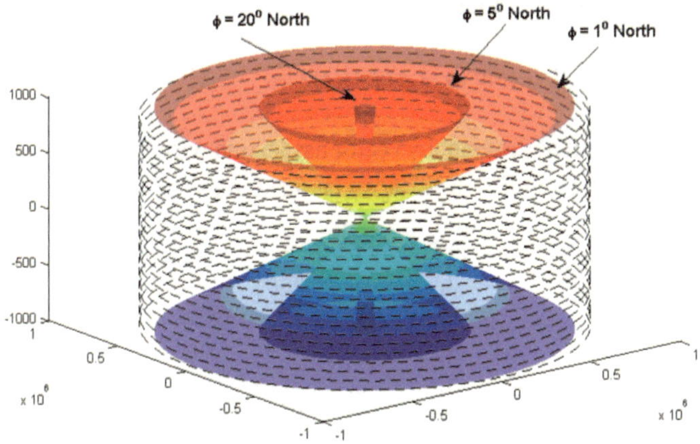

Figure 8.15: Effects of rotation on the nonlinear whirlpool for different latitudes: $\phi = 1°$, $\phi = 5°$, and $\phi = 20°$ North.

8.6 Weakly nonlinear model

We now expand the regular perturbation (i. e., Born's approximation) in a power series of ϵ, i. e.,

$$u_r = \epsilon u_r', \quad u_\theta = u_\theta^0 + \epsilon u_\theta', \quad w = \epsilon w', \quad p = p^0 + \epsilon p', \quad \rho = \rho^0 + \epsilon \rho', \tag{8.151}$$

where $\epsilon \ll 1$ characterizes the size of the perturbation due to the waves and the basic state $(u_\theta^0, \overline{p}, \overline{\rho})$. By neglecting the quadratic terms of the perturbation, we linearize the Boussinesq model (8.9)–(8.13) with the expansion (8.151) to the form (the primes are omitted)

$$Du_r = -\frac{\partial p}{\partial r}, \quad Du_\theta = -\frac{1}{r}\frac{\partial p}{\partial \theta}, \quad Dw = -\frac{\partial p}{\partial z} - \rho g, \quad D\rho = \frac{N^2}{g} w, \tag{8.152}$$

$$\frac{1}{r}\frac{\partial (r u_r)}{\partial r} + \frac{1}{r}\frac{\partial u_\theta}{\partial \theta} + \frac{\partial w}{\partial z} = 0, \tag{8.153}$$

where

$$D = \frac{\partial}{\partial t} - \frac{f}{2}\frac{\partial}{\partial \theta}. \tag{8.154}$$

In view of the presence of the operator D in system (8.152)–(8.153), it is natural to introduce the new independent variable

$$\tau = \theta + \frac{f}{2} t \tag{8.155}$$

and regard dependent variables as a function of τ, e. g.,

$$w(r, z, \theta, t) = w^*(r, z, \tau[\theta, t], t), \tag{8.156}$$

where w^* is the vertical component of the velocity vector associated with the coordinate τ. Thus, the weakly nonlinear model (8.152)–(8.153) can be written as

$$\frac{\partial u_r^*}{\partial t} = -\frac{\partial p^*}{\partial r}, \tag{8.157}$$

$$\frac{\partial u_\theta^*}{\partial t} = -\frac{1}{r}\frac{\partial p^*}{\partial \tau}, \tag{8.158}$$

$$\frac{\partial w^*}{\partial t} = -\frac{\partial p^*}{\partial z} - \rho^* g, \tag{8.159}$$

$$\frac{\partial \rho^*}{\partial t} = \frac{N^2}{g} w^*, \tag{8.160}$$

$$\frac{1}{r}\frac{\partial (r u_r^*)}{\partial r} + \frac{1}{r}\frac{\partial u_\theta^*}{\partial \tau} + \frac{\partial w^*}{\partial z} = 0. \tag{8.161}$$

The system of linearized equations (8.157)–(8.161) reduces to the linear equation for the pressure perturbation, which is commonly used in dynamical meteorology. The resulting equation is

$$\frac{\partial^2}{\partial t^2}\Delta p^* + N^2\Delta_h p^* = 0, \tag{8.162}$$

where

$$\Delta_h = \frac{1}{r}\frac{\partial}{\partial r}\left(r\frac{\partial}{\partial r}\right) + \frac{1}{r^2}\frac{\partial^2}{\partial \theta^2} \tag{8.163}$$

is the two-dimensional Laplace–Beltrami operator in polar angles and

$$\Delta = \Delta_h + \frac{\partial^2}{\partial z^2}. \tag{8.164}$$

For example, returning to the original coordinates, equation (8.162) is written as

$$\frac{\partial^2}{\partial t^2}\Delta p + N^2\Delta_h p + \frac{f^2}{4}\frac{\partial^2}{\partial \theta^2}\Delta p - f\frac{\partial^2}{\partial t\partial \theta}\Delta p = 0. \tag{8.165}$$

The corresponding boundary conditions at the bottom and the surface for equation (8.162) are

$$\left.\frac{\partial p^*}{\partial z}\right|_{z=0} = \left.\frac{\partial p^*}{\partial z}\right|_{z=H} = 0, \tag{8.166}$$

whereas at the lateral boundary we have

$$\left.\frac{\partial p^*}{\partial r}\right|_{r=r_0} = 0. \tag{8.167}$$

We employ the usual normal mode analysis and allow solutions to be found as expansions of vertical modes $\psi_n(z)$ of the form

$$p_n^* = \sum_n p_n^*(r,\theta,t)\psi_n(z), \tag{8.168}$$

where the uniform mean vertical stratification $\overline{p}^*(z)$ is absorbed into the definition of the pressure and density so that p^* has units of (potential) energy per unit mass and ρ^* represents (dimensionless) density anomaly [57].

In view of the boundary conditions (8.166), it is more convenient to look for a specific example of the idealized vertical profile $\psi_n(z)$, e. g.,

$$p_n^* = \sum_n \widehat{p}_n^*(r)\cos(n\theta - \omega_n t)\cos(m_n z), \tag{8.169}$$

where m_n is the vertical wave number vector. The first boundary condition in (8.166) is then automatically satisfied with the choice of the sinusoidal vertical structure (8.169), whereas the second boundary condition in (8.166) implies that there exist an infinite number of internal wave modes satisfying the condition

$$m_n H = n\pi, \quad n = \pm 1, \pm 2, .., \tag{8.170}$$

which also agrees with the asymptotic expression obtained in Ref. [45], where the internal gravity waves in rectangular geometry with zero rotation were studied.

Substitution of the presentation (8.169) into equation (8.162) yields an ODE for $\widehat{p}_n^*(r)$:

$$\frac{d^2\widehat{p}_n^*}{dr^2} + \frac{1}{r}\frac{d\widehat{p}_n^*}{dr} + \left[\frac{m_n^2\,\widehat{\omega}_n^2}{N^2 - \widehat{\omega}_n^2} - \frac{n^2}{r^2}\right]\widehat{p}_n^* = 0, \tag{8.171}$$

where we denote

$$\widehat{\omega}_n = \omega_n + \frac{fn}{2}. \tag{8.172}$$

We denote

$$\frac{m_n^2\,\widehat{\omega}_n^2}{N^2 - \widehat{\omega}_n^2} = \alpha_n^2 > 0. \tag{8.173}$$

We next introduce a new independent variable ξ defined via

$$\xi = \alpha_n r \tag{8.174}$$

and regard $\widehat{p}$ as a function of ξ. Thus, equation (8.171) is written as the Bessel equation of order n with the single singular point at $\xi = 0$, i. e.,

$$\xi^2\frac{d^2\widehat{p}_n}{d\xi^2} + \xi\frac{d\widehat{p}_n}{d\xi} + (\xi^2 - n^2)\widehat{p}_n = 0. \tag{8.175}$$

Correspondingly, the solution of equation (8.175) is given by the Bessel function of the first kind of order n:

$$\widehat{p}_n^*(\xi) = J_n(\xi) = \left(\frac{\xi}{2}\right)^n \sum_l \frac{(-1)^l}{l!(l+n)!}\left(\frac{\xi}{2}\right)^{2l}. \tag{8.176}$$

Thus, the solution of equation (8.171) is written as

$$p^*(r,\tau,z,t) = \sum_n J_n(\alpha_n r)\cos\left(n\left[\tau - \frac{f}{2}t\right] - \omega_n t\right)\cos(m_n z), \tag{8.177}$$

where the infinite discrete set of the frequencies ω_n is determined from the boundary condition (8.167) at $r = r_0$.

8.6.1 Evanescent modes

In a greater generality, instead of the notation (8.173) we can write

$$\frac{m_n^2 \widehat{\omega}_n^2}{N^2 - \widehat{\omega}_n^2} = s\alpha_n^2, \tag{8.178}$$

where

$$\alpha_n^2 > 0 \quad \text{and} \quad s = \pm 1 \text{ or } s = 0. \tag{8.179}$$

The case $s = 0$ corresponds to the flow that can be realized at frequencies

$$\omega_n = -\frac{fn}{2} \tag{8.180}$$

and equation (8.171) becomes

$$r^2 \frac{d^2 \widehat{p}_n^*}{dr^2} + r\frac{d\widehat{p}_n^*}{dr} - n^2 \widehat{p}_n^* = 0, \tag{8.181}$$

with the particular solutions of the form $\widehat{p}_n^* = r^n$ or $\widehat{p}_n^* = r^{-n}$. In this case, the boundary condition (8.167) does not provide the oscillatory modes.

Similarly, for the case $s = -1$, we obtain

$$\omega_n > N - \frac{fn}{2}, \tag{8.182}$$

which provides a reasonable range of admissible frequencies. However, in this case, transformation (8.174) reduces equation (8.171) to the modified Bessel equation

$$\xi^2 \frac{d^2 \widehat{p}_n}{d\xi^2} + \xi \frac{d\widehat{p}_n}{d\xi} - (\xi^2 + n^2)\widehat{p}_n = 0 \tag{8.183}$$

with two particular solutions $\widehat{p}_n^* = I_n(\xi)$ and $\widehat{p}_n^* = K_n(\xi)$, where $I_n(\xi)$ and $K_n(\xi)$ are the modified Bessel functions of the first and second kind, respectively. Since these two functions do not oscillate, the boundary condition (8.167) does not provide the oscillatory modes as well.

8.6.2 Neutral stability

We now consider the case when the frequency can be complex, e. g., we set

$$\omega = \omega_r + i\omega_i, \tag{8.184}$$

where $\omega_r = \mathrm{Re}\{\omega\}$ and $\omega_i = \mathrm{Im}\{\omega\}$.

First, we consider the condition (8.178) with $s = +1$ under assumption (8.184). If $\omega_i \neq 0$, one can see that the condition (8.178) is met if $n^2\pi^2/H^2 = -\alpha_{n,k}^2 > 0$, which is not valid.

Now, we consider the condition (8.178) with $s = -1$. Similarly, we can find that ω_i can be expressed in terms of ω_r by the relation

$$\omega_i = \pm\sqrt{\widehat{\omega}_r^2 - \frac{\frac{\alpha_{n,k}^2}{m_n^2}N^2}{\frac{\alpha_{n,k}^2}{m_n^2} - 1}}, \tag{8.185}$$

in which $\widehat{\omega}_r = \omega_r + fn/2$. In case when $\omega_i \neq 0$, one can see that the relation (8.178) is constrained by the condition that either $\alpha_{n,k}^2 = m_n^2$ or $N = 0$ (which is not valid for stratified media). However, it is possible to investigate the dynamic stability of the solution within the asymptotic approximations in the case when $\alpha_{n,k}^2 \approx m_n^2$. The results of these studies will be reported elsewhere.

8.7 Existence of nonlinear whirlpools

A general presentation of solutions in the form (8.177) is valid only if α_n given by (8.173) is real. This leads to the condition

$$\omega_n < \gamma_{n,\phi} = N - \frac{fn}{2}, \tag{8.186}$$

in which we shall refer to $\gamma_{n,\phi}$ as the latitude-dependent *cut-off frequency* of a nonlinear whirlpool. Alternatively, for given values ω_n, this condition can be used to identify the modes n of the nonevanescent whirlpool motion, i. e., the whirlpools exist for such modes that

$$n < \frac{2(N - \omega_n)}{f}. \tag{8.187}$$

For example, at latitude $\phi = 20°$ North and at the value $N = 10^{-3}\,\mathrm{s}^{-1}$, the condition (8.187) yields $n < 38$. For example, at latitude $\phi = 20°$ North and at the value $N = 10^{-3}\,\mathrm{s}^{-1}$, the condition (8.187) yields $n < 38$.

Since the frequency band of internal waves is limited, we have

$$|f| < |\omega| < |N|. \tag{8.188}$$

In view of notation (8.173), the condition (8.186) can also be written as

$$\delta_{n,H} = \frac{\alpha_n}{\sqrt{m_n^2 + \alpha_n}} < 1, \tag{8.189}$$

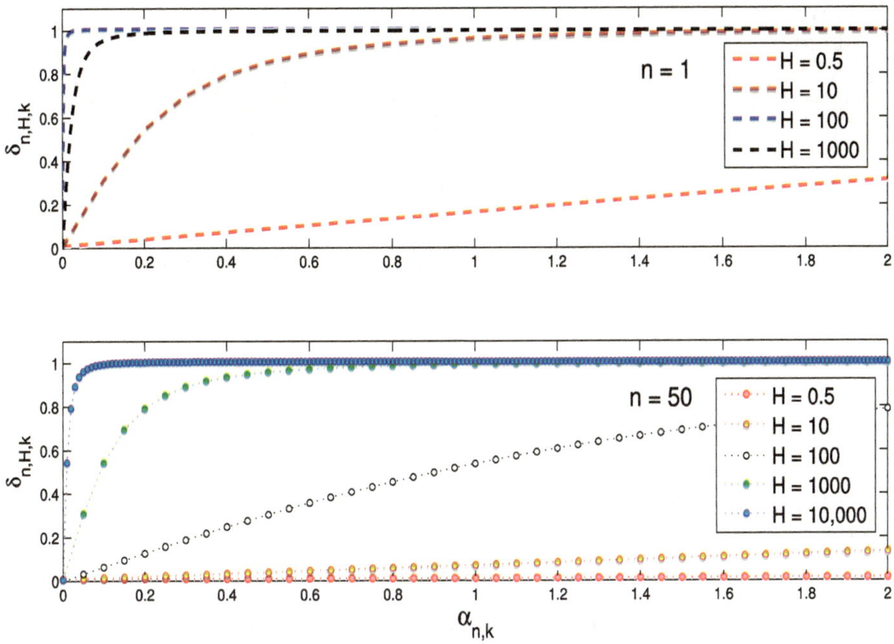

Figure 8.16: The nonlinear whirlpool exists for such values of H that $\delta_{n,H} < 1$.

which is always true, as also illustrated in Figure 8.16, showing the effects of increasing H on the behavior of $\delta_{n,H}$ versus α_n. Particularly, since $\alpha_n \in (0, \infty)$ for any latitude ϕ, the limitation (8.189) is independent of latitude, as expected.

8.7.1 Two examples

Here we consider two illustrative examples for the “localized whirlpools” whose persistence does not depend on the boundary condition (8.167), e. g., it does not depend on the temperature differences of the water column, as, e. g., the whirlpool shown in Figure 8.15.

Example 8.1. Since under oceanic conditions, as a rule, whirlpools are observed in the presence of external flows, which, for instance, can be connected with tides, we consider the flows governed by the tidal M_2 frequency, i. e.,

$$\omega_0 = \omega_{M_2} \sim 1.4 \times 10^{-4}\,\mathrm{s}^{-1}, \tag{8.190}$$

where ω_{M_2} denotes the tidal frequency.

Figure 8.17 compares the frequency ω_{M_2} given by (1.8) with the *critical frequency* $\sigma_{n,\phi}$ defined by (8.186). The simulations were done using the values $N = 10^{-3}\,\mathrm{s}^{-1}$ and

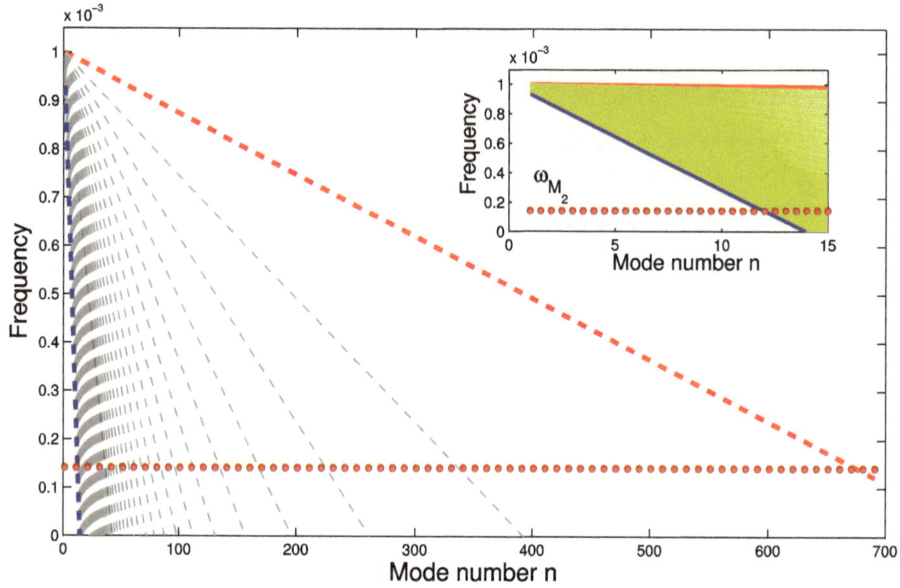

Figure 8.17: Existence of oscillatory flows far away for lateral circular boundary $r = r_0$ forced at M_2 tidal frequency $\omega_{M_2} \sim 1.4 \times 10^{-4}\,s^{-1}$.

$r_0 = 10^4$ m. In Figure 8.17 both frequencies are plotted versus the mode number n. The red circles are used to show the constant frequency ω_{M_2}, and the critical frequency $\sigma_{n,\theta}$ is plotted as a red dashed line for latitude $\theta = 1°$ North and as a blue dashed red line for latitude $\theta = 80°$ North. The frequencies for the values of latitude $\theta \in [2°, 79°]$ North are plotted as gray dashed lines, with a "latitude step size" of $\Delta\theta = 1°$. We observe that for $\theta = 1°$ North, the tidally generated whirlpool exists for sufficiently large values of mode numbers $n \leqslant 670$. However, at latitude $\theta = 80°$ North, the tidally generated whirlpool exists for smaller values of n. In particular, the perturbed whirlpool solution exists for $n \leqslant 11$, which is illustrated in more detail for a smaller latitude range in the upper right corner of Figure 8.17. □

Example 8.2. Since waves can be generated not only at the fundamental frequency but also at all of its harmonics $\omega_n = n\omega_0$, we next consider the case of an infinite number of discrete internal wave modes satisfying the dispersion relationship. Here, again, we use the values $N = 10^{-3}\,s^{-1}$ and $\omega_0 = \omega_{M_2}$, given by (8.190). In the upper panel of Figure 8.18, the frequency ω_{M_2} is compared with the *critical frequency* $\sigma_{n,\phi}$. Both frequencies are plotted versus the mode number n. The blue and yellow circles are used to show the constant values of the frequency $\omega = \omega_{M_2}$ and $\omega = n\omega_{M_2}$, respectively, for $n \in [1, 8]$, whereas the critical frequency $\sigma_{n,\phi}$ is plotted as red and blue lines for latitude $\theta = 1°$ North and $\theta = 80°$ North, respectively, and the frequencies $\sigma_{n,\phi}$ for the values of latitude $\theta \in [2°, 79°]$ North are plotted as gray lines with a "latitude step size" of $\Delta\theta = 1°$. For

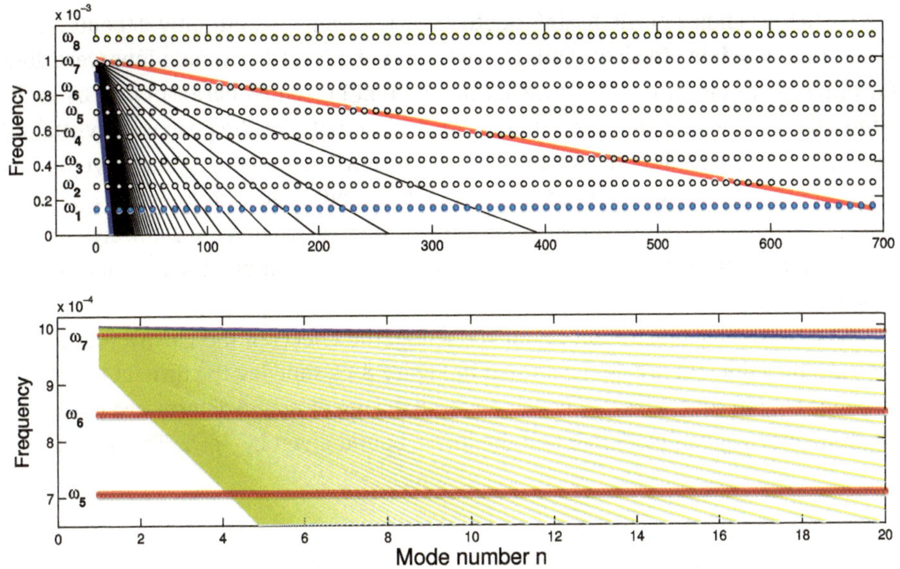

Figure 8.18: Existence of perturbed flows far away for lateral circular boundary $r = r_0$ forced by harmonic waves $\omega = n\omega_{M_2}$.

example, the lower panel of Figure 8.18 shows a more detailed plot for $n \in [5, 7]$, from which we observe that the harmonic waves exist only if $n \leqslant 4$ for $\theta = 80°$ North. □

8.7.2 Oscillatory solution

Substitution of the solution (8.177) into the condition (8.167) gives the relation

$$G_n := J_n'(\xi_n^0) = 0, \tag{8.191}$$

in which

$$\xi_n^0 = \alpha_n r_0, \tag{8.192}$$

where α_n is defined by (8.173). Equation (8.191) guarantees that an infinite set of oscillations exists provided that ω_n and ξ_n^0 are related by the equation

$$\omega_{n,k} = \pm\left(\frac{N^2[\xi_{n,k}^0]^2/r_0^2}{m_n^2 + [\xi_{n,k}^0]^2/r_0^2}\right)^{\frac{1}{2}} - \frac{fn}{2}, \tag{8.193}$$

for each solution number k of equation (8.191). The eigenfrequencies $\omega_{n,k}$ are determined as follows. First, we find numerically the roots $\xi_{n,k}^0$ ($k = 1, 2, 3, \ldots$) of equation (8.191). We remark that the numerical values of the roots $\xi_{n,k}^0$ do not depend on the particular choice of the basin's depth H or the radius r_0, but only depend on the aspect ratio $\sigma = H/r_0$. In view of this remark, most of our further simulations were done for different values of $\sigma < 1$. However, the method presented here can also be easily adjusted for the aspect ratios $\sigma > 1$ or $\sigma = 1$. Additionally, for the purpose of visualization of the analytic solutions, we consider the case when $\xi \in [0, \xi_{\max}]$, and we further set $\xi_{\max} = 100$. Next, for each root $\xi^0_{n,k}$, the corresponding eigenfrequencies $\omega_{n,k}$ were obtained from the dispersion relation (8.193). In the case when $N^2 = 0$, the eigenvalues of the similar eigenvalue problem (8.192)–(8.193) have been obtained numerically in Kudlick [46].

As a particular illustration to our modeling, the effects of the Earth's rotation on the existence domain are demonstrated in Figure 8.19, showing the distribution of the obtained eigenfrequencies $\omega_{n,k}$ versus ξ within the upper bound determined by the cut-off frequency γ given by (8.186) and the lower bound given by $-fn/2$. The aspect ratio

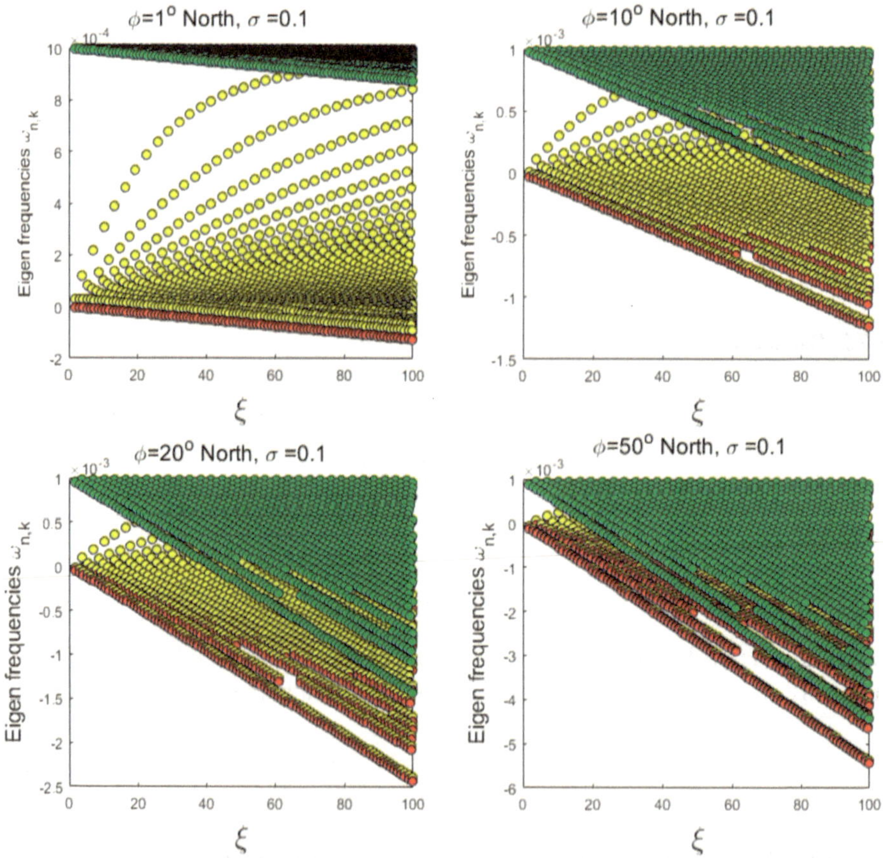

Figure 8.19: The eigenfrequencies $\omega_{n,k}$ of oscillatory solutions bounded by $-fn/2$ and $\gamma_{n,\phi}$ at latitudes ϕ = 1° North, ϕ = 10° North, ϕ = 20° North, and ϕ = 50° North.

is taken as $\sigma = 0.1$. As we observe from Figure 8.19, the domain of the existence of oscillatory solutions becomes smaller at larger values of latitude ϕ.

Returning to the original variables, we can write the solution for small perturbations in the following form:

$$p(r,\theta,z,t) = \sum_{n=1}^{n_{\max}} \sum_{k=1}^{k_{\max}} J_n(\xi_{n,k}) \cos(n\theta - \omega_{n,k}t) \cos(m_n z), \tag{8.194}$$

$$u_r(r,\theta,z,t) = \sum_{n=1}^{n_{\max}} \sum_{k=1}^{k_{\max}} \frac{a_{n,k}}{\widehat{\omega}_{n,k}} J_n'(\xi_{n,k}) \sin(n\theta - \omega_{n,k}t) \cos(m_n z), \tag{8.195}$$

$$u_\theta(r,\theta,z,t) = \sum_{n=1}^{n_{\max}} \sum_{k=1}^{k_{\max}} \frac{n}{r\widehat{\omega}_{n,k}} J_n(\xi_{n,k}) \cos(n\theta - \omega_{n,k}t) \cos(m_n z), \tag{8.196}$$

$$w(r,\theta,z,t) = \sum_{n=1}^{n_{\max}} \sum_{k=1}^{k_{\max}} \frac{m_n \widehat{\omega}_{n,k}}{N^2 - \widehat{\omega}_{n,k}^2} J_n(\xi_{n,k}) \sin(n\theta - \omega_{n,k}t) \sin(m_n z), \tag{8.197}$$

$$\rho(r,\theta,z,t) = \sum_{n=1}^{n_{\max}} \sum_{k=1}^{k_{\max}} \frac{m_n N^2}{g(N^2 - \widehat{\omega}_{n,k}^2)} J_n(\xi_{n,k}) \cos(n\theta - \omega_{n,k}t) \sin(m_n z), \tag{8.198}$$

where $\omega_{n,k}$ are determined from (8.193).

The solution set (8.194)–(8.198) provides insight into the distribution of pressure, density, and velocity within a perturbed whirlpool. The occurrence of such a whirlpool is contingent upon the existence of frequencies ω, which are only found in a limited set of frequencies known as eigenfrequencies and denoted by $\omega_{n,k}$. In simpler terms, these are frequencies that satisfy the dispersion relation (8.38) and correspond to the frequencies in which the solution of the model (8.14)–(8.18) exists, provided that the boundary conditions (8.19) are met. Eigenfrequencies for specific parameters of the model are depicted in Figures 8.4 and 8.19. In nature, whirlpools can exist within cylindrical domains, such as the Guatavita lagoon (Figure 8.2) or other domains that restrict the whirlpool to a cylindrical domain due to natural constraints, as exhibited in Figure 8.1, showing two massive whirlpools in the Atlantic Ocean off the coast of Guyana and Suriname, or in Figure 8.3, showcasing a colossal whirlpool in the atmosphere. Mathematically, the whirlpool can be visualized as demonstrated in Figure 8.13.

The primary purpose of this chapter was to study the effects of the Earth's rotation on small-amplitude oscillatory motions in density-stratified fluid in a cylindrical basin. The new features "*cut-off frequency*" and "*geometric aspect ratio of the basin*" were identified in the present modeling scenario. One of the possible directions for further studies would be to perform asymptotic analysis for the case $a_{n,k}^2 \approx m_n^2$ in formula (8.185) in order to investigate the stability of the resulting nonlinear whirlpools with the goal to relate them with the whirlpools observed in nature. For example, under oceanic conditions, as a rule, whirlpools are observed in the presence of external flows, which, for instance, can be connected with tides.

8.8 Exercises

Exercise 8.1. Can you show that the operator is admitted by the system (8.47)–(8.51)?

Exercise 8.2. Can you show that the operator

$$X_4 = r\frac{\partial}{\partial r} + z\frac{\partial}{\partial z} + u_\theta\frac{\partial}{\partial u_\theta} + w\frac{\partial}{\partial w} + 2p\frac{\partial}{\partial p} + \rho\frac{\partial}{\partial \rho}$$

is admitted by the system (8.47)–(8.51)?

Exercise 8.3. Check that a general solution of the system reducing to the form

$$\begin{aligned}
\frac{\partial u_\theta}{\partial t} &= 0,\\
\frac{\partial p}{\partial r} &= \frac{u_\theta^2}{r} + fu_\theta,\\
\frac{\partial w}{\partial t} &= -\rho g,\\
\frac{\partial \rho}{\partial t} &= \frac{N^2}{g}w
\end{aligned}$$

is given by

$$\begin{aligned}
&u_\theta = U(r), \quad p = \int\left[\frac{1}{r}U^2(r) + fU(r)\right]dr + V(t - t_0),\\
&w = W_1(r)\cos N(t - t_0) + W_2(r)\sin N(t - t_0),\\
&\rho = \frac{N^2}{g}[W_1(r)\sin N(t - t_0) - W_2(r)\cos N(t - t_0)],
\end{aligned}$$

where $U(r)$, $W_1(r)$, $W_2(r)$, and $V(t)$ are arbitrary functions and t_0 is an arbitrary constant.

Exercise 8.4. Check if the operators

$$\begin{aligned}
X_1 &= \frac{\partial}{\partial t}, \quad X_2 = \frac{\partial}{\partial \theta}, \quad X_3 = \frac{\partial}{\partial z}, \quad X_\varphi = g\varphi(z)\frac{\partial}{\partial p} - \varphi'(z)\frac{\partial}{\partial \rho}, \quad X_\psi = \psi(t)\frac{\partial}{\partial p},\\
X_4 &= r\frac{\partial}{\partial r} + z\frac{\partial}{\partial z} + u_\theta\frac{\partial}{\partial u_\theta} + w\frac{\partial}{\partial w} + 2p\frac{\partial}{\partial p} + \rho\frac{\partial}{\partial \rho},\\
X_5 &= 2(ft + 2\theta)\frac{\partial}{\partial \theta} - 4r\frac{\partial}{\partial r} + 2fr\frac{\partial}{\partial u_\theta} + f^2r^2\frac{\partial}{\partial p}.
\end{aligned}$$

form a Lie algebra.

9 Modeling scenario 6: invariant solutions of internal waves in the ocean

In this chapter, the main objective is to derive the exact solutions of the nonlinear two-dimensional Boussinesq model for internal gravity waves, which cannot be guessed from the anisotropic and other physical properties. The new class of exact solutions were deduced by means of dilations and rotation symmetries and these exact solutions possess the unique property of spinning, which has not been reported in previous studies.

9.1 Mathematical modeling

Here we consider the model corresponding to two-dimensional (y-independent) internal wave propagation in the ocean. In two dimensions, the nonlinear Boussinesq equations (8.1)–(8.3) are written as

$$u_t + uu_x + wu_z - fv = -p_x, \tag{9.1}$$

$$v_t + uv_x + wv_z + fu = 0, \tag{9.2}$$

$$w_t + uw_x + ww_z = -p_z - \rho g, \tag{9.3}$$

$$\rho_t + u\rho_x + w\rho_z = 0, \tag{9.4}$$

$$u_x + w_z = 0, \tag{9.5}$$

where all variables depend on x, z, and t. Note that although the motion is two-dimensional, the term $-fv_z$ is presented in equations because of nonzero Earth rotation.

We look for a solution of system (9.1)–(9.5) in the form

$$\vec{u} = \vec{0} + \vec{u}', \quad \rho = \rho_0 + \overline{\rho}(z) + \rho', \quad p = \overline{p}(z) + p'. \tag{9.6}$$

Since $\overline{\rho}(z)$ in (9.6) can be expressed in terms of the buoyancy frequency N defined by (8.5), we can express the total fluid density as

$$\rho = \rho_0\left(1 - \frac{N^2}{g}z + \frac{\rho'}{\rho_0}\right). \tag{9.7}$$

Using the expansion (9.7), we can write the right side term in equation (9.3) as

$$-p_z - \rho g = \rho_0 g + g\overline{\rho} - p_z' - \rho_0 g + \rho_0 N^2 z - \rho' g = p_z' - \rho' g.$$

Thus, substituting the presentation (9.6) in the governing equations (9.1)–(9.5), we arrive at the following nonlinear equations for perturbed quantities:

$$u_t' + u'u_x' + w'u_z' - fv' = -p_x', \tag{9.8}$$

https://doi.org/10.1515/9783111387499-009

$$v'_t + u'v'_x + w'v'_z + fu' = 0, \tag{9.9}$$

$$w'_t + u'w'_x + w'w'_z = -p'_z - \rho' g, \tag{9.10}$$

$$\rho'_t + u'\rho'_x + w'\rho'_z - \frac{N^2}{g} w' = 0, \tag{9.11}$$

$$u'_x + w'_z = 0. \tag{9.12}$$

The incompressibility condition (9.12) can be taken into account by introducing a *stream function* ψ which is related to the two velocity components u' and w' by

$$u' = \psi_z, \quad w' = -\psi_x. \tag{9.13}$$

We next substitute (9.13) into equations (9.8)–(9.12) and exclude the pressure p' by subtracting equation (9.11) differentiated with respect to x from equation (9.8) differentiated with respect to z to arrive at the following three-equation system for ψ, ρ', and v':

$$\Delta\psi_t - g\rho_x - fv_z = \psi_x \Delta\psi_z - \psi_z \Delta\psi_x, \tag{9.14}$$

$$v_t + f\psi_z = \psi_x v_z - \psi_z v_x, \tag{9.15}$$

$$\rho_t + \frac{N^2}{g}\psi_x = \psi_x \rho_z - \psi_z \rho_x, \tag{9.16}$$

where the primes are omitted, the constant density value ρ_0 is normalized to one, and Δ denotes the two-dimensional Laplacian:

$$\Delta = \frac{\partial^2}{\partial x^2} + \frac{\partial^2}{\partial z^2}, \quad \text{e.g., } \Delta\psi_t = \frac{\partial^2 \psi_t}{\partial x^2} + \frac{\partial^2 \psi_t}{\partial z^2}. \tag{9.17}$$

In the linear approximation, equations (9.14)–(9.16) can be written as the single equation

$$(\Delta\psi)_{tt} + N^2 \psi_{xx} + f^2 \psi_{zz} = 0. \tag{9.18}$$

Since the coefficients of equation (9.18) are independent of the horizontal and vertical coordinates x and z, respectively, this equation admits solutions that are trigonometric in x and z. We can therefore assume the solution has the form of a wave packet, i. e.,

$$\psi(x, z, t) = \Psi e^{i(kx + mz - \omega t)}, \tag{9.19}$$

where Ψ is a constant. Substitution of the assumed solution (9.19) into the resulting linear equation (9.18) shows that this equation is satisfied identically provided that the following *dispersion relation* between the frequency ω and the wave number $\vec{k} = (k, m)$ is satisfied:

$$\omega^2 = \frac{N^2 k^2 + f^2 m^2}{k^2 + m^2}. \tag{9.20}$$

The nonlinear model (9.1)–(9.5) is constrained by the anisotropic nature of the internal wave motion, which means that the frequency ω of internal waves depends only on the orientation of the wave number vector $\vec{k} = (k, m)$ but not on its magnitude, which is also apparent from the dispersion relation (9.20).

The anisotropic property (9.20) for the nonlinear internal wave model (9.14)–(9.16) can also be written in the trigonometric form

$$\omega^2 = N^2 \sin^2 \alpha + f^2 \cos^2 \alpha, \tag{9.21}$$

where α is the beam inclination relative to the horizontal.

In terms of Lie group analysis, it can be shown that unidirectional internal wave beams, i. e., the nonlinear solution obeying the dispersion relation (9.20), can also be obtained as invariant solutions of nonlinear equations of motion. Namely, the invariant solution based on the dilation and translation transformations can be represented by unidirectional internal wave beams of the form

$$\psi(\eta, t) = e^{-i\omega t} \int_{-\infty}^{0} \hat{Q}(s) e^{is\eta} ds + e^{i\omega t} \int_{0}^{\infty} \hat{Q}^*(s) e^{-is\eta} ds, \tag{9.22}$$

where the asterisk denotes the complex conjugate and $\eta = kx + mz = x \sin\alpha - z \cos\alpha$ is the invariant of the translational symmetry $X = m\frac{\partial}{\partial x} - k\frac{\partial}{\partial z}$. Physically, η represents the direction of internal wave beam propagation. The latter invariant solution coincides with the exact solutions deduced from the anisotropic property of internal waves in the previous studies in Tabaei & Akylas [75]. In our notation

$$\hat{Q}(s) = \int_{-\infty}^{\infty} Q(\eta) e^{-is\eta} ds \tag{9.23}$$

is the Fourier transform of the initial condition for $\psi(\eta, 0)$, which represents an amount of energy radiated along the internal plane-wave beam at the angle α in the positive or negative (depending on the sign of s) direction. In fact, the anisotropic property expressed by (9.20) makes it possible to construct, via superposition of sinusoidal plane waves with wave numbers inclined at the same angle α to the vertical, general plane-wave disturbances of frequency ω in the form of beams that are uniform along the transverse direction $\xi = x \cos\alpha + z \sin\alpha$ and have a general profile along the invariant η. Furthermore, since $Q(\eta)$ is an arbitrary wave amplitude of a superposition of plane waves, the direction of beam propagation satisfying the initial condition $\psi(\eta, 0) = Q(\eta)$ is determined uniquely, since for any sinusoidal wave of the form $h(x) = a \sin(kx - \omega t)$ the phase velocity $\vec{C}$ can be written as $(\omega/k^2)\vec{k}$ and the sign of s in formula (9.22) is determined uniquely by the sign of the frequency. In particular, the direction of energy propagation is along the group velocity $\vec{C}_g = \nabla_{\vec{k}}\omega$, which, according to the anisotropic property (9.21), is orthogonal to the phase velocity, so that $\vec{C}$ and $\vec{C}_g$ have opposite vertical components. While the invariant solution (9.22) describes plane-wave beams for any choice of amplitude $Q(\eta)$, unidirectional beams, in which energy propagates in one direction, involve plane waves with wave numbers of the same sign only. For example, in view of four possible configurations of the group $\vec{C}_g$ and phase $\vec{C}$ velocities, for the energy to be radiated upward to the right, the radiation condition requires that $k > 0$ and $m < 0$ (i. e., the wave number vector $\vec{k}$ is along η and thus directed downward to the right), and so forth.

It is worth noting that in 2003, Tabaei and Akylas [75] pointed out that uniform plane-wave beams that obey the linear dispersion relation (9.20) with $f = 0$ in a uniformly stratified Boussinesq fluid are also solutions of the nonlinear inviscid equations of motion (9.14)–(9.16). Namely, the latter invariant solution (9.22) coincides with the exact solutions deduced from the anisotropic property of internal waves in previous studies [75]. However, all existing theoretical models and previous discussions of internal gravity wave beams in general are based either on the linearized equations of motion or on the anisotropic properties of internal waves. On the other hand, it so happens that more general forms of uniform plane-wave beams that obey the dispersion relation (9.20) can be obtained as invariant solutions of the nonlinear equations. This observation, which apparently had passed unnoticed in previous studies, proves most useful here in developing a group theoretic model presented in the first part of this book. Since this invariant solution can be deduced from the anisotropic property of internal waves, we will not focus on its derivation here. The primary focus in this section is on finding the more general forms of invariant solutions of the nonlinear two-dimensional Boussinesq model for internal gravity waves, which cannot be guessed from the anisotropic property and correspondingly were not reported in previous studies. In particular, the invariant solutions considered here will be deduced from the infinite-dimensional Lie algebra spanned by the infinitesimal symmetries.

9.2 Rotationally invariant solutions and comparison with linear theory

The maximal symmetry group admitted by equations (9.14)–(9.16) with two parameters f and N is generated by the infinite-dimensional Lie algebra spanned by nine operators:

$$X_1 = \frac{\partial}{\partial v}, \quad X_2 = \frac{\partial}{\partial \rho}, \quad X_3 = a(t)\frac{\partial}{\partial \psi}, \quad X_4 = \frac{\partial}{\partial t}, \tag{9.24}$$

$$X_5 = b(t)\left[\frac{\partial}{\partial x} - f\frac{\partial}{\partial v}\right] + zb'(t)\frac{\partial}{\partial \psi}, \tag{9.25}$$

$$X_6 = c(t)\left[\frac{\partial}{\partial z} + \frac{N^2}{g}\frac{\partial}{\partial \rho}\right] - xc'(t)\frac{\partial}{\partial \psi}, \tag{9.26}$$

$$X_7 = x\frac{\partial}{\partial x} + z\frac{\partial}{\partial z} + v\frac{\partial}{\partial v} + \rho\frac{\partial}{\partial \rho} + 2\psi\frac{\partial}{\partial \psi}, \tag{9.27}$$

$$X_8 = t\frac{\partial}{\partial t} + 2x\frac{\partial}{\partial x} + 2z\frac{\partial}{\partial z} + 3\psi\frac{\partial}{\partial \psi} - 2fx\frac{\partial}{\partial v} + \frac{2zN^2}{g}\frac{\partial}{\partial \rho}, \tag{9.28}$$

$$X_9 = z\frac{\partial}{\partial x} - x\frac{\partial}{\partial z} - \frac{1}{f}[g\rho - zN_f]\frac{\partial}{\partial v} + \frac{1}{g}[fv - xN_f]\frac{\partial}{\partial \rho}, \tag{9.29}$$

where $a(t)$, $b(t)$, and $c(t)$ are arbitrary functions of time t and $N_f = N^2 - f^2$ is a notation.

There are no group extensions in the case of homogeneous fluid ($N = 0$) and in the physically irrelevant case for the ocean $f = N$. In general, as is clear from the dispersion relation, propagating internal waves only exist for excitation frequencies in the range $|f| < |\omega| < |N|$. Since the frequency band of internal waves is limited, hereafter we limit our analysis to the case when

$$\frac{N^2 - \omega^2}{\omega^2 - f^2} > 0. \tag{9.30}$$

In agreement with Buchnev [15], the presence of the arbitrary functions $a(t)$, $b(t)$, and $c(t)$ in the symmetry Lie algebra is a characteristic property of incompressible fluids. Namely, the operator X_3 generates the group transformation $\overline{\psi} = \psi + \varepsilon_3 a(t)$ of the stream function ψ, where ε_3 is the group parameter. The invariance of fluid flows under this transformation is quite obvious because the velocity vector $(\psi_z, v, -\psi_x)$ is invariant under this transformation. The operators X_5, X_6 express the invariance under the generalization $\overline{x} = x + \varepsilon_5 b(t)$, $\overline{z} = z + \varepsilon_6 c(t)$ of the coordinate translations and the Galilean transformations.

For example, the operator X_9 represents the rotation symmetry given by the following transformations:

$$\widetilde{x} = x\cos\varepsilon + z\sin\varepsilon, \quad \widetilde{z} = z\cos\varepsilon - x\sin\varepsilon, \quad \widetilde{t} = t, \quad \widetilde{\psi} = \psi, \tag{9.31}$$

$$g\widetilde{\rho} = g\rho\cos\varepsilon + fv\sin\varepsilon - \frac{N^2 + f^2}{2}x\sin\varepsilon, \tag{9.32}$$

$$f\widetilde{v} = fv\cos\varepsilon - g\rho\sin\varepsilon + \frac{N^2 + f^2}{2}z\sin\varepsilon, \tag{9.33}$$

which depend on a continuous parameter ε. The latter transformations represent a *symmetry group* (or admitted group) for equations (9.14)–(9.16) in the sense that the latter equations have the same form when they are written in the new variables $\widetilde{x}$, $\widetilde{z}$, and so forth. The latter fact can be checked by direct differentiation and substitution.

Following the standard methodology of finding invariant solutions outlined in the first part of the book, we arrive at the following result. The nontrivial exact solution of the Boussinesq equations that is invariant with respect to the symmetries X_7 and X_9 is written as follows:

$$v = \frac{1}{f}\left[\left(2\phi^2(t) + A + \frac{N^2 - f^2}{2}\right)x + 2z\phi'(t)\right], \tag{9.34}$$

$$\rho = \frac{1}{g}\left[2x\phi'(t) - \left(2\phi^2(t) + A - \frac{N^2 - f^2}{2}\right)z\right], \tag{9.35}$$

$$\psi = (x^2 + z^2)\phi(t), \tag{9.36}$$

where A is a constant and $\phi(t)$ is the solution of the nonlinear boundary value problem

$$\phi'' + 2\phi^3 + \left(A + \frac{f^2 + N^2}{2}\right)\phi = 0, \quad \phi(0) = \phi_0, \quad \left.\frac{d\phi}{dt}\right|_{t=0} = \phi_0'. \tag{9.37}$$

In this notation, ϕ_0 is an arbitrarily chosen parameter and ϕ_0' is determined from the condition

$$\phi' = \pm\sqrt{B^2 - \phi^4 - K\phi^2}, \tag{9.38}$$

where B is a constant resulting from the integration of equation (9.37) and K is another constant given by

$$K \equiv A + \frac{f^2 + N^2}{2}. \tag{9.39}$$

Furthermore, the solution of equation (9.37) is a bounded oscillating function, satisfying the condition

$$-C_* \leqslant \phi(t) \leqslant C_*, \tag{9.40}$$

where C_* is the positive constant defined explicitly in terms of the parameters A, B, f, and N.

We remark that the invariant solution (9.34)–(9.36) includes terms independent of time and they are also a solution of the nonlinear model (9.14)–(9.16) by themselves with $\phi(t) = 0$. Thus, the resulting wave motion can also be discussed in terms of the perturbation to the changed background state. Hereafter, without loss of generality we set $A = 0$. The corresponding equations (9.14)–(9.16) linearized on the steady terms

$$v_0 = \frac{x(N^2 - f^2)}{2f}$$

and

$$\rho_0 = \frac{z(N^2 - f^2)}{2g}$$

are written as follows:

$$\Delta\psi_{tt} + \frac{N^2 + f^2}{2}\Delta\psi = 0. \tag{9.41}$$

As is clear from (9.41), the invariant solution (9.36) for $\psi(x, z, t)$ is also the solution of the linear model (9.18) provided that $\phi(t)$ solves the linear equation

$$\phi'' + \frac{f^2 + N^2}{2}\phi = 0. \tag{9.42}$$

Under the assumption of smallness of $\phi(t)$ expressed by (9.40), the resulting linear equation (9.42) coincides with the linearized equation (9.37) (in which the term $2\phi^3$ is dropped). We use the latter fact to illustrate the result (9.40) on the boundness of the invariant solution (9.34)–(9.36) by comparing the general solution of the linear model (9.42) with the numerical solution of the nonlinear boundary value problem (9.37) under the condition that both solutions satisfy the same initial conditions in (9.37) with $A = 0$. The general solution of the linear model (9.42) is given by

$$\phi(t) = c_1 \cos\left(\sqrt{\frac{f^2 + N^2}{2}}t\right) + c_2 \sin\left(\sqrt{\frac{f^2 + N^2}{2}}t\right), \tag{9.43}$$

where, according to (9.37), we set

$$c_1 = \phi_0, \quad c_2 = \pm\sqrt{\frac{2(B^2 - \phi_0^4)}{f^2 + N^2} - \phi_0^2}. \tag{9.44}$$

We next write the nonlinear model (9.37) in the equivalent form of the first-order coupled system

$$\frac{d\phi}{dt} = \varphi(t), \tag{9.45}$$

$$\frac{d\varphi}{dt} = -2\phi^3 - \frac{f^2 + N^2}{2}\phi, \tag{9.46}$$

where, according to (9.38), the initial conditions in (9.37) are written as

$$\phi(0) = \phi_0, \quad \varphi(0) = \pm\sqrt{B^2 - \phi_0^4 - \frac{f^2 + N^2}{2}\phi_0^2}. \tag{9.47}$$

The invariance under rotation means that the solution is valid at any circle by stretching the radius, and due to the boundness criterion (9.40), the linear approximation (9.43) for $\phi(t)$ predicts the spherically pulsating behavior. The conservation laws for the model (9.14)–(9.16) can also be written in the following integral form:

$$\frac{d}{dt}\iint v\,dxdz = 0, \quad \frac{d}{dt}\iint \rho\,dxdz = 0, \quad \frac{d}{dt}\iint\left[v^2 + \frac{g^2}{N^2}\rho^2 + |\nabla\psi|^2\right]dxdz = 0, \tag{9.48}$$

$$\frac{d}{dt}\iint\left[fxv - gz\rho - \frac{1}{2}|\nabla\psi|^2\right]dxdz = 0, \quad \frac{d}{dt}\iint\left[v\rho + fx\rho - \frac{N^2}{g}zv\right]dxdz = 0. \tag{9.49}$$

We remark that similar conservation laws were also obtained in Cho et al. [17], in which the classical and nonlinear stability of baroclinic jets was compared to symmetric disturbances for nonhydrostatic, adiabatic Boussinesq equations. Within the context of Fjortoft's theorem from *energy-Casimir* (or *pseudoenergy*) stability analysis [24], we associate the first conservation law in (9.49) with a Hamiltonian, which differs from the energy by Casimir invariants which constitute "hidden" conservation laws of the Eulerian representation of fluid flows. Thus, the second conservation law in (9.49) is another Casimir (see also [68]). However, it follows directly from Ref. [68] that the Casimir invariants are related to the symmetries of the Lagrangian representation.

The values for the densities of the conservation laws in (9.49) can be obtained by direct substitution of the expressions for v, ρ, and ψ given by (9.34)–(9.36) into the corresponding conservation laws. For example, the density corresponding to the first conservation form in (9.49) is written as

$$fxv - gz\rho - \frac{1}{2}|\nabla\psi|^2 = A(x^2 + z^2) + \left(\frac{N^2 - f^2}{2}\right)(x^2 - z^2). \tag{9.50}$$

The conservation of energy for the model (9.14)–(9.16) follows directly from the linear theory of internal waves. On the other hand, we remark that the energy conservation law can also be recovered for the nonlinear model in question by means of group theoretical modeling. In terms of the rotationally invariant solution (9.34)–(9.36), the *energy density* for the third conservation law in (9.48) is given by

$$\begin{aligned} E = {} & 4\left(\frac{1}{f^2} - \frac{1}{N^2}\right)(x^2 - z^2)[\phi^2(t) + K]\phi^2(t) \\ & + \left(N - \frac{K}{N}\right)^2 z^2 + 4\left(\frac{1}{f^2} - \frac{1}{N^2}\right)xz[2\phi^2(t) + K]\phi'(t), \end{aligned} \tag{9.51}$$

where we denote

$$K = A + \frac{f^2 + N^2}{2}.$$

It is remarkable that the conserved density (9.51), unlike the energy density, does not depend on the function $\phi(t)$. Another remarkable property is that the conserved density (9.51) is not rotationally invariant even though the solution is rotationally symmetric, while the energy density is both rotationally invariant and symmetric.

The effect of nonlinearity on the spinning phenomenon for rotationally invariant solutions (9.34)–(9.36) is demonstrated for the energy density E at latitude $\theta = 20°$ North in Figure 9.1. Namely, the first two columns in Figure 9.1 show the equipotential curves of the energy density E for the linear approximation (9.43) of $\phi(t)$, whereas the last

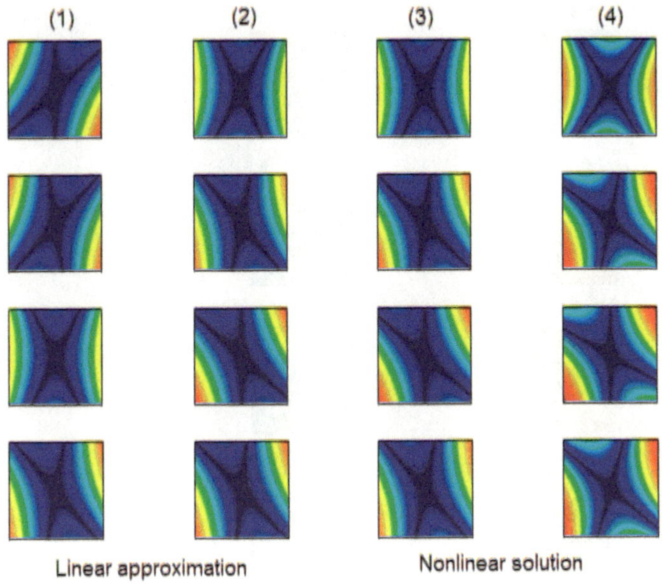

Figure 9.1: Effects of nonlinearity on the spinning phenomenon for the energy density $E = v^2 + \frac{g^2}{N^2}\rho^2 + |\nabla\psi|^2$ at $\theta = 20°$ North. The first two columns (1 and 2) show the equipotential curves of the energy density E for the linear approximation (9.43) of $\phi(t)$, whereas the last two columns (3 and 4) show the results of numerical simulations for the corresponding energy density for the case when $\phi(t)$ is given by the nonlinear problem (9.45)–(9.47). The average time step is 0.0088 inertial periods.

two columns (3 and 4) show the results of numerical simulations for the corresponding energy density for the case when $\phi(t)$ is given by the nonlinear problem (9.45)–(9.47). The comparison (based on the numerical estimates) between the results presented in columns 2, 3, and 4 shows that the spinning is more intensive at larger values of latitude. The first column shows the difference between the initial equilibrium position of the energy density for the linear and nonlinear cases. The effect of the Earth's rotation on the spinning phenomenon for rotationally invariant solutions (9.34)–(9.36) is demonstrated for the energy density E in Figure 9.2, which shows the results of numerical simulations for the corresponding energy density for the case when $\phi(t)$ is given by the nonlinear problem (9.45)–(9.47) at different latitudes. Namely, the first column shows the time series of the equipotential curves for the energy density at $\theta = 20°$ North. The second column shows the corresponding time series of the energy density at latitude $\theta = 60°$, and the third column shows the equipotential curves of the energy density E close to the critical latitude at $\theta = 74.2°$ North. The spinning intensity does not visibly appear to change significantly with changes in latitude (although a slight variation can be observed in a log scale), but there is change in the numerical values.

The next plot shown in Figure 9.3 is used to demonstrate the phenomenon of spinning for the corresponding conserved density $v\rho + fx\rho - \frac{N^2}{g} zv$. We observe the coun-

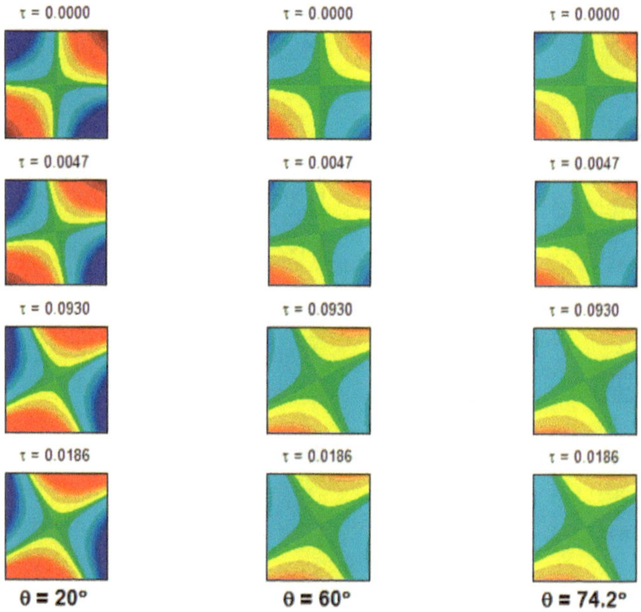

Figure 9.2: Effects of the Earth's rotation on the spinning phenomenon for the energy density $E = v^2 + \frac{g^2}{N^2}\rho^2 + |\nabla\psi|^2$ at different latitudes. The first column shows the time series of the equipotential curves for the energy density at $\theta = 20°$ North. The second and third columns show the corresponding time series of the energy density at latitudes $\theta = 60°$ and $\theta = 74.2°$ North, respectively.

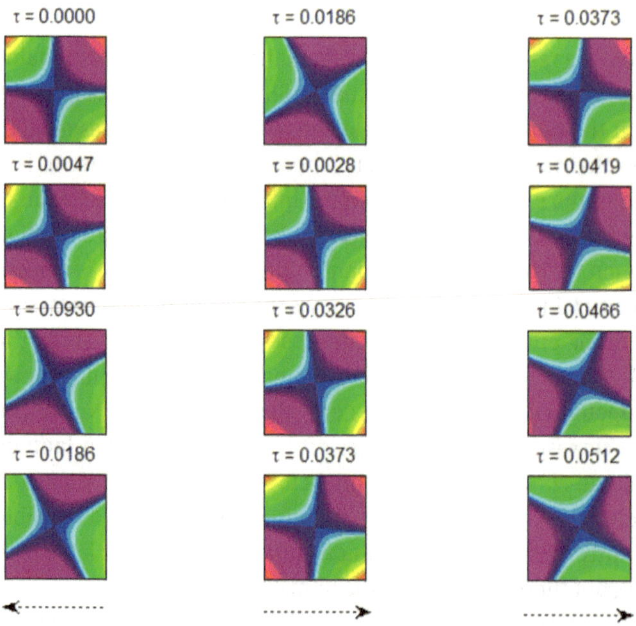

Figure 9.3: Spinning phenomena for the conserved quantity $v\rho + fx\rho - \frac{N^2}{g}zv$.

terclockwise rotation (in the first column) from the equilibrium position (defined as a vertical location of the minimum value of the conserved density indicated by the darkest blue color) at initial time $\tau = 0$ to the "left" maximum azimuth deviation attained at later time $\tau = 0.0186$ inertial periods. The second column shows the clockwise rotation from the left maximum azimuth deviation toward the equilibrium position, whereas the third column shows further clockwise rotation from the equilibrium position to the "right" maximum azimuth deviation attained at $\tau = 0.0512$ inertial periods. Since we are only interested in qualitative analysis, the quantitative results have been omitted from the present analysis.

9.3 Lagrangian, conservation laws, and exact solutions of the nonlinear internal waves

In some calculations, it is convenient to write equations (9.14)–(9.16) by using the Jacobians $J(\psi, v) = \psi_x v_z - \psi_z v_x$, etc., in the following form (hereafter the symbol "prime" is omitted):

$$\Delta\psi_t - g\rho_x - fv_z = J(\psi, \Delta\psi), \tag{9.52}$$

$$v_t + f\psi_z = J(\psi, v), \tag{9.53}$$

$$\rho_t + \frac{N^2}{g}\psi_x = J(\psi, \rho). \tag{9.54}$$

The variational derivative (the *Euler–Lagrange operator*) is defined by

$$\frac{\delta}{\delta u^\alpha} = \frac{\partial}{\partial u^\alpha} + \sum_{s=1}^{\infty}(-1)^s D_{i_1}\cdots D_{i_s}\frac{\partial}{\partial u^\alpha_{i_1\cdots i_s}}, \quad \alpha = 1,\dots,m, \tag{9.55}$$

where the summation over the repeated indices $i_1 \dots i_s$ runs from 1 to n.

Definition 9.1. The adjoint equations to nonlinear PDEs

$$F_\alpha(x, u, \dots, u_{(s)}) = 0, \quad \alpha = 1,\dots,m, \tag{9.56}$$

are given by

$$F^*_\alpha(x, u, \mu, \dots, u_{(s)}, \mu_{(s)}) = 0, \quad \alpha = 1,\dots,m, \tag{9.57}$$

where $\mu = (\mu^1, \dots, \mu^n)$ are new dependent variables and F^*_α are defined by

$$F^*_\alpha(x, u, \mu, \dots, u_{(s)}, \mu_{(s)}) = \frac{\delta(\mu^\beta F_\beta)}{\delta u^\alpha}. \tag{9.58}$$

In the case of linear equations, Definition 9.1 is equivalent to the classical definition of the adjoint equation.

Consider the function

$$\mathcal{L} = \mu^\beta F_\beta(x, u, \dots, u_{(s)}) \tag{9.59}$$

involved in (9.58). Equations (9.56) and their adjoint equations (9.57) can be obtained from (9.58) by taking the variational derivatives (9.55) with respect to the dependent variables u and the similar variational derivatives with respect to the new dependent variables μ:

$$\frac{\delta}{\delta\mu^\alpha} = \frac{\partial}{\partial\mu^\alpha} + \sum_{s=1}^{\infty}(-1)^s D_{i_1}\cdots D_{i_s}\frac{\partial}{\partial\mu^\alpha_{i_1\cdots i_s}}, \quad \alpha = 1,\dots,m. \tag{9.60}$$

That is,

$$\frac{\delta\mathcal{L}}{\delta\mu^\alpha} = F_\alpha(x, u, \dots, u_{(s)}), \tag{9.61}$$

$$\frac{\delta\mathcal{L}}{\delta u^\alpha} = F^*_\alpha(x, u, \mu, \dots, u_{(s)}, \mu_{(s)}). \tag{9.62}$$

This circumstance justifies the following definition.

Definition 9.2. The differential function (9.59) is called a formal Lagrangian for the differential equation (9.56). For the sake of brevity, formal Lagrangians are also referred to as Lagrangians.

If the variables u are known, the new variables μ are obtained by solving equations (9.57), which are, according to (9.58), linear PDEs with respect to μ^α. We will call μ^α nonlocal variables.

Nonlocal variables can be excluded from physical quantities such as conservation laws if equations (9.56) are *self-adjoint*, or, in general, *quasi-self-adjoint* in the following sense.

Definition 9.3. Equations (9.56) are said to be self-adjoint if the system is obtained from the adjoint equations (9.57) by substituting $\mu = u$:

$$F^*_\alpha(x, u, u, \dots, u_{(s)}, u_{(s)}) = \Phi^\beta_\alpha F_\beta(x, u, \dots, u_{(s)}), \quad \alpha = 1,\dots,m, \tag{9.63}$$

with regular (in general, variable) coefficients Φ^β_α.

Definition 9.4. Equations (9.56) are said to be quasi-self-adjoint if the system of adjoint equations (9.57) becomes equivalent to the original system (9.56) upon substituting

$$\mu = h(u) \tag{9.64}$$

with a certain function $h(u)$ such that $h'(u) \neq 0$.

9.3.1 Adjoint system to the Boussinesq model (9.52)–(9.54)

Let us apply the methods from the first part of the book to the Boussinesq model (9.52)–(9.54). For the latter equations of motion, the formal Lagrangian (9.59) is written

$$\mathcal{L} = \varphi[\Delta\psi_t - g\rho_x - fv_z - J(\psi,\Delta\psi)] + \mu[v_t + f\psi_z - J(\psi,v)] \tag{9.65}$$

$$+ r\left[\rho_t + \frac{N^2}{g}\psi_x - J(\psi,\rho)\right], \tag{9.66}$$

where φ, μ, and r are new dependent variables. The adjoint equations to equations (9.52)–(9.54) are obtained by taking the variational derivative of $\mathcal{L}$, namely,

$$\frac{\delta\mathcal{L}}{\delta\psi} = 0, \quad \frac{\delta\mathcal{L}}{\delta v} = 0, \quad \frac{\delta\mathcal{L}}{\delta\rho} = 0, \tag{9.67}$$

where (see (9.55))

$$\begin{aligned}
\frac{\delta}{\delta v} &= \frac{\partial}{\partial v} - D_x\frac{\partial}{\partial v_x} - D_z\frac{\partial}{\partial v_z},\\
\frac{\delta}{\delta\rho} &= \frac{\partial}{\partial\rho} - D_x\frac{\partial}{\partial\rho_x} - D_z\frac{\partial}{\partial\rho_z},\\
\frac{\delta}{\delta\psi} &= \frac{\partial}{\partial\psi} - D_x\frac{\partial}{\partial\psi_x} - D_z\frac{\partial}{\partial\psi_z} + D_xD_t\frac{\partial}{\partial\psi_{xt}} + D_zD_t\frac{\partial}{\partial\psi_{zt}} + \cdots.
\end{aligned}$$

Taking into account the special form (9.65) of $\mathcal{L}$, we have

$$\begin{aligned}
\frac{\delta\mathcal{L}}{\delta\psi} &= -D_x\frac{\partial\mathcal{L}}{\partial\psi_x} - D_z\frac{\partial\mathcal{L}}{\partial\psi_z} - (D_x^2 + D_z^2)\left[D_t\frac{\partial\mathcal{L}}{\partial\Delta\psi_t} + D_x\frac{\partial\mathcal{L}}{\partial\Delta\psi_x} + D_z\frac{\partial\mathcal{L}}{\partial\Delta\psi_z}\right]\\
&= D_x\left(\varphi\Delta\psi_t + \mu v_z - \frac{N^2}{g}r + r\rho_z\right) - D_z(\varphi\Delta\psi_x + f\mu + \mu v_x + r\rho_x)\\
&\quad - (D_x^2 + D_z^2)[D_t(\varphi) + D_x(\varphi\psi_z) - D(\varphi\psi_x)]\\
&= J(\varphi,\Delta\psi) + \mu_x v_z - \frac{N^2}{g}r_x + J(r,\rho) - f\mu_z - \mu_z v_x\\
&\quad - \Delta\varphi_t + 2[\varphi_{xz}\psi_{xx} + \varphi_{zz}\psi_{xz} - \varphi_{xx}\psi_{xz} - \varphi_{xz}\psi_{zz}],\\
\frac{\delta\mathcal{L}}{\delta v} &= -D_t\frac{\partial\mathcal{L}}{\partial v_t} - D_x\frac{\partial\mathcal{L}}{\partial v_x} - D_z\frac{\partial\mathcal{L}}{\partial v_z} = -\mu_t + f\varphi_z - J(\mu,\psi),\\
\frac{\delta\mathcal{L}}{\delta\rho} &= -D_t\frac{\partial\mathcal{L}}{\partial\rho_t} - D_x\frac{\partial\mathcal{L}}{\partial\rho_x} - D_z\frac{\partial\mathcal{L}}{\partial\rho_z} = -r_t + g\varphi_x - J(r,\psi).
\end{aligned}$$

Hence, the adjoint equations (9.67) can be written as follows:

$$\Delta\varphi_t + \frac{N^2}{g}r_x + f\mu_z - J(\varphi,\Delta\psi) - \Theta = 0, \tag{9.68}$$

$$-\mu_t + f\varphi_z - J(\mu, \psi) = 0, \tag{9.69}$$

$$-r_t + g\varphi_x - J(r, \psi) = 0, \tag{9.70}$$

where

$$\Theta = J(\mu, \nu) + J(r, \rho) - 2\left[\frac{\partial}{\partial x}J(\varphi, \psi) + \frac{\partial}{\partial z}J(\varphi, \psi)\right]. \tag{9.71}$$

9.3.2 Self-adjointness of the Boussinesq model (9.52)–(9.54)

Theorem 9.1. *Equations* (9.52)–(9.54) *are quasi-self-adjoint.*

Proof. Looking for substitution (9.64) in the form of a general scaling transformation, one can readily obtain that after the transformation

$$\varphi = \psi, \quad \mu = -\nu, \quad r = -\frac{g^2}{N^2}\rho, \tag{9.72}$$

the quantity Θ given by equation (9.71) vanishes. Therefore, the adjoint equations (9.68)–(9.70) become identical with equations (9.52)–(9.54) after the substitution (9.72). Hence, according to Definition 9.4, equations (9.52)–(9.54) are *quasi-self-adjoint*. Since equations (9.72) are obtained just by simple scaling of the equations $\varphi = \psi$, $\mu = \nu$, and $r = \rho$ required for the self-adjointness, we say that equations (9.52)–(9.54) are *self-adjoint*. □

9.3.3 Conservation laws

General discussion of conservation equations

Along with the individual notation t, x, and z for the independent variables and ν, ρ, and ψ for the dependent variables, we will also use the index notation $x^1 = t$, $x^2 = x$, $x^3 = z$ and $u^1 = \nu$, $x^2 = \rho$, $u^3 = \psi$, respectively. We will write the conservation laws both in the differential form

$$D_t(C^1) + D_x(C^2) + D_z(C^3) = 0 \tag{9.73}$$

and the integral form

$$\frac{d}{dt}\iint C^1 dxdz = 0, \tag{9.74}$$

where the double integral is taken over the (x, z)-plane $\mathbb{R}^2$. Equations (9.73) and (9.74) provide a conservation law for equations (9.52)–(9.54) if they hold for solutions of the same equations. The vector $\vec{C} = (C^1, C^2, C^3)$ satisfying the conservation equation (9.73) is termed a *conserved vector*. Its component C^1 is called the *density* of the conservation law due to equation (9.74). The two-dimensional vector (C^2, C^3) defines the *flux* of the conservation law.

The integral form (9.74) of a conservation law follows from the differential form (9.73) provided that the solutions of equations (9.52)–(9.54) vanish or rapidly decrease at infinity on $\mathbb{R}^2$. Indeed, integrating equation (9.73) over an arbitrary region $\Omega \subset \mathbb{R}^2$, we have

$$\frac{d}{dt}\iint_{\Omega} C^1 dxdz = -\iint_{\Omega} [D_x(C^2) + D_z(C^3)]dxdz.$$

According to Green's theorem, the integral on the right-hand side reduces to the integral along the boundary $\partial\Omega$ of Ω,

$$-\iint_{\Omega} [D_x(C^2) + D_z(C^3)]dxdz = \int_{\partial\Omega} C^3 dx - C^2 dz,$$

and hence vanishes as Ω expands and becomes the plane $\mathbb{R}^2$.

Remark 9.1. It is manifest from this discussion that one can ignore in C^1 "divergent type" terms because they do not change the integral in the conservation equation (9.74). Specifically, if C^1 evaluated on the solutions of equations (9.52)–(9.54) has the form

$$C^1 = \widetilde{C}^1 + D_x(h^2) + D_z(h^3) \tag{9.75}$$

with some functions h^2 and h^3, then the conservation equation (9.73) can be equivalently rewritten in the form

$$D_t(\widetilde{C}^1) + D_x(\widetilde{C}^2) + D_z(\widetilde{C}^3) = 0,$$

where

$$\widetilde{C}^2 = C^2 + D_t(h^2), \quad \widetilde{C}^3 = C^3 + D_t(h^3).$$

Accordingly, we have

$$\iint C^1 dxdz = \iint \widetilde{C}^1 dxdz,$$

and hence the integral conservation equation (9.74) provided by the conservation density C^1 of the form (9.75) coincides with that provided by the density $\widetilde{C}^1$.

In particular, if $\widetilde{C}^1 = 0$, the integral in equation (9.74) vanishes. This kind of conservation laws are trivial from a physical point of view. Therefore we single out physically useless conservation laws by the following definition.

Definition 9.5. The conservation law is said to be trivial if its density C^1 evaluated on the solutions of equations (9.52)–(9.54) is the divergence

$$C^1 = D_x(h^2) + D_z(h^3).$$

The following statement simplifies calculations while dealing with conservation equations.

Lemma 9.1. *A function $F(v, \rho, \psi, v_x, v_z, \rho_x, \rho_z, \psi_x, \psi_z, \psi_{xt}, \psi_{zt}, \ldots)$ is the divergence*

$$F = D_x(C^1) + D_z(C^2) \tag{9.76}$$

if and only if it satisfies the following equations:

$$\frac{\delta F}{\delta v} = 0, \quad \frac{\delta F}{\delta \rho} = 0, \quad \frac{\delta F}{\delta \psi} = 0. \tag{9.77}$$

Here the variational derivatives act on F as usual:

$$\begin{aligned}
\frac{\delta F}{\delta v} &= \frac{\partial F}{\partial v} - D_x\left(\frac{\partial F}{\partial v_x}\right) - D_z\left(\frac{\partial F}{\partial v_z}\right),\\
\frac{\delta F}{\delta \rho} &= \frac{\partial F}{\partial \rho} - D_x\left(\frac{\partial F}{\partial \rho_x}\right) - D_z\left(\frac{\partial F}{\partial \rho_z}\right),\\
\frac{\delta F}{\delta \psi} &= \frac{\partial F}{\partial \psi} - D_x\left(\frac{\partial F}{\partial \psi_x}\right) - D_z\left(\frac{\partial F}{\partial \psi_z}\right) + D_x D_t\left(\frac{\partial F}{\partial \psi_{xt}}\right) + D_z D_t\left(\frac{\partial F}{\partial \psi_{zt}}\right) + \cdots .
\end{aligned} \tag{43}$$

Corollary 9.1. *A function C^1 is the density of a conservation law* (9.73) *if and only if the function*

$$F = D_t(C^1)|_{(9.52)\text{–}(9.54)} \tag{9.78}$$

satisfies equations (9.77). *Here $|_{(9.52)\text{–}(9.54)}$ means that the quantity $D_t(C^1)$ is evaluated on the solutions of equations* (9.52)–(9.54).

In particular, Lemma 9.1 allows one to single out trivial conservation laws as follows.

Corollary 9.2. *The conservation law* (9.73) *is trivial if and only if its density C^1 evaluated on the solutions of equations* (9.52)–(9.54), *i. e., the quantity*

$$C^1_* = C^1|_{(9.52)\text{–}(9.54)}, \tag{9.79}$$

satisfies equations (9.77),

$$\frac{\delta C^1_*}{\delta v} = 0, \quad \frac{\delta C^1_*}{\delta \rho} = 0, \quad \frac{\delta C^1_*}{\delta \psi} = 0, \tag{9.80}$$

on the solutions of equations (9.52)–(9.54).

9.3.4 Variational derivatives of expressions with Jacobians

We will use in our calculations the following statement on the behavior of certain expressions with Jacobians under the action of the variational derivatives (43).

Proposition 9.1. *The following equations hold:*

$$\frac{\delta J(\psi, v)}{\delta v} = 0, \quad \frac{\delta J(\psi, v)}{\delta \psi} = 0, \tag{9.81}$$

$$\frac{\delta [vJ(\psi, v)]}{\delta v} = 0, \quad \frac{\delta J[vJ(\psi, v)]}{\delta \psi} = 0, \tag{9.82}$$

$$\frac{\delta [\rho J(\psi, \rho)]}{\delta \rho} = 0, \quad \frac{\delta [\rho J(\psi, \rho)]}{\delta \psi} = 0, \tag{9.83}$$

$$\frac{\delta J(\psi, \Delta\psi)}{\delta \psi} = 0, \quad \frac{\delta [\psi J(\psi, \Delta\psi)]}{\delta \psi} = 0. \tag{9.84}$$

Proof. Let us verify that the first equation in (9.81) holds. We have (see (43))

$$\frac{\delta J(\psi, v)}{\delta v} = \frac{\delta(\psi_x v_z - \psi_z v_x)}{\delta v} = -D_z(\psi_x) + D_x(\psi_z) = -\psi_{xz} + \psi_{zx} = 0.$$

Replacing v by ψ, one obtains the second equation in (9.81). Let us verify now that equations (9.82) are satisfied. We have

$$\frac{\delta [vJ(\psi, v)]}{\delta v} = \frac{\partial [v(\psi_x v_z - \psi_z v_x)]}{\partial v} - D_z(v\psi_x) + D_x(v\psi_z),$$

$$J(\psi, v) - D_z(v\psi_x) + D_x(v\psi_z) = J(\psi, v) - J(\psi, v) - v\psi_{xz} + v\psi_{zx} = 0,$$

and

$$\frac{\delta [vJ(\psi, v)]}{\delta \psi} = -D_x(vv_z) + D_z(vv_x) = -v_x v_z - vv_{xz} + v_z v_x + vv_{zx} = 0.$$

Replacing v by ρ, one obtains equations (9.83). Equations (9.84) are derived likewise even though they involve higher-order derivatives. We have

$$\begin{aligned}\frac{\delta J(\psi, \Delta\psi)}{\delta \psi} &= \frac{\delta [\psi_x(\psi_{xxz} + \psi_{zzz}) - \psi_z(\psi_{xxx} + \psi_{xzz})]}{\delta \psi} \\ &= -D_x(\Delta\psi_z) + D_z(\Delta\psi_x) - D_z(D_x^2 + D_z^2)(\psi_x) + D_x(D_x^2 + D_z^2)(\psi_z) \\ &= -D_x(\Delta\psi_z) + D_z(\Delta\psi_x) - D_z(\Delta\psi_x) + D_x(\Delta\psi_z) = 0.\end{aligned}$$

Derivation of the second equation in (9.84) requires only a simple modification of the previous calculations. That is,

$$\frac{\delta [\psi J(\psi, \Delta\psi)]}{\delta \psi} = \frac{\delta [\psi(\psi_x \Delta\psi_z - \psi_z \Delta\psi_x)]}{\delta \psi}$$

$$
\begin{aligned}
&= \psi_x \Delta\psi_z - \psi_z \Delta\psi_x - D_x(\psi\Delta\psi_z) + D_z(\psi\Delta\psi_x) \\
&\quad - \Delta D_z(\psi\psi_x) + \Delta D_x(\psi\psi_z) \\
&= \psi_x \Delta\psi_z - \psi_z \Delta\psi_x - \psi_x \Delta\psi_z - \psi\Delta\psi_{zx} + \psi_z \Delta\psi_x + \psi\Delta\psi_{zx} \\
&\quad - \frac{1}{2}\Delta D_z D_x(\psi^2) + \frac{1}{2}\Delta D_x D_z(\psi^2) = 0. \qquad \square
\end{aligned}
$$

9.3.5 Nonlocal conserved vectors

It can be shown that for any operator

$$X = \xi^i \frac{\partial}{\partial x^i} + \eta^\alpha \frac{\partial}{\partial u^\alpha}, \quad i = 1, 2, 3, \tag{9.85}$$

admitted by the system (9.52)–(9.54), the quantities

$$
\begin{aligned}
C^i &= \xi^i \mathcal{L} + W^\alpha \left[\frac{\partial \mathcal{L}}{\partial u_i^\alpha} - D_j\left(\frac{\partial \mathcal{L}}{\partial u_{ij}^\alpha} \right) + D_j D_k \left(\frac{\partial \mathcal{L}}{\partial u_{ijk}^\alpha} \right) \right] \\
&\quad + D_j(W^\alpha) \left[\frac{\partial \mathcal{L}}{\partial u_{ij}^\alpha} - D_k\left(\frac{\partial \mathcal{L}}{\partial u_{ijk}^\alpha} \right) + D_j D_k \left(\frac{\partial \mathcal{L}}{\partial u_{ijk}^\alpha} \right) \right] + D_j D_k(W^\alpha) \left[\frac{\partial \mathcal{L}}{\partial u_{ijk}^\alpha} \right]
\end{aligned} \tag{9.86}
$$

define the components of a conserved vector for equations (9.52)–(9.54) considered together with the adjoint equations (9.68)–(9.70). Here

$$W^\alpha = \eta^\alpha - \xi^j u_j^\alpha, \quad \alpha = 1, 2, 3. \tag{9.87}$$

Formula (9.86) is written by taking into account that the Lagrangian (9.65) involves the derivatives up to the third order. Moreover, noting that the Lagrangian (9.65) vanishes on the solutions of equations (9.52)–(9.54), we can drop the first term in (9.86) and use the conserved vector in the abbreviated form (9.87).

For computing the conserved vectors (9.86), the Lagrangian (9.65) containing the mixed derivatives should be written in the symmetric form

$$
\begin{aligned}
\mathcal{L} &= \frac{1}{3}\varphi[\psi_{txx} + \psi_{xtx} + \psi_{xxt} + \psi_{tzz} + \psi_{ztz} + \psi_{zzt} - 3g\rho_x - 3fv_z \\
&\quad - \psi_x(\psi_{zxx} + \psi_{xzx} + \psi_{xxz} + 3\psi_{zzz}) + \psi_z(3\psi_{xxx} + \psi_{xzz} + \psi_{zxz} + \psi_{zzx})] \\
&\quad + \mu[v_t + f\psi_z - \psi_x v_z + \psi_z v_x] + r\left[\rho_t + \frac{N^2}{g}\psi_x - \psi_x\rho_z + \psi_z\rho_x \right].
\end{aligned} \tag{9.88}
$$

Since the Lagrangian $\mathcal{L}$ and hence the components (9.86) of a conserved vector contain the nonlocal variables φ, μ, and r, we obtain in this way *nonlocal conserved vectors.*

9.3.6 Computation of nonlocal conserved vectors

The substitution (9.88) into the expression for the conserved vector (9.86) yields

$$
\begin{aligned}
C^1 = {} & W^1 \frac{\partial \mathcal{L}}{\partial v_t} + W^2 \frac{\partial \mathcal{L}}{\partial \rho_t} + W^3 \left[D_x^2 \left(\frac{\partial \mathcal{L}}{\partial \psi_{txx}} \right) + D_z^2 \left(\frac{\partial \mathcal{L}}{\partial \psi_{tzz}} \right) \right] \\
& - \left[D_x(W^3) D_x \left(\frac{\partial \mathcal{L}}{\partial \psi_{txx}} \right) + D_z(W^3) D_z \left(\frac{\partial \mathcal{L}}{\partial \psi_{tzz}} \right) \right] \\
& + D_x^2(W^3) \frac{\partial \mathcal{L}}{\partial \psi_{txx}} + D_z^2(W^3) \frac{\partial \mathcal{L}}{\partial \psi_{tzz}},
\end{aligned}
$$

or

$$
\begin{aligned}
C^1 = {} & W^1 \mu + W^2 r + \frac{1}{3} W^3 [D_x^2(\varphi) + D_z^2(\varphi)] \\
& - \frac{1}{3} [\varphi_x D_x(W^3) + \varphi_z D_z(W^3)] + \frac{1}{3} \varphi [D_x^2(W^3) + D_z^2(W^3)].
\end{aligned} \tag{9.89}
$$

Furthermore, using the same procedure, we obtain

$$
\begin{aligned}
C^2 = {} & W^1 \frac{\partial \mathcal{L}}{\partial v_x} + W^2 \frac{\partial \mathcal{L}}{\partial \rho_x} + W^3 \left[\frac{\partial \mathcal{L}}{\partial \psi_x} + D_x^2 \left(\frac{\partial \mathcal{L}}{\partial \psi_{xxx}} \right) + D_z^2 \left(\frac{\partial \mathcal{L}}{\partial \psi_{xzz}} \right) \right. \\
& \left. + D_t D_x \left(\frac{\partial \mathcal{L}}{\partial \psi_{xtx}} + \frac{\partial \mathcal{L}}{\partial \psi_{xxt}} \right) + D_x D_z \left(\frac{\partial \mathcal{L}}{\partial \psi_{xxz}} + \frac{\partial \mathcal{L}}{\partial \psi_{xzx}} \right) \right] \\
& - D_x(W^3) D_x \left(\frac{\partial \mathcal{L}}{\partial \psi_{xxx}} \right) - D_t(W^3) D_x \left(\frac{\partial \mathcal{L}}{\partial \psi_{xtx}} \right) - D_x(W^3) D_t \left(\frac{\partial \mathcal{L}}{\partial \psi_{xxt}} \right) \\
& - D_z(W^3) D_z \left(\frac{\partial \mathcal{L}}{\partial \psi_{xzz}} \right) - D_z(W^3) D_x \left(\frac{\partial \mathcal{L}}{\partial \psi_{xzx}} \right) - D_x(W^3) D_z \left(\frac{\partial \mathcal{L}}{\partial \psi_{xxz}} \right) \\
& - D_x^2(W^3) \left(\frac{\partial \mathcal{L}}{\partial \psi_{xxx}} \right) + D_z^2(W^3) \left(\frac{\partial \mathcal{L}}{\partial \psi_{xzz}} \right) + D_t D_x(W^3) \left(\frac{\partial \mathcal{L}}{\partial \psi_{xtx}} + \frac{\partial \mathcal{L}}{\partial \psi_{xxt}} \right) \\
& + D_x D_z(W^3) D_z \left(\frac{\partial \mathcal{L}}{\partial \psi_{xxz}} + \frac{\partial \mathcal{L}}{\partial \psi_{xzx}} \right),
\end{aligned}
$$

or

$$
\begin{aligned}
C^2 = {} & W^1 \mu \psi_z + W^2 (r \psi_z - g\varphi) + W^3 \left[-\Delta \psi_z - \mu v_z + \frac{N^2}{g} r - r \rho_z \right. \\
& \left. + D_x^2(\phi \psi_z) + \frac{1}{3} D_z^2(\varphi \psi_z) + \frac{2}{3} \varphi_{xt} - \frac{2}{3} D_x D_z(\varphi \psi_x) \right] \\
& - D_x(W^3) \left[D_x(\varphi \psi_z) + \frac{1}{3} \varphi_t - \frac{1}{3} D_z(\varphi \psi_x) \right] - \frac{1}{3} D_t(W^3) \varphi_x \\
& - \frac{1}{3} D_z(W^3) [D_z(\varphi \psi_z) - D_x(\varphi \psi_x)] + \left[D_x^2(W^3) + \frac{1}{3} D_z^2(W^3) \right] \varphi \psi_z \\
& + \frac{2}{3} \varphi D_t D_x(W^3) - \frac{2}{3} \varphi \psi_x D_z D_x(W^3).
\end{aligned} \tag{9.90}
$$

Likewise, we get

$$
\begin{aligned}
C^3 = &-W^1(\mu\psi_x + f\varphi) - W^2 r\psi_x + W^3\Big[-\Delta\psi_x + (f + v_x)\mu + r\rho_x \\
&- \frac{1}{3}D_x^2(\phi\psi_x) - D_z^2(\varphi\psi_x) + \frac{1}{3}D_z^2(\varphi\psi_z) + \frac{2}{3}\varphi_{xt} + \frac{2}{3}D_xD_z(\varphi\psi_z)\Big] \\
&+ \frac{1}{3}D_x(W^3)[D_x(\varphi\psi_x) - D_z(\varphi\psi_z)] - \frac{1}{3}D_t(W^3)\varphi_z \\
&- D_z(W^3)\Big[\frac{1}{3}\varphi_t - D_z(\varphi\psi_x) + \frac{1}{3}D_x(\varphi\psi_z)\Big] \\
&- \Big[\frac{1}{3}D_x^2(W^3) + D_z^2(W^3)\Big]\varphi\psi_x + \frac{2}{3}\varphi D_tD_z(W^3) + \frac{2}{3}\varphi\psi_z D_zD_x(W^3).
\end{aligned}
\tag{9.91}
$$

9.3.7 Local conserved vectors

The quantities (9.89)–(9.91) define a nonlocal conserved vector because they contain the nonlocal variables φ, μ, and r. In consequence, the conservation equation (9.73) requires not only the basic equations (9.52)–(9.54), but also the adjoint equations (9.68)–(9.70).

However, we can eliminate the nonlocal variables using the self-adjointness of equations (9.52)–(9.54), thus transforming the nonlocal conserved vector into a local one. Namely, we substitute in equations (9.89)–(9.91) the expressions (9.72) for φ, μ and r:

$$\varphi = \psi, \quad \mu = -v, \quad r = -\frac{g^2}{N^2}\rho. \tag{9.92}$$

Then the adjoint equations (9.68)–(9.70) are satisfied for any solutions of the basic equations (9.52)–(9.54), and hence the quantities (9.89)–(9.91) satisfy the conservation equation (9.73) on all solutions of equations (9.52)–(9.54).

Let us apply the procedure to C^1. We eliminate the nonlocal variables in (9.89) by substituting there the expressions (9.72) and write C^1 in the following form:

$$C^1 = -vW^1 - \frac{g^2}{N^2}\rho W^2 + \frac{1}{3}[W^3\Delta\psi - \psi_x D_x(W^3) - \psi_z D_z(W^3) + \psi\Delta W^3],$$

where

$$\Delta\psi = D_x^2(\psi) + D_z^2(\psi), \quad \Delta W^3 = D_x^2(W^3) + D_z^2(W^3).$$

We further simplify the expression for C^1 by using the identities

$$
\begin{aligned}
W^3D_x^2(\psi) &= D[_x W^3D_x(\psi)] - \psi_x D_x(W^3), \\
W^3D_z^2(\psi) &= D_z[W^3D_z(\psi)] - \psi_z D_z(W^3)
\end{aligned}
$$

and

$$\psi D_x^2(W^3) = D_x[\psi D_x(W^3)] - \psi_x D_x(W^3),$$
$$\psi D_z^2(W^3) = D_z[\psi D_z(W^3)] - \psi_z D_z(W^3).$$

Then we have

$$C^1 = -\nu W^1 - \frac{g^2}{N^2}\rho W^2 - \psi_x D_x(W^3) - \psi_z D_z(W^3) + \frac{1}{3}\Delta(\psi W^3). \tag{9.93}$$

Dropping in (9.93) the divergent type term

$$\frac{1}{3}\Delta(\psi W^3) = D_x\left[\frac{1}{3}D_x(\psi W^3)\right] + D_z\left[\frac{1}{3}D_z(\psi W^3)\right]$$

in accordance with Remark 9.1, we finally obtain

$$C^1 = -\nu W^1 - \frac{g^2}{N^2}\rho W^2 - \psi_x D_x(W^3) - \psi_z D_z(W^3). \tag{9.94}$$

We will not dwell on a similar modification of the expressions (9.90) and (9.91) for the components C^2 and C^3 of conserved vectors. We will see further that they can be found by a simpler method when density C^1 is known.

Utilization of obvious symmetries

Equations (9.52)–(9.54) do not contain the dependent and independent variables explicitly and therefore they are invariant with respect to addition of arbitrary constants to all these variables. This means that equations (9.52)–(9.54) admit the one-parameter groups of translations in all variables,

$$\overline{\nu} = \nu + a_1, \quad \overline{\rho} = \rho + a_2, \quad \overline{\psi} = \psi + a_3, \quad \overline{t} = t + a_4, \quad \overline{x} = x + a_5, \quad \overline{z} = z + a_6,$$

with generators

$$X_1 = \frac{\partial}{\partial \nu}, \quad X_2 = \frac{\partial}{\partial \rho}, \quad X_3 = \frac{\partial}{\partial \psi}, \quad X_4 = \frac{\partial}{\partial t}, \quad X_5 = \frac{\partial}{\partial x}, \quad X_6 = \frac{\partial}{\partial z}. \tag{9.95}$$

One can also find by simple calculations the dilations (scaling transformations)

$$\overline{\nu} = a\nu, \quad \overline{\rho} = b\rho, \quad \overline{\psi} = c\psi, \quad \overline{t} = dt, \quad \overline{x} = sx, \quad \overline{z} = sz \tag{9.96}$$

admitted by equations (9.52)–(9.54). These transformations are defined near the identity transformation if the parameters $a, \dots, s$ are positive. The dilations of x and z are taken by the same parameter s in order to keep invariant the Laplacian Δ. In order to find the parameters $a, \dots, s$, we use the invariance condition of equations (9.52)–(9.54).

Namely, under transformations (9.96), the derivatives involved in equations (9.52)–(9.54) are changed as follows:

$$\begin{aligned}\overline{v}_{\overline{t}} &= adv_t, \quad \overline{v}_{\overline{x}} = asv_x, \quad \overline{v}_{\overline{z}} = asv_z,\\ \overline{\rho}_{\overline{t}} &= bd\rho_t, \quad \overline{\rho}_{\overline{x}} = bs\rho_x, \quad \overline{\rho}_{\overline{z}} = bs\rho_z,\\ \overline{\psi}_{\overline{t}} &= cd\psi_t, \quad \overline{\psi}_{\overline{x}} = cs\psi_x, \quad \overline{\psi}_{\overline{z}} = cs\psi_z,\\ \overline{\Delta}\overline{\psi}_{\overline{t}} &= cds^2\Delta\psi_t, \quad \overline{\Delta}\overline{\psi}_{\overline{x}} = cs^3\Delta\psi_x, \quad \overline{\Delta}\overline{\psi}_{\overline{z}} = cs^3\Delta\psi_z,\end{aligned} \tag{9.97}$$

where $\overline{\Delta}$ is the Laplacian written in the variables $\overline{x}$ and $\overline{z}$. The invariance of equations (9.52)–(9.54) under the transformations (9.96) means that the following equations are satisfied:

$$\begin{aligned}\overline{\Delta\psi_{\overline{t}}} - g\overline{\rho}_{\overline{x}} - f\overline{v}_{\overline{z}} - \overline{\psi}_{\overline{x}}\overline{\Delta\psi_{\overline{z}}} + \overline{\psi}_{\overline{z}}\overline{\Delta\psi_{\overline{x}}} &= 0,\\ \overline{v}_{\overline{t}} + f\overline{\psi}_{\overline{z}} - \overline{\psi}_{\overline{x}}\overline{v}_{\overline{z}} + \overline{\psi}_{\overline{z}}\overline{v}_{\overline{x}} &= 0,\\ \overline{\rho}_{\overline{t}} + \frac{N^2}{g}\overline{\psi}_{\overline{x}} - \overline{\psi}_{\overline{x}}\overline{\rho}_{\overline{z}} + \overline{\psi}_{\overline{z}}\overline{\rho}_{\overline{x}} &= 0,\end{aligned}$$

where equations (9.52)–(9.54) hold. Substituting here the expressions (9.97), we have

$$\begin{aligned}\overline{\Delta\psi_{\overline{t}}} - g\overline{\rho}_{\overline{x}} - f\overline{v}_{\overline{z}} - \overline{\psi}_{\overline{x}}\overline{\Delta\psi_{\overline{z}}} + \overline{\psi}_{\overline{z}}\overline{\Delta\psi_{\overline{x}}} &= cds^{22}\Delta\psi_t - bs\rho_x - cs^4(\psi_x\Delta\psi_z - \psi_z\Delta\psi_x),\\ \overline{v}_{\overline{t}} + f\overline{\psi}_{\overline{z}} - \overline{\psi}_{\overline{x}}\overline{v}_{\overline{z}} + \overline{\psi}_{\overline{z}}\overline{v}_{\overline{x}} &= adv_t + cs\psi_z - acs^2(\psi_x v_z - \psi_z v_x),\\ \overline{\rho}_{\overline{t}} + \frac{N^2}{g}\overline{\psi}_{\overline{x}} - \overline{\psi}_{\overline{x}}\overline{\rho}_{\overline{z}} + \overline{\psi}_{\overline{z}}\overline{\rho}_{\overline{x}} &= bd\rho_t + cs\frac{N^2}{g}\psi_x - bcs^2(\psi_x\rho_z - \psi_z\rho_x).\end{aligned}$$

These equations show that the invariance of equations (9.52)–(9.54) is guaranteed by the following six equations for five undetermined parameters a, b, c, d, and s:

$$cd = bs = c^2s^4, \quad ad = cs = acs^2, \quad bd = cs = bcs^2. \tag{9.98}$$

It can be verified by simple calculations that equations (9.98) yield

$$d = 1, \quad b = a, \quad c = a^2, \quad s = \frac{1}{a},$$

where a is an arbitrary parameter. We substitute these values of the parameters in the dilations determined by (9.96), denote the positive parameter a by $\exp(\tilde{a})$, and drop the tilde to conclude that equations (9.52)–(9.54) admit that one-parameter nonuniform dilation group

$$\overline{v} = ve^a, \quad \overline{\rho} = \rho e^a, \quad \overline{\psi} = \psi e^{2a}, \quad \overline{t} = t, \quad \overline{x} = xe^a, \quad \overline{z} = ze^a$$

with the following generator:

$$X_7 = x\frac{\partial}{\partial x} + z\frac{\partial}{\partial z} + v\frac{\partial}{\partial v} + \rho\frac{\partial}{\partial \rho} + 2\psi\frac{\partial}{\partial \psi}. \tag{9.99}$$

We will consider the operators (9.95) and (9.99) as obvious symmetries of equations (9.52)–(9.54) and will find the local conservation laws provided by these symmetries.

Translations of the northward velocity component *v*

For the operator X_1 from (9.95), equations (9.87) yield

$$W^1 = 1, \quad W^2 = 0, \quad W^3 = 0.$$

Substituting the latter expressions in equation (9.94), we obtain

$$C^1 = -v.$$

In this case, equations (9.90) and (9.91) are also simple. They are written

$$C^2 = \mu\psi_z, \quad C^3 = -(\mu\psi_x + f\varphi)$$

and upon using transformations (9.72) yield

$$C^2 = -v\psi_z, \quad C^3 = v\psi_x - f\psi.$$

Since any conserved vector is defined up to multiplication by an arbitrary constant, we change the sign of C^i $(i = 1, 2, 3)$ and obtain the following conserved vector:

$$C^1 = v, \quad C^2 = v\psi_z, \quad C^3 = f\psi - v\psi_x. \tag{9.100}$$

We have

$$D_t(C^1) + D_x(C^2) + D_z(C^3) = v_t + (v_x + f)\psi_z - v_z\psi_x.$$

Hence, the conservation equation (9.73) coincides with equation (9.53).

Translation of the density *ρ*

For the operator X_2 from (9.95), equations (9.87) yield

$$W^1 = 0, \quad W^2 = 1, \quad W^3 = 0.$$

Substituting the latter expressions in equation (9.94), we obtain

$$C^1 = -\frac{g^2}{N^2}\rho.$$

Furthermore, equations (9.90) and (9.91) and transformations (9.72) yield

$$C^2 = -g\psi - \frac{g^2}{N^2}\rho\psi_z, \quad C^3 = \frac{g^2}{N^2}\rho\psi_x.$$

Multiplying C^i $(i = 1, 2, 3)$ by $-N^2/g$, we arrive at the following conserved vector:

$$C^1 = \rho, \quad C^2 = \frac{N^2}{g}\psi, \quad C^3 = -\rho\psi_x. \tag{9.101}$$

One can readily verify that the conservation equation (9.73) for the vector (9.101) is also satisfied. Namely, it coincides with equation (9.54).

Translation of the stream function ψ

For the operator X_3 from (9.95), equations (9.87) yield

$$W^1 = 0, \quad W^2 = 0, \quad W^3 = 1.$$

Substituting the latter expressions in equation (9.94), we obtain

$$C^1 = 0.$$

Hence, the invariance of equations (9.52)–(9.54) under the translation of ψ furnishes only a trivial conservation law (see Definition 9.5).

Derivation of the flux of conserved vectors with known densities

We will show here how to find the components C^2 and C^3 of the conserved vector (9.100) without employing equations (9.90) and (9.91), provided that the conserved density $C^1 = v$ is known. To this end, let us verify first that $C^1 = v$ satisfies Corollary 9.1. Indeed, in this case, $D_t(C^1) = v_t$, and hence the invariance criterion (9.78) yields

$$F = D_t(C^1)\big|_{(9.52)-(9.54)} = J(\psi, v) - f\psi_z. \tag{9.102}$$

Using Proposition 9.1, we see that equations (9.77) are satisfied, i. e.,

$$\frac{\delta F}{\delta v} = \frac{\delta F}{\delta \rho} = 0, \quad \frac{\delta F}{\delta \psi} = D_z(f) = 0. \tag{9.103}$$

Therefore, Corollary 9.2 guarantees that F defined by equation (9.102) satisfies equation (9.76):

$$\psi_x v_z - \psi_z v_x - f\psi_z = D_x(H^1) + D_z(H^2) \tag{9.104}$$

with certain unknown functions H^j $(j = 1, 2)$.

In order to find H^j, we write

$$\psi_x v_z - f\psi_z = D_z(v\psi_x - f\psi) - v\psi_{xz}, \quad -\psi_z v_x = D_x(-v\psi_z) + v\psi_{zx}$$

and obtain

$$\psi_x v_z - \psi_z v_x - f\psi_z = D_x(-v\psi_z) + D_z(v\psi_x - f\psi).$$

Thus, $H^1 = -v\psi_z$ and $H^2 = v\psi_x - f\psi$. Denoting $C^2 = -H^1$, $C^3 = -H^2$, i. e.,

$$C^1 = v\psi_z, \quad C^2 = f\psi - v\psi_x,$$

and invoking equation (9.102), we write equation (9.104) in the form

$$D_t(C^1)\big|_{(9.52)-(9.54)} = D_x(C^1) + D_z(C^2) = 0.$$

This is precisely the conservation equation (9.73) for the vector (9.100). Thus, we have obtained the components C^2 and C^3 of the conserved vector (9.100) without using equations (9.90) and (9.91).

The components C^2 and C^3 of the conserved vector (9.101) can be derived likewise.

Translation of the eastward coordinate *x*

For the operator X_5 from (9.95), equations (9.87) yield

$$W^1 = -v_x, \quad W^2 = -\rho_x, \quad W^3 = -\psi_x.$$

Substituting these expressions in equation (9.94), we obtain

$$\begin{aligned} C^1 &= vv_x + \frac{g^2}{N^2}\rho\rho_x + \psi_x\psi_{xx} + \psi_z\psi_{xz} \\ &= D\left[\frac{v^2}{2} + \frac{g^2}{2N^2}\rho^2 + \frac{1}{2}(\psi_x^2 + \psi_z^2)\right]. \end{aligned}$$

Hence, the invariance of equations (9.52)–(9.54) under the transformation of *x* furnishes only a trivial conservation law (see Definition 9.5). Similar calculations show that the invariance under the translation of *z* provides also a trivial conservation law.

Time translation

For the operator X_4 from (9.95), equations (9.87) yield

$$W^1 = -v_t, \quad W^2 = -\rho_t, \quad W^3 = -\psi_t.$$

Substituting these expressions in equation (9.94), we obtain

$$C^1 = vv_t + \frac{g^2}{N^2}\rho\rho_t + \psi_x\psi_{xt} + \psi_z\psi_{zt}. \tag{9.105}$$

Changing the last two terms of C^1 by using the identity

$$\begin{aligned}\psi_x\psi_{xt} + \psi_z\psi_{zt} &= D_x(\psi\psi_{xt}) - \psi\psi_{xxt} + D_z(\psi\psi_{zt}) - \psi\psi_{zzt}\\ &= D_x(\psi\psi_{xt}) + D_z(\psi\psi_{zt}) - \psi\Delta\psi_t\end{aligned} \tag{9.106}$$

and dropping the divergent type terms, we rewrite C^1 given by equation (9.105) in the form

$$C^1 = \nu\nu_t + \frac{g^2}{N^2}\rho\rho_t - \psi\Delta\psi_t. \tag{9.107}$$

Let us clarify if the conservation law with the density (9.107) is trivial or nontrivial. To do so, according to Definition 9.5, we have to evaluate the density (9.107) on the solutions of equations (9.52)–(9.54). Then, from the invariance criterion (9.79), $C^1_* = C^1|_{(9.52)-(9.54)}$, we have

$$C^1_* = \left\{\nu J(\psi,\nu) + \frac{g^2}{N^2}\rho J(\psi,\rho) - \psi J(\psi,\Delta\psi)\right\} - fD_z(\nu\psi) - gD_x(\rho\psi)$$

and Corollary 9.2 shows that the conservation law is trivial. Indeed, the last two terms of C^1_* have the divergent form. The expression in braces for C^1_* satisfies equations (9.80) according to Proposition 9.1, and hence it also has the divergent form.

Thus, the invariance of equations (9.52)–(9.54) under the time translation furnishes only a trivial conservation law.

Use of the dilation and conservation of energy

Consider the generator (9.99) of the dilation group,

$$X_7 = x\frac{\partial}{\partial x} + z\frac{\partial}{\partial z} + \nu\frac{\partial}{\partial \nu} + \rho\frac{\partial}{\partial \rho} + 2\psi\frac{\partial}{\partial \psi}.$$

In this case the quantities (9.87) have the form

$$W^1 = \nu - x\nu_x - z\nu_z, \quad W^2 = \rho - x\rho_x - z\rho_z, \quad W^3 = 2\psi - x\psi_x - z\psi_z. \tag{9.108}$$

Substitution of (9.108) in (9.94) yields

$$\begin{aligned}C^1 = &-\nu^2 + x\nu\nu_x + z\nu\nu_z + \frac{g^2}{N^2}(-\rho^2 + x\rho\rho_x + z\rho\rho_z)\\ &- \psi_x^2 + x\psi_x\psi_{xx} + z\psi_x\psi_{xz} - \psi_z^2 + x\psi_z\psi_{xz} + z\psi_z\psi_{zz}.\end{aligned} \tag{9.109}$$

We modify (9.109) by using the identities

$$\begin{aligned}x\nu\nu_x + z\nu\nu_z &= \frac{1}{2}D_x(x\nu^2) + \frac{1}{2}D_z(z\nu^2) - \nu^2,\\ x\rho\rho_x + z\rho\rho_z &= \frac{1}{2}D_x(x\rho^2) + \frac{1}{2}D_z(z\rho^2) - \rho^2,\end{aligned}$$

$$x\psi_x\psi_{xx} + x\psi_z\psi_{xz} = \frac{1}{2}D_x[x(\psi_x^2 + \psi_x^2)] - \frac{1}{2}(\psi_x^2 + \psi_x^2),$$
$$z\psi_x\psi_{xz} + x\psi_z\psi_{zz} = \frac{1}{2}D_z[z(\psi_x^2 + \psi_x^2)] - \frac{1}{2}(\psi_x^2 + \psi_x^2).$$

Substituting these in (9.109) and dropping the divergent type terms, we have

$$C^1 = -2\left(v^2 + \frac{g^2}{N^2}\rho^2 + |\nabla\psi|^2\right),$$

where

$$|\nabla\psi|^2 = \psi_x^2 + \psi_z^2.$$

Dividing C^1 by the essential coefficient (-2) we finally obtain the following conservation law in the integral form (9.74):

$$\frac{d}{dt}\iint\left[v^2 + \frac{g^2}{N^2}\rho^2 + |\nabla\psi|^2\right]dxdz = 0. \tag{9.110}$$

Equation (9.110) represents the *conservation of the energy* with the *density*

$$E = v^2 + \frac{g^2}{N^2}\rho^2 + |\nabla\psi|^2. \tag{9.111}$$

Let us find the components C^2 and C^3 of this conservation law written in the differential form (9.73). Now, we first verify that E defined by equation (9.111) satisfies Corollary 9.1 for densities of conservation laws. We have

$$D_t(E) = 2\left(vv_t + \frac{g^2}{N^2}\rho\rho_t + \psi_x\psi_{xt} + \psi_z\psi_{zt}\right). \tag{9.112}$$

Since the expression in the brackets in equation (9.112) is identical with C^1 in (9.105), it can be rewritten in the form (9.107), and hence satisfies equations (9.77). Thus, Corollary 9.2 guarantees that E is the density of a conservation law. Furthermore, it is manifested from equation (9.111) that this conservation law is nontrivial.

According to Corollary 9.1, $D_t(E)$ defined by equation (9.112) and evaluated on the solutions of equations (9.52)–(9.54) satisfies equation (9.76),

$$D_t(E)|_{(9.52)-(9.54)} = D_x(H^1) + D_z(H^2), \tag{9.113}$$

with certain functions H^1 and H^2. In order to find these functions, we use identity (9.106), i. e.,

$$2(\psi_x\psi_{xt} + \psi_z\psi_{zt}) = D_x(2\psi\psi_{xt}) + D_z(2\psi\psi_{zt}) - 2\psi\Delta\psi_t, \tag{9.114}$$

to obtain

$$
\begin{aligned}
2vv_t &= 2\psi_x vv_z - \psi_z vv_x - 2fv\psi_z \\
&= D_x(v^2\psi_x) - D_x(v^2\psi_z) - 2fD_z(v\psi) + 2f\psi v_z, \qquad (9.115)\\
2\frac{g^2}{N^2}\rho\rho_t &= 2\frac{g^2}{N^2}(\psi_x\rho\rho_z - \psi_z\rho\rho_x) - 2g\rho\psi_x \\
&= \frac{g^2}{N^2}[D_z(\rho^2\psi_x) - D_x(\rho^2\psi_z)] - 2gD_x(\rho\psi) + 2g\psi\rho_x, \qquad (9.116)\\
-2\psi\Delta\psi_t &= -2g\psi\rho_x - 2f\psi v_z - 2\psi\psi_x\Delta\psi_z + 2\psi\psi_z\Delta\psi_x \\
&= -2g\psi\rho_x - 2f\psi v_z - D_x(\psi^2\Delta\psi_z) + D_z(\psi^2\Delta\psi_x). \qquad (9.117)
\end{aligned}
$$

Substituting the expressions (9.115), (9.116), (9.114), and (9.117) in the right-hand side of equation (9.112), we arrive at equation (9.113) with

$$
\begin{aligned}
H^1 &= -v^2\psi_z - \frac{g^2}{N^2}\rho^2\psi_z - 2g\rho\psi + 2\psi\psi_{xt} - \psi^2\Delta\psi_z, \\
H^2 &= v^2\psi_z + \frac{g^2}{N^2}\rho^2\psi_x - 2fv\psi + 2\psi\psi_{zt} + \psi^2\Delta\psi_x.
\end{aligned}
$$

Thus, denoting $C^2 = -H^1$ and $C^3 = -H^2$, we arrive at the following differential form (9.73) of the conservation of energy for equations (9.52)–(9.54):

$$
D_t(E) + D_x(C^2) + D_x(C^3) = 0 \qquad (9.118)
$$

with the density E given by equation (9.111) and the flux given by the equations

$$
\begin{aligned}
C^1 &= v^2\psi_z + \frac{g^2}{N^2}\rho^2\psi_z + 2g\rho\psi - 2\psi\psi_{xt} + \psi^2\Delta\psi_z, \\
C^2 &= -v^2\psi_z - \frac{g^2}{N^2}\rho^2\psi_x + 2fv\psi - 2\psi\psi_{zt} - \psi^2\Delta\psi_x.
\end{aligned}
$$

9.4 Concluding remarks

In this modeling scenario, the focus was on demonstrating how Lie group analysis can provide general and effective methods for constructing exact solutions of nonlinear equations analytically. One aspect of Lie group analysis that was explored was the use of differential equations to obtain exact solutions of two-dimensional stratified rotating nonlinear Boussinesq equations. The obtained invariant solution is representative of the exact solution of the nonlinear Boussinesq equations of stratified fluid in the ocean and was obtained from the rotation and dilation symmetries. Qualitative and numerical analysis of the obtained invariant solutions revealed that they remain bounded oscillating functions of time. The analysis also revealed that the orthogonality of the group and the phase velocity vectors implies the beam energy is transported along rather than perpendicularly to the wave crests. As a result, cylindrical wavefronts cannot be expected from a two-dimensional oscillating source, unlike in an isotropic medium where the phase and group velocity vectors are collinear. However, the form of the rotationally invariant solution suggests some similarity to the three-dimensional case, where mechanical oscillators act as a point rather than line sources and generated gravity wave disturbances have a conical shape, with nearly circular phase lines in planes of constant height. Such circular patterns have been identified in satellite images of the upper stratosphere, and there are ongoing studies on the possibility of modeling thunderstorms through rotationally invariant solutions. In particular, Dewan et al. [19] were able to link these gravity waves to isolated thunderstorms. The possibility for modeling of thunderstorms by means of rotationally invariant solutions is the question of further studies that will be reported elsewhere.

The modeling scenario also explored the derivation of conservation laws from the symmetries of differential equations. While Noether's theorem is commonly used for this purpose, it requires the existence of a classical Lagrangian and is not applicable to equations (8.1)–(8.3). The scenario described an editorialized method for deriving conservation laws from symmetries of nonlinear differential equations, which involves new variables connected with the physical variables in question by adjoint equations and termed nonlocal variables. The corresponding conservation laws are also nonlocal. However, it was demonstrated that the difficulty of clarifying the physical significance of nonlocal conservation laws due to the presence of nonlocal variables can be avoided if the system of nonlinear differential equations under consideration is self-adjoint or quasi-self-adjoint. The system (8.1)–(8.3) was shown to be self-adjoint, which allowed for the association of physically significant conservation laws with the symmetries of the system. One of the conservation laws obtained was the conservation of energy, which was also obtained by using the invariance of equations (8.1)–(8.3) under dilations (scaling transformations).

Bibliography

[1] Aguiar, A., Read, P., Wordsworth, R., Salter, T., Yamazaki, Y. *A laboratory model of Saturn's North Polar hexagon*, Icarus, 206 (2), 755–763, 2010.

[2] Anderson, R. F., Ali, S., Brandtmiller, L. L., Nielsen, S. H., Fleisher, M. Q. *Wind-driven upwelling in the Southern Ocean and the deglacial rise in atmospheric* CO_2, Science, 323, 1443, 2009.

[3] Bachelor, G. *An Introduction to Fluid Dynamics*, Cambridge University Press, Cambridge. 1967.

[4] Balasuriya, S. *Vanishing viscosity in the barotropic β-plane*, J. Math. Anal. Appl., 214, 128–150, 1966.

[5] Baldwin, M., Dunkerton, T., *Stratospheric harbingers of anomalous weather regimes*, Science V., 80 (294), 581–584, 2001. doi: https://doi.org/10.1126/science.1063315.

[6] Belotserkovskii, O., Mingalev, I., Mingalev, O. *Formation of large-scale vortices in shear flows of the lower atmosphere of the earth in the region of tropical latitudes*, Cosm. Res., 47 (6), 466–479, 2009.

[7] Biebricher, A. *Planetary waves: The biggest coherent pressure structures in the Earth's atmosphere*, Space Sci., 2015.

[8] Blinova, E. *A method of solution of the nonlinear problem of atmospheric motions on a planetary scale*, Dokl. Acad. Nauk USSR, 110, 975–977, 1956.

[9] Bluman, G. W. and Anco, S. C., *Symmetry and Integration Methods for Differential Equations*, Springer-Verlag, New York, 2002.

[10] G. Bluman, S. Kumei, *Symmetries and Differential Equations*, Springer-Verlag, New-York, 1989.

[11] G. Bluman, J. Cole, *Similarity Methods for Differential Equations*, Applied Mathematical Sciences (AMS), volume 13, Springer, New York, 1974.

[12] Boussinesq, J. *Essai sur la theorie des eaux courantes*, Memoires presentes par divers savants 'l'Acad. des Sci. Inst. Nat. France, XXIII, pp. 1–680, 1877.

[13] Brand, A., Avrahami, I., Einav, S., Ryvkin, M. *Numerical models of net-structure Stents inserted into arteries*, Comput. Biol. Med., 52, 102–110, 2014.

[14] Bransky, A., Korin, N., Nemirovski, Y., Dinnar, U. *Correlation between erythrocytes deformability and size: A study using a microchannel based cell analyzer*, Microvasc. Res., 73, 713, 2007.

[15] Buchnev A, *Lie Group Admitted by the Equations of Motion of an Ideal Incompressible Fluid*, Continuum Dynamics 7, pp. 212–214. Institute of Hydrodynamics, USSR Acad. Sci., Siberian Branch, Novosibirsk (Russian), 1971.

[16] Cohen, A. *An Introduction to the Lie Theory of One-Parameter Groups with Applications to the Solution of Differential Equations*, D.C. Heath, New York, 1911.

[17] Cho, H., Shepherd, T., Vladimirov, V., *Application of the direct Liapunov method to the problem of symmetric stability in the atmosphere*, J. Atmos. Sci., 50 (6), 822–836, 1993.

[18] Cenedese, C., Linden, P. *Cyclone and anticyclone formation in a rotating stratified fluid over a sloping bottom*, J. Fluid Mech., 381, 199–223, 1999.

[19] Dewan E, Picard R, OíNeil R, Gardiner H, Gibson J, *MSX satellite observations of thunderstorm-generated gravity waves in mid-wave infrared images of the upper stratosphere*, Geophys. Res. Lett., 25, 939–942, 1998.

[20] Dickson, L. *Differential equations from the group standpoint*, Ann. Math., 25, 287, 1924.

[21] Dolgov, D., Zakharov, Y. *Modeling of viscous inhomogeneous fluid flow in large blood vessels*, Vestnik Kemerovo State Univ., 2 (62) T.1, 30–35, 2015 (in Russian).

[22] Einav, S., Stolero, D. *Pulsatile blood flow in an arterial bifurcation, numerical solution*, Med. Biol. Eng. Comput., 25, 12–17, 1987.

[23] Elad, D., Einav, S. *Simulation of airway closure during forced vital capacity*, Ann. Biomed. Eng., 17, 617–631, 1989.

[24] Fjortoft R, *Application of integral theorems in deriving criteria of stability for laminar flows and for the baroclinic circular vortex*, Geophys. Publ., 17 (6), 1–52, 1950.

[25] Formaggia, L., Lamponi, D., Quarteroni, A. *One dimensional models for blood flow in arteries*, J. Eng. Math., 47 (3–4), 251–276, 2003.

https://doi.org/10.1515/9783111387499-010

[26] Formaggia, L., Gerbeau, J., Nobile, F., Quarteroni, A. *On the coupling of 3D and 1D Navier–Stokes equations for flow problems in compliant vessels*, Comput. Methods Appl. Mech. Eng., 191 (6–7), 561–582, 2001.

[27] Formaggia, L., Nobile F., Gerbeau, J., Quarteroni, A. *Numerical treatment of defective boundary conditions for the Navier-Stokes equations*, SIAM J. Numer. Anal., 40, 376–401, 2002.

[28] Friedlander, S., Siegmann, W. *Internal waves in a contained rotating stratified fluid*, J. Fluid Mech., 114, 123–156, 1982.

[29] Fushchych, W., Shtelen, W. *On approximate symmetry and approximate solution of the non-linear wave equation with a small parameter*, J. Phys. A, Math. Gen., 22, 887–890, 1989.

[30] Geidarov, N., Zakharov, Y., Shokin, Y. *Solution of the problem of viscous fluid with a given pressure differential*, Russ. J. Numer. Anal. M., 26 (1), 39–48, 2011.

[31] Gill, A., *Atmosphere-Ocean Dynamics*, Academic Press, New York, etc., 1983.

[32] Golovkin, H. *Vanishing viscosity in Cauchy's problem for hydromechanics equation*, Proc. Steklov Inst. Math., 92, 33–53, 1966.

[33] Gummel, E., Milosevic, H., Ragulin, V., Zakharov, V., Zimin, Y. *Motion of Viscous Inhomogeneous Incompressible Fluid of Variable Viscosity*, Zbornik rabova konferencije MIT 2013, Beograd, pp. 267–274, 2014.

[34] Herrmann, E. *The motions of the atmosphere and especially its waves*, Bull. Am. Math. Soc., 2 (9), 285–296, 2896, 1896.

[35] Holton, J., Haynes, P., McIntyre, M., Douglass, A., Rood, C., Pfister, L. *Stratosphere–troposphere exchange*, Rev. Geophys., 33, 403–439, 1995.

[36] Hsieh, P., *Application of modflow for oil reservoir simulation during the Deepwater Horizon crisis*, Ground Water, 49 (3), 319–323, 2011.

[37] Ibragimov, R. *Oscillatory nature and dissipation of the internal waves energy spectrum in the deep ocean*, Eur. Phys. J. Appl. Phys., 40, 315–334, 2007.

[38] Ibragimov, R., Pelinovsky, D. *Incompressible viscous fluid flows in a thin spherical shell*, J. Math. Fluid Mech., 11, 60–90, 2009.

[39] Ibragimov, N., Ibragimov R. *Integration by quadratures of the nonlinear Euler equations modeling atmospheric flows in a thin rotating spherical shell*, Phys. Lett. A, 375, 3858, 2011.

[40] Ibragimov, R., Dameron, M., *Spinning phenomena and energetics of spherically pulsating patterns in stratified fluids*, Phys. Scr., 84, 015402, 2011.

[41] Ibragimov, N. *A Practical Course in Differential Equations and Mathematical Modeling*, 2nd ed., ALGA Publications, 2005.

[42] Jefferson, G., Carminati, J. *ASP: Approximate symbolic computation of approximate symmetries of differential equations*, Comput. Phys. Commun., 185, 430, 2014.

[43] Kim, J., HoYoon, L., Sehyun, S., *Advances in the measurement of red blood cell deformability: A brief review*, J. Cell. Biotechnol., 1 (1), 63–79, 2015.

[44] Korteweg, D., de Vries, G. *On the change of form of long waves advancing in a rectangular canal, and on a new type of long stationary waves*, Philos. Mag., 39, (1895), 422–443.

[45] Krauss, W. *Methoden und Ergebnisse der Theoretischen Oceanographie II: Interne Wellen*, Gebrueder-Borntraeger, Berlin, 1966.

[46] Kudlick, M. *On Transient Motions in a Contained Rotating Fluid*, Ph.D. dissertation, Massachusetts Institute of Technology, 1966.

[47] Lamb, H. *Hydrodynamics*, Cambridge University Press, 5th edition, 1924.

[48] Lee, C., Peng, H., Yuan, H., Wu, J., Zhou, M., Hussain, F. *Experimental studies of surface waves inside a cylindrical container*, J. Fluid Mech., 677, 39–62, 2011.

[49] Lie, S., *Vorlesungen über Differentialgleichungen mit bekannten infinitesimalen Transformationen* (Bearbeited und herausgegeben von Dr. G. Scheffers), Teubner, B.G., Leipzig, 1891.

[50] Lions, J., Teman, R., Wang, S. *On the equations of the large-scale ocean*, Nonlinearity, 5, 1007–1053, 1992.

[51] Lions, J., Teman, R., Wang, S. *New formulations of the primitive equations of atmosphere and applications*, Nonlinearity, 5, 237–288, 1992.

[52] McCreary, J. *A linear stratified ocean model of the equatorial undercurrent*, Philos. Trans. R. Soc. Lond., 302, 385–413, 1981.

[53] Milosevic, H., Gaydarov, N., Zakharov, Y. *Model of incompressible viscous fluid flow driven by pressure difference in a given channel*, Int. J. Heat Mass Transf., 62, 242–246, 2013.

[54] Musak, L. *Topographically trapped waves*, Annu. Rev. Fluid Mech., 12, 45–76, 1980.

[55] Newman, P. *Chemistry and dynamics of the Antarctic ozone hole*, in The Stratosphere: Dynamics, Transport, and Chemistry, Geophys. Monogr. Ser., 2010, doi:https://doi.org/10.1029/2009GM000873.

[56] Noether, E. *Invariante Variationsprobleme. Konigliche Gessellschaft der Wissenschaften*, Gottingen Math. Phys. K1., (1918), English transl.: Transp. Theory Stat. Phys., 1 (3), 186–207, 1971.

[57] Nethery, D., Shankar, D. *Vertical propagation of baroclinic Kelvin waves along the west coast of India*, J. Earth Syst. Sci., 116 (4), 331–339, 2007.

[58] Ovsyannikov, L. V. *Group Analysis of Differential Equations*, Nauka, Moscow, 1978. English transl., ed. W. F. Ames, Academic Press, New York, 1982.

[59] Ovsyannikov, L. *Lectures on the Theory of Group Properties of Differential Equations*, Novosibirsk University Press, Novosibirsk, 1966. English transl., ed. Ibragimov, N., ALGA Publications, Karlskrona, 2009.

[60] Phythoud, F., Stergiopulos, N., Meister, J. *Forward and backward waves in the arterial system: nonlinear separation using Riemann invariants*, Technol. Health Care, 201–207, 1995.

[61] Quarteroni, A., Tuveri, M., Veneziani, A. *Computational vascular fluid dynamics problems, models and methods*, Comput. Vis. Sci., 2, 163–197, 2000.

[62] Quemada, D. *Rheology of concentrated disperse systems and minimum energy dissipation principle*, Rheol. Acta, 16 (1), 82–94, 1997.

[63] Toggweiler, J., Russel, J. *Ocean circulation on a warming climate*, Nature, 451, 286–288, 2008.

[64] Rideout, V., Dick, D. *Difference-differential equations for fluid flow in distensible tubes*, IEEE Trans. Biomed. Eng., 14, 171–177, 1967.

[65] Riemann, G., *Uber die Fortpáanzung ebener Luftwellen von endlicher Schwingungsweite*, Abhandl. Konigl. Ges. Wiss. Gottingen, 8, 43–65, 1860.

[66] Rotman, O., Shav, D., Raz, S., Zaretsky, U., Einav, S. *Biomechanical aspects of catheter-related thrombophlebitis*, J Biomed. Sci. Eng., 6, 6–13, 2013.

[67] Shen, J. *On Pressure Stabilization Method and Projection Method for Unsteady Navier-Stokes Equations*, Advances in Computer Methods for Partial Differential Equations, 658–662, IMACS, New Brunswick, NJ, 1992.

[68] Sheshadri, A., Plumb, R., Lindgren, E., Domeisen, D., *The vertical structure of annular modes*, J. Atmos. Sci., 2018, doi: https://doi.org/10.1175/JAS-D-17-0399.1.

[69] Shindell, D., Schmidt, G. *Southern Hemisphere climate response to ozone changes and greenhouse gas increases*, Res. Lett., 31, L18209, 2004.

[70] Summerhayes, C., Thorpe, S., *Oceanography*, An Illustrative Guide, New York: John Willey & Sons, 1996.

[71] Streeter, V., Keitzer, W., Bohr, D. *Pulsatile pressure and flow through distensible vessels*, Circ. Res., 13, 3–20, 1963.

[72] Su, C. H. and Gardner, C. S., *Derivation of the Korteweg-de Vries equations and Burgers equations*, J. Math. Phys., 10, 3, 1968.

[73] Summerhayes, C. P., Thorpe, S. A. *Oceanography: An Illustrative Guide*, Wiley, New York, 1996.

[74] Swarztrauber, D. *Shallow water flow on the sphere*, Mon. Weather Rev., 132, 3010–3018, 2004.

[75] Tabaei, A., Akylas, T. R., *Nonlinear internal gravity wave beams*, J. Fluid Mech., vol. 482, pp. 141–161, 2003.

[76] Temam, R., Ziane, M. *Navier-Stokes equations in thin spherical domains*, Contemp. Math., 209, 281–314, 1997.

[77] Vu, K., Jefferson, G., Carminati, J. *Finding generalised symmetries of differential equations using the MAPLE package DESOLVII*, Comput. Phys. Commun., 183, 1044–1054, 2012.

[78] Wittberg, L., Wyk, S., Fuchs, L., Gutmark, E., Bakeljauw, P., Little-Gutmark, I. *Effects of aortic irregularities on blood flow*, Biomech. Model. Mechanobiol., 15, 345–360, 2016.

[79] Westerhof, N., Bosman, F., Vries, C., Noordergraaf, A., *Analog studies of the human systemic arterial tree*, J. Biomech., 2, 121–143, 1969.

Index

https://doi.org/10.1515/9783111387499-011

De Gruyter Series in Nonlinear Analysis and Applications

Volume 42
Catherine Bandle, Alfred Wagner
Shape Optimization. Variations of Domains and Applications, 2023
ISBN 978-3-11-102526-1, e-ISBN (PDF) 978-3-11-102543-8, e-ISBN (EPUB) 978-3-11-102581-0

Volume 41
Jesús Garcia-Falset, Khalid Latrach
Nonlinear Functional Analysis and Applications, 2023
ISBN 978-3-11-103096-8, e-ISBN (PDF) 978-3-11-103181-1, e-ISBN (EPUB) 978-3-11-103208-5

Volume 40
Dung Le
Cross Diffusion Systems. Dynamics, Coexistence and Persistence, 2022
ISBN 978-3-11-079498-4, e-ISBN (PDF) 978-3-11-079513-4, e-ISBN (EPUB) 978-3-11-079517-2

Volume 39
Jesús Ildefonso Díaz, David Gómez-Castro, Tatiana A. Shaposhnikova
Nonlinear Reaction-Diffusion Processes for Nanocomposites. Anomalous Improved Homogenization, 2021
ISBN 978-3-11-064727-3, e-ISBN (PDF) 978-3-11-064899-7, e-ISBN (EPUB) 978-3-11-064751-8

Volume 38
Ireneo Peral Alonso, Fernando Soria de Diego
Elliptic and Parabolic Equations Involving the Hardy-Leray Potential, 2021
ISBN 978-3-11-060346-0, e-ISBN (PDF) 978-3-11-060627-0, e-ISBN (EPUB) 978-3-11-060560-0

Volume 37
Gershon Wolansky
Optimal Transport. A Semi-Discrete Approach, 2021
ISBN 978-3-11-063312-2, e-ISBN (PDF) 978-3-11-063548-5, e-ISBN (EPUB) 978-3-11-063317-7

Volume 36
Giovanni Molica Bisci, Patrizia Pucci
Nonlinear Problems with Lack of Compactness, 2021
ISBN 978-3-11-064886-7, e-ISBN (PDF) 978-3-11-065201-7, e-ISBN (EPUB) 978-3-11-064893-5

Volume 35
Takashi Suzuki
Semilinear Elliptic Equations. Classical and Modern Theories, 2020
ISBN 978-3-11-055535-6, e-ISBN (PDF) 978-3-11-055628-5, e-ISBN (EPUB) 978-3-11-055545-5

www.degruyter.com